Rural History in Europe 8

COST
Action A 35 PROGRESSORE

COST – the acronym for European COoperation in the field of Scientific and Technical Research – is the oldest and widest European intergovernmental network for cooperation in research. Established by the Ministerial Conference in November 1971, COST is presently used by the scientific communities of 35 European countries to cooperate in common research projects supported by national funds. The funds provided by COST – less than 1% of the total value of the projects – support the COST cooperation networks (COST Actions) through which, with EUR 30 million per year, more than 30,000 European scientists are involved in research having a total value which exceeds EUR 2 billion per year. This is the financial worth of the European added value which COST achieves.

A 'bottom up approach' (the initiative of launching a COST Action comes from the European scientists themselves), *à la carte participation* (only countries interested in the Action participate), 'equality of access' (participation is open also to the scientific communities of countries not belonging to the European Union) and 'flexible structure' (easy implementation and light management of the research initiatives) are the main characteristics of COST (Web: www.cost.esf.org).

As precursor of advanced multidisciplinary research COST has a very important role for the realization of the European Research Area (ERA) anticipating and complementing the activities of the Framework Programmes, constituting a 'bridge' towards the scientific communities of emerging countries, increasing the mobility of researchers across Europe and fostering the establishment of 'Networks of Excellence' in many key scientific domains such as: Biomedicine and Molecular Biosciences; Food and Agriculture; Forests, their Products and Services; Materials, Physical and Nanosciences; Chemistry and Molecular Sciences and Technologies; Earth System Science and Environmental Management; Information and Communication Technologies; Transport and Urban Development; Individuals, Societies, Cultures and Health. It covers basic and more applied research and also addresses issues of pre normative nature or of societal importance.

Integration through Subordination. The Politics of Agricultural Modernisation in Industrial Europe

Edited by

Peter Moser and Tony Varley

BREPOLS

This publication is supported by COST (European COoperation in the field of Scientific and Technical Research), the Archives of Rural History (Bern, Switzerland) and the Centre de Recherches Historiques (CNRS/EHESS, Paris, France). It is the result of the work undertaken in the working group 4 'State and rural societies' of the COST Action A35 "Progressore".

Cover: ©. Science in practice: the introduction of a new harvester in the 1950s. ©. Archives of Rural History, Bern (http://www.agrararchiv.ch)

This is a highly fragile area, hit hard by erosion, desertification, salinization of agricultural land and fires which destroy the vegetation and result in further soil erosion.

© D/2013/0095/154 ISBN 978-2-503-54529-5

Printed in the E.U. on acid free paper.

CONTENTS

LIST OF CONTRIBUTORS

Paul Brassley	University of Exeter, United Kingdom
Daniel Flückiger	Archives of Rural History/University of Bern, Switzerland
Dulce Freire	Universidade de Lisboa, Portugal
Fritz Georg Von Graevenitz	European University Institute, Florence, Italy
Jonathan Harwood	University of Manchester, United Kingdom
Christian Hoffmann	European Academy of Bozen/Bolzano, Italy
Niek Koning	Wageningen University, Netherlands
Daniel Lanero Táboas	Universidade de Santiago de Compostela, Spain
Ernst Langthaler	Institute of Rural History, St. Pölten, Austria
Peter Moser	Archives of Rural History, Bern, Switzerland
Socrates D. Petmezas	Institute for Mediterranean Studies/FORTH and University of Crete, Rethymno, Greece
Kathrin Renner	European Academy of Bozen/Bolzano, Italy
Flavio V. Ruffini	Landesagentur für Umwelt, Autonome Provinz Bozen, Italy
Gloria Sanz Lafuente	Universidad Pública de Navarra, Spain
Serge Schmitz	University of Liege, Belgium
Anton Schuurman	Wageningen University, Netherlands
Thomas Streifeneder	European Academy of Bozen/Bolzano, Italy
Zsuzsanna Varga	Eötvös Loránd University, Hungary
Tony Varley	National University of Ireland, Galway

LIST OF FIGURES

LIST OF TABLES

The Series Rural History in Europe

Sometimes viewed as the producer of foodstuffs and raw materials, as a garden or a space for the production of energy; sometimes seen as a reservoir of manpower destined to be subverted to the needs of cities and industries or as a peaceful haven which welcomes weary city-dwellers; also regarded as a place of easily incensed 'backward' societies or as an idyllic spot close to nature where societies act as guardians of a certain conception of quality of life, the countryside has not ceased to be at the centre of economic, social, political and now environmental concern of governments and public opinion. Pulled between contradictory demands, the countryside of today presents an image of a place and a society in clear transformation, which, in itself, is difficult to decipher.

In Europe, the difficulty is that the countryside, which forms a complex and evolving universe, is experiencing ruptures and also exhibits inertias. These complex transformations of the rural world can only be fully understood if they are viewed in a manner which transcends national boundaries and if the discrepancies, which can only be observed by adopting a broad view, in fact at the continental scale, are taken into account. These changes cannot be explained without reference to their past. The question is how to create such a dialogue between researchers which goes beyond national boundaries, crosses chronological barriers and breaks disciplinary boundaries.

The main objective of the 'Rural History in Europe' collection is to provide just such keys to unlock the changes experienced by present-day European rural societies in the light of their historical experience. It is to produce the historical knowledge which will allow us to conceptualize the future of European country-dwellers, as they face the kinds of problems which historians have grappled with in examining societies in the past: under employment and multiple occupations, migrations and rural depopulation, distortion of competition by the marketplace or by the policies established by political authorities, competition with non-European producers, problems of resource allocation (land and water), distribution and redistribution of heritages and land-holding, the future of owner-occupied farms and the function of agriculture as employer, the tensions between private ownership and public access to land, the changes induced by settlement of urban migrants, environment and sustainable development, and so on.

In most European countries, rural society and long term change have been the subject of intensive research. Over the last half-century, a lively historiography has developed around questions concerning the factors of growth and systems of agricultural production, as well as the periodization of growth and the impact of agriculture on industrial development. As a result, a body of knowledge has built up on techniques, forms of land-use, levels of production, the differing conditions and nature of the peasantry and the strategies of economic agents. More recently environmental concerns have made a dramatic impact on the consciousness of historians in the field, while previous findings have been questioned in a critical way and new perspectives have been drawn.

The field known as Rural History remains very much alive, despite the decline of the agricultural sector over the last several decades. Rural History remains fundamental in a Europe which has for so long been kept together by its Common Agricultural Policy and which now has to face the heavy impact of Brussels-based initiatives upon its rural regions, and an unprecedented revision of current agrarian systems and systems of production in the countries which have recently joined the EU. How can the changes taking place in present-day Europe be understood without taking into account a past which is still very present, and which determines both structures and behaviour?

The volumes of this collection are the result of several workshops which have been held during the past years and which have mainly been supported by funds of the European Action COST A 35 and by other institutions like CORN, GDR CNRS Sociétés Rurales Européennes, Universities... The COST Action, which was initiated three years ago, intends to extend the historical analysis of rural society over the last millenium in order to envisage the problems of the countryside in an extended timeframe and also to draw upon the commentary and expertise of specialists from other disciplines (sociology, economics, anthropology). This will enable the first real comparison of Europe from historians from all over the continent, from Scandinavia to the Mediterranean and from the western frontier to the eastern limits. Papers from these meetings are published after a peer-review process, supported by the editorial board and, of course, the work of the authors and editors of the volumes. This collection is constructed upon four main pillars: *Landed property; The management of rural land; Peasant societies; The state, government, politics and peasants*. Each of these brings new perspectives and produce new tools to better understand the changes which are taking place today. In order to ask the relevant questions about the future of these peasantries and rural spaces in transformation, the books of this collection will deal with the *longue durée* and will present either research in progress or a synthesis on a regional or national scale. The cumulative effect of this approach will be to produce volumes, the geographical coverage of which will be the whole of Europe.

The volumes also, of course, take into account the role of history as an explanatory factor for contemporary European Societies. If rural societies have been overthrown, if rural landscapes have been profoundly transformed and if the intervention of the State has considerably strengthened the regulation of production and trade, the contrast between a contemporary rural world in rapid transformation and a traditional rural world with frozen landscapes, petrified societies, immobile economies and lethargic political contexts will be an illusion. It is important, therefore, to detect, to measure and to interpret the range of recent changes by illuminating those that have taken place in past centuries in a European context.

1. The state and agricultural modernisation in the nineteenth and twentieth centuries in Europe

Peter MOSER & Tony VARLEY

I. Introduction

The aim of this book is to deepen our understanding of the causes, nature and implications of agricultural change in Europe since the eighteenth century[1]. We examine the political contexts in which agriculture was embedded by considering the appearance of modernising agricultural policies in industrial societies and some of their more consequential impacts. To set the scene – and to explore how notions of broad agricultural change have shifted over time – the first part of this introduction considers the historical development of agriculture in Western Europe. In view of the state's centrality to agricultural policy, an analytical discussion of the capacities of states to induce agricultural modernisation will then follow. The third part of our discussion will compare the variety of agricultural projects pursued in the specific political contexts of concern to the contributors to this volume.

II. Historical contexts

Attempts to tie instances of broad agricultural change to specific historical periods are always potentially contestable as regards their starting and ending points as well as their sources of dynamism. At the same time, an attempt at periodisation is indispensable if we are to impose some order on realities that otherwise threaten to become overly diffuse. While political reasons have often featured in attempts to account for broad agricultural change, by no means have they been the only sorts of reasons deemed consequential. Access to new resources as well as economic, social and technological developments and the creation of new 'useful knowledge' have often been regarded as of equal or greater importance as drivers of change (Hobsbawm, 1977; Crotty, 2001; Mokyr 2002; Sieferle et. al., 2006; Wrigley, 2010).

[1] This volume has its origins in a workshop, the third in a series organised under the auspices of Working Group 4 of the COST Action A 35, held in Möschberg, Switzerland in June 2008. We are thankful to the Swiss National Research Fund for its financial support of this workshop, and to Vicente Pinilla and especially to Jürgen Schlumbohm for their careful reading of the manuscript.

In what sense can periods of broad agricultural change in Europe be described as 'revolutionary'? Although the term 'agricultural revolution' is not universally accepted as accurately describing the broad changes within agriculture (Thirsk, 1987), the term remains viable as long as it is handled carefully. Paul Bairoch (1989) provides an ambitious and – especially if it is combined with an analysis of the issue of agriculture's widening resource base – useful periodisation of transformative change in western European agriculture. Bairoch distinguishes three agricultural revolutions occurring over the last 250 years or so. The first of these, extending from the seventeenth to the middle of the nineteenth century, saw change in agriculture as a matter of a broad re-organisation of the long-established system of using biotic resources on the land. What became definitive in the second agricultural revolution, which extended from the mid-nineteenth century to the 1940s, was the ever-expanding interaction between the new industrial societies and their agricultural sectors. The third agricultural revolution, commencing in the 1940s and continuing to the present day, was by far the most comprehensive and mould-breaking of all three – and therefore qualifies as the most unequivocally revolutionary of the three periods. It is characterised by the sharp drop in demand for human and animal labour and the unprecedented increases in production and productivity as well as a rapidly decreasing biodiversity. Critical to the third agricultural revolution was the newly gained access to the vast, but finite fuel reserves of the lithosphere. Since the 1950s food production, processing and transport have been largely, albeit never fully, based on the consumption of the same fossil fuel reserves that manufacturing industry has relied on since the early days of the thermo-industrial revolution when for the first time in history it became possible to decouple the processes of production and reproduction in the industrial sector (Georgescu-Roegen, 1960; Sieferle *et. al.*, 2006; Wrigley 2010).

II.1. The period of the first agricultural revolution 1750-1850: Growth within the limits set by the use of biotic resources

The first of the three revolutionary periods illustrates how change that eventually works to transform the *status quo* can be a long drawn out affair that can be at once continuous with the past while breaking sharply with it. Selective and incremental change aimed at practical improvements has been a feature of agrarian societies for millennia, but a significant awakening to the possibilities of transformative change in the sphere of agriculture can be observed in the last quarter of the eighteenth century in many parts of northern and western – but not southern (Clar & Pinilla, 2009) – Europe. Agricultural change was evident not only in Britain but also in countries like France, Germany, the Netherlands, Sweden and Switzerland, where the interest in changing agricultural matters was stimulated by the rapid pace at which knowledge was developing and organisational innovations were appearing. The interest in improving food and fodder production was strengthened by a combination of political and economic crisis conditions born out of the 'double experience of war and famine in the 1750/60s and early 1770s (Behrisch, 2008: 51; Stuber *et al.*, 2009).

In France and Germany the desire to achieve fundamental change in agriculture was influenced by the close relationship that developed between the 'absolute' monarchies and the Enlightenment rationalism associated with Physiocracy and Cameralism (Behrisch, 2008). For both schools it was essential that the enlightened absolutist rulers, as a means of enhancing their own political power and relative standing in the emerging competitive state system, supported the project of agriculture-led economic growth. In breaking with the earlier mercantilist ideas both the more theoretical Physiocrats and the rather pragmatic Cameralists shared the view that the growth of national economies would depend on agricultural production undergoing a fundamental transformation. Progress in achieving more output on the land by using the existing resources primarily required the creation and transformation of new knowledge, the application of technical innovations and the cultivation of new plants.

The state – whether organised along monarchical lines or constituted as a city republic – played a crucial role in stimulating agricultural change in order to achieve growth. Political initiatives, in other words, were pivotal in giving impetus to the movement for fundamental agricultural change. But how transformative in practice were the changes occurring during the period of the first agricultural revolution? The growing of nitrogen-fixing plants like clover, the re-organisation of husbandry methods such as stall-feeding and the extension of relatively newly discovered plants (such as the potato) were not revolutionary in themselves. However, when combined they enabled a substantial increase in livestock keeping which provided the essential manure for increased plant production and in turn this generated the basis for expanded livestock keeping (Pfister, 1995a). The closure of the age-old fertilizer gap, a characteristic of all solar-based agricultural societies, not only created the material background for unprecedented economic growth in agricultural production but, probably even more significantly, it opened up a radical new developmental perspective, inducing people to look forwards in their economic expectations (Koselleck, 1979). The preference given to *production* over *trade* by the Physiocrats in France, the Cameralists in Germany and the economic patriots as well as other elite-dominated agricultural improvement societies in many other countries challenged the hitherto dominant perception of agriculture as being an intrinsically static or stagnant form of economic activity.

A crucial feature of this development was the stimulus given to the complex process of creating and applying knowledge with a view to transforming agriculture. Not only did a substantial segment of the educated elite become interested in questions of how farming methods could be improved, but a number of practical farmers began to read and sometimes even write about the new production methods which they were developing, either together or in parallel with the academically educated agriculturalists of the new agricultural improvement societies (Stuber *et al.*, 2009). 'Practice with Science', was therefore a very suitable motto for the Royal Agricultural Society of England (established in 1837) to choose as its rallying cry (Brassley, 2008). Out of the first agricultural revolution arose not only significantly increased production but an institutionalisation of actors and discourses of knowledge within the agricultural departments of universities, agricultural colleges and farmers' clubs as well as within the print media of numerous farming-related publications and periodicals.

The most significant outcome of the first agricultural revolution was a swiftly emerging consciousness of the possibility of transforming agriculture into an increasingly dynamic sector capable of continuous growth (albeit a growth still inherently limited by its use of biotic resources which by nature are bound to cyclical processes of reproduction). In the longer term another, initially unintended, consequence of the first agricultural revolution was its contribution to the disruption of the established social and political order. Not alone did this contribute to hastening the end of the *Ancien Régime* but it facilitated the breakthrough to an industrial society that would bring forth new state elites and private and civil society actors with new interests and ideas as to how transformative change was to be conceived and pursued in the agricultural sphere. In other words, the changes during the first agricultural revolution are of an endogenous character, induced by actors shaping and re-shaping their environment (Boserup, 1965).

II.2. The period of the second agricultural revolution c. 1850-1950: integration of the agricultural sector into the industrial society

In the period of the second agricultural revolution the terms *modern* and *industrial* became increasingly regarded as one and the same. From now on modernising farming primarily meant seeking to emulate the new mode of industrial production in re-organising agricultural production. But since the resource base of agriculture in this period remained primarily the use of living animals and plants, the results of the modernising measures introduced in farming not surprisingly were often significantly different from those in industry. While, for example, the combustion engine rapidly displaced draught animals in industrial production, this source of power became even more important in agriculture as can be observed in the sharp rise of horses on farms from the middle of the nineteenth century on (Moser, 2012). The same is to be said about wage labour: while it became a constitutive element of industry in the second half of the nineteenth century, it began to shrink in the agricultural sector (Koning, 1994).

The persistence and sometimes the new appearance of elements perceived as surviving from the pre-industrial area, evoked bewilderment in industrial and progressive agricultural circles. Agriculture and quite often those engaged in it, were increasingly seen as deviating from what was now more and more the industrially defined norm. This perception strengthened the pre-existing resolve to modernise farming and those working the land. One way of doing this was to convert peasants into farmers who conformed to the standard set by manufacturing industry. Such a normative ideal was succinctly formulated by the Swiss politician (and farmer) Zacharis Gysel who in 1854 published an important book with the revealing title: *Der Schaffhauser Bauer, wie er sein sollte, und wie er nicht ist, wie er ist, und wie er nicht sein sollte* (literally: 'The Schaffhauser peasant as he should be, and is not, and how he is and should not be') (Gysel, 1854). A growing number of civil servants, scientists and individual farmers became dedicated to modernising agriculture according to the norms followed in manufacturing industry (Sombart, 1902).

Although state interventions took a variety of different forms, as time passed practically every measure strengthened the tendency to integrate agriculture into industrial society by trying to make it look and function more and more like industry (Fitzgerald, 2003). This underlying tendency was often obscured by discourses of the late nineteenth century in which pressure groups, political parties and administrative and scientific interests demanded special treatment for the agricultural sector. So far historians have tended to reproduce rather than analyse these contemporary discourses. Historiography has overwhelmingly interpreted the wide-ranging state intervention in agriculture before World War I simply as attempts to protect the farming population from the negative impacts of the world markets at the cost of the consumers. While the protectionist rhetoric may support such a reading, a careful scrutiny of the actual policies being adopted and the outcomes of these policies suggest a rather different interpretation.

Even the most consciously protectionist measures did not inhibit the integration of agricultural production into the world market. In virtually all European countries the agricultural sector was integrated into the new division of labour dictated by the new global trading regime. It is commonly accepted that farmers in Holland, Denmark and Britain, countries strongly committed to free trade until the 1930s, changed their production patterns from grain to animal husbandry. Yet the very same process is evident in countries like Switzerland, Germany and France, where governments had opted for a so-called 'protectionist' policy (Aldenhoff-Hübinger, 2002; Baumann, 1993; van Molle, 2008). Based on broadly converging outcomes, there are probably more similarities than differences between Britain after the repeal of the corn laws in 1846 and those Continental states with formally 'protectionist' trade policies in the late nineteenth century. In Switzerland, an early industrialised country like Britain or Belgium, grain-growing decreased just as dramatically as it did in free-trading countries. And productivity growth in Germany (with its agricultural protection) was among the highest in Europe (see Koning, below, Figure 2.3).

The crucial challenge, therefore, is not so much to describe the supposed success of lobby organisations but to analyze and explain why and how the different policies produced broadly similar outcomes in this period. One way forward is to scrutinise more critically the so-called protectionist policies. In Switzerland as well as in Spain, for example, it is much more precise to speak of a 'selective protectionism' than a general 'protectionist' policy (David, Mach & Straumann, 2008; Gallego Martinez & Pinilla, 1996). Such a policy empowered farmers to adapt to changing circumstances more than it protected them in their existing practices. Furthermore, agricultural policy reliant on selective protectionist measures was in its outcomes – as against its justifications or declarations of intent – far more dynamic than static in character. Usually it was accompanied by educational measures as well as by prohibitions and monetary incentives. Those engaged in agriculture were thus enabled to adapt to the new realities created by the transport revolution while simultaneously preserving the potential to produce at least a substantial portion of the food so badly needed when international trade became seriously disrupted during World War I.

In the last quarter of the nineteenth century agriculture had become – along with increasing imports – virtually the sole provider of food for the ever-expanding urban-industrial population whose members were increasingly losing the knowledge, skill and capacity to produce even a small portion of their own nourishment. Simultaneously agriculture also became an essential provider of many of the raw materials such as leather in shoemaking, or plants and wool in the expanding textile and clothes-making industries. And, equally importantly, the demand for industrial goods in farming circles itself grew rapidly in the same period. Agriculture before World War I, in other words, became more rather than less important within the industrial societies of western Europe. It is, therefore, not surprising that it was regarded more as a valuable resource than as a burden. Agriculture not only provided food for the expanding urban population, it also 'contributed in an essential way to modern economic growth' (Federico, 2005). It made as much economic as political sense for the European industrial states to invest in, regulate and protect this increasingly important sector by pursuing modernising policies that were by no means synonymous with protecting individual farmers.

It is noteworthy that official measures taken in this period still usually took into account the distinctive organising principles (*Eigenlogik)* of the agricultural sphere where the process of production was still closely tied to the process of reproduction. Production in agriculture had therefore to be organised cyclically and seasonally rather than along linear and continuous lines as was now customary in industrial production. Educational institutions and labour policies of the industrial states allowed for the idiosyncrasies of agricultural production by, for instance, establishing so-called winter schools for young farmers or fixing variable summer and winter working hours and minimum wages for farm labourers (Flückiger, below; Collins, 2000).

The purpose of these 'special' measures, deviating from the industrial norm, was (as with selective protectionism) by no means to preserve a pre-industrial peasantry. On the contrary, it was meant to help to bridge the widening gap between agriculture and industry and effect a transition from a peasant to a modern agriculture in practical ways. Instead of turning 'peasants into Frenchmen' (Weber, 1976) the agricultural policies were aimed at transforming 'peasants into farmers', as Ernst Laur, the internationally acclaimed agricultural economist and director of the Swiss Farmers Union expressed it. Laur maintained that one of the most efficient measures to achieve such a transformation was the adoption of systematic book-keeping even on small and medium sized farms. He developed, on a scientific basis, a book-keeping system for peasant farms (Laur, 1911). Laur was convinced, that the practice of book-keeping 'will turn the peasant into a farmer' (Laur, 1907). Right from its inception in 1898 the *Swiss Bauernsekretariat*, the scientific department of the Swiss Farmers' Union, organised book-keeping courses for hundreds of peasants with the intention to transform them into small and medium-sized commercial farmers who closely aligned their cultivation of the land with the demands on the markets and the regulations of the state. Farmers with such abilities were, in many ways, more successful in adopting the new modernising measures than the big estates heavily dependent on wage labour. All over western Europe, as the coal-based transport revolution and the advent of industrial fertilizer broke the Malthusian connection between population and prices (Koning, 1994), these big estates found themselves unable to compete successfully with the medium-sized farms depending mainly on family labour.

For the first time since the beginning of the second agricultural revolution the hazards of relying on global food markets became obvious in the later stages of World War I. In many European societies under the impact of the ensuing food crisis a wide variety of actors (including hitherto liberal-minded free traders and substantial sections of the political left who had long fought more for cheap food prices than for higher wages) came to realise the agricultural sector's crucial importance for industrial society's continued well-being. Under the new circumstances even societies that had adopted free trade policies in the second half of the nineteenth century now began to opt for an agricultural policy that favoured the production of food over trading in foodstuffs. A whole range of measures – such as compulsory production, educational offensives and price incentives – was implemented in pursuit of this goal. In Switzerland this policy went so far that agriculture became organised along the lines of a public service (Moser, 2012). Even in hitherto free trade strongholds like the United Kingdom or the Netherlands the state intervened heavily in the agricultural sector during the interwar period (Schuurman, below).

The desire to integrate the agricultural sector seamlessly into industrial society was carried to such lengths in the period of the second agricultural revolution that, in light of the difficulties of saying definitively where agriculture ended and manufacturing industry began, contemporary historians have commenced to question the adequacy of the very term 'agricultural sector'. For the very same reason the term agrarian knowledge society (Uekötter, 2010) has been extended and reformulated as 'agrarian-industrial knowledge society'. This is a much more precise (if less elegant) term for analysing the processes of knowledge production, transformation and implementation in agriculture since the middle of the nineteenth century (Auderset, Bächi & Moser, 2012).

In summary then a far-reaching integration of the agricultural sector into the industrial society was achieved during the period of the second agricultural revolution. At the same time it should not be forgotten that agricultural production remained not entirely, but to a large degree, tied to the reproduction cycles of its biotic resources and thereby remained capable of reproducing a substantial part of its own resource base within the process of production.

II.3. The third agricultural revolution: a partially successful attempt to industrialise agricultural production

'The relationship with the environment has always been a distinctive feature of agriculture, but its nature has changed deeply in the recent decades'. To formulate this important observation by Giovanni Federico (2005: 5) more precisely: a *fundamental* change in the 'colonisation of nature and the societal metabolism' (Landsteiner & Langthaler, 2010) occurred in the post-war decades with the widespread adoption of the combustion engine in agriculture. Access to the enormous fuel reserves of the lithosphere – hitherto a resource almost exclusively consumed in industrial production – substantially, albeit never fully, transformed the *cultivation* of soil, plants and animals into an activity of *converting* mineral resources into food for human consumption. This was the single most important factor that, in the ensuing 'productivist' phase (Ilbery & Bowler, 1998), enabled the historically unprecedented growth in the volume of food production as well as in the productivity of agricultural labour, the latter even beginning to surpass that of the industrial and service sectors (Bairoch, 1989; Halbeisen, Müller & Veyrassat, 2012; Federico 2012).

From the late 1940s onwards it was the versatile, multi-purpose oil-fuelled tractor that truly revolutionised work in Europe's fields. Together with the ready availability of other fossil fuel-based inputs and machinery – mineral fertilizer, pesticides, fungicides and motorised mowing, milking and harvesting machines – a historically unprecedented replacement of people and draft animals was set in train. What is called the replacement of labour by capital in neo-classical economics can more precisely be described as the substitution of living resources by mineral ones, thus permitting the partial decoupling of production and reproduction in agriculture for the first time in history.

But the spectacular increase in food production of the post-war era is the result of a complex process. It cannot be adequately explained by one factor alone. The ready availability of cheap oil (Pfister, 1995b) would hardly have produced these momentous results on its own; a variety of new and old actors were active in intervening and shaping the unfolding process. New knowledge and skills were created, transmitted, acquired and applied to agricultural practice. The outcome would have been quite different had the space left by the declining number of farmers, farm women and farm labourers not been simultaneously filled by the swelling numbers of agronomists, technicians, veterinary surgeons, inseminators, advisers, accountants, salesmen and other actors specialised in interpreting the new (as well as sometimes the old) knowledge needed to exploit the new production possibilities (Uekötter, 2010; Flückiger, below). In other words, the approach of the farming population towards the process of cultivating the land had to be substantially changed. Agricultural institutions and individuals sceptical, reluctant or even hostile towards this form of modernisation found themselves increasingly marginalised. It is only since the late 1980s with the advent of post-productivism, often co-existing with productivism within the same farm, that state institutions and scientists began to take seriously alternative notions of how agriculture should be conceived and practiced.

Largely because of the newly acquired access to the fossil fuel-based resources the same broad project of agricultural modernisation as pursued in the two preceding periods now produced strikingly different results. Alongside the spectacularly growing volume of food and labour productivity stands the loss of social- and biodiversity in the countryside. The extent and severity of ecological degradations in the rural world in the second half of the twentieth century illustrates how similar (or even bigger) problems can arise in farming as in industry when it comes to be based on the same resources and is organised according to the same principles and practices as manufacturing industry. The animal welfare issue aptly illustrates in its turn what becomes unavoidable when what is considered 'normal' and desirable in industry (i.e. production on a large scale and the transportation of goods over long distances for cost and market reasons) becomes morally highly objectionable in agriculture on the grounds that it often harms living creatures. Another example of the different results produced in the third agricultural revolution compared to its two predecessors is the loss of energy efficiency and biodiversity. The eighteenth and nineteenth century improvements such as stall-feeding, the introduction of human and animal-operated machinery and improved crop rotations actually increased the energy efficiency of production along with biodiversity. Yet, in the second half of the twentieth century, when agricultural production began to rely increasingly on the consumption of fossil fuel, the efficiency ratio declined dramatically alongside rapidly shrinking biodiversity (Krausmann, 2006).

The policy of integrating the agricultural sector into industrial society gained strong support from the breakthrough that saw the 'social state' being virtually universally accepted by all the relevant political forces. From the 1940s on agricultural policy in almost all European states contained a social dimension in favour of the still numerous segment of the working population toiling on the land. It enabled those surviving in agriculture to realise an income comparable to that earned by those employed in industry and services. The extension of the 'social-democratic' welfare state idea to the rural areas was supported in principle, if not in every detail, by European liberal, social-democratic and conservative governments alike. Similarly the Common Agriculture Policy of what became the European Union contained social, regional as well as economic elements. Individual farmers were encouraged and indeed obliged to increase the volume of their production as well as changing it even more towards animal products in order to meet increasing consumer demand. In some areas production ultimately exceeded demand, thus leading at first to the introduction of costly export-subsidies to dispose of the surplus (Spoerer, 2010; Patel, 2009) and then to production restrictions at the level of the individual farm.

The agricultural policy of practically all European states, within and outside the EU, not only offered incentives to stimulate increased production but also put in place mechanisms to achieve the most radical structural reform ever seen in agriculture. The modernising policy, as Yves Tavernier (1972: 22) had already realised by the early 1970s, literally provided fuel for the machinery of displacing producers. Even economists fundamentally critical of the European 'high-price' policy stressed the point that it was specifically this policy that forced the farmers to achieve the highest possible productivity (Kleinewefers, 1972). Not surprisingly, it was precisely during the period of this policy that the vast majority of all farms were given up within a generation or two, thus reducing the ever smaller fraction gainfully occupied in agriculture to a tiny minority even in most rural areas. The widespread, endless but ultimately futile demonstrations and protests of European farmers against their displacement from the land in the post-war area are a telling indication that this policy was a much more complex one than has been often depicted in the literature which tends to misread this defensive collective action as an instance of successful 'rent seeking' behaviour (Hofreither, 2009).

Not surprisingly, the numerical, social and spatial contraction of agriculture significantly influenced the agricultural organisations. The once relatively powerful farmers' organisations, which more and more had acquired the role of implementing state agricultural policies during the latter years of the second agricultural revolution, now tended to become the ever more co-opted 'machines of integration', ready to adapt industrial and urban values to a degree that frequently even went far beyond that of sections of the urban elites. This can be seen, for instance, in their hostility towards peasant and organic farming-friendly policy measures in the 1970/80s (Moser, 1994; Vogt, 2000). While serving in the corporatist era to strengthen the policy of integrating agriculture into industrial society almost uncritically, the main farmers' organisations more and more departed from their equally important role of representing the diversity of the agricultural world. The result of this departure has been a weakening of agriculture's position vis-à-vis the agri-industrial world, as is illustrated revealingly by the sharp decline in the prices producers realise since the 1990s even though consumer prices continue to increase (Moser, 2012). The above-mentioned farmer demonstrations directed against state agricultural policy have been a notable consequence of this process of co-optation. Not surprisingly too, they were often aimed in part at their own official representatives. Parallel to the integration of the mainstream farmers' organisations into the agri-industrial complex and the state administration, we can observe the appearance of a great variety of non-governmental organisations (including a number of newly established farmers' organisations) tied to the emergence of different farming styles and production methods (Van der Ploeg, 2003).

Notwithstanding this diversification of agricultural representation in the last three decades, agriculture as a whole has experienced a dramatic weakening on the cultural and symbolic level. Today, even the very term 'agriculture' is fast disappearing from the lexicon of everyday life, as is seen for example when departments of agriculture are re-named departments of food and rural development or political parties drop any reference to their rural origins so as to remain – or to become – part of the (post)modern world. Agriculture today is more and more incapable of intellectually explaining and politically defending and upholding its own distinctive organising principles (*Eigenlogik*), principles that despite everything still retain their relevance. Even the most advanced forms of industrialised agriculture continue to depend on using plants and animals whose production is tied to the reproduction of its own basis and therefore remains subject to cyclical and seasonal constraints. The abandonment of customised education for agronomists and the way agricultural colleges tend to ignore the seasonality dimension in organising farmers' education today are only two examples of the tendency to ignore agriculture's distinctive peculiarities[2]. At the same time agriculture remains, primarily for food and recreation related reasons, an important factor in European societies even in the early twenty-first century – a fact that is partly mirrored in its continued significance within the European Union's policy regime and also in the attractions it continues to have for young people in spite of the long working hours and relatively low incomes awaiting those who opt to work in agriculture and live from it.

When Wrigley (2010) argued that the energy revolution in manufacturing at the end of the eighteenth century industry opened a Pandora's box, the same must even more forcefully be said of what happened in the attempt to industrialise agricultural production during the period of the third agricultural revolution. It is therefore appropriate that we now turn to look more closely at the forces responsible for giving shape to these changes and for driving them forward.

III. Strong states and agricultural modernisation viewed analytically: four ideal-typical scenarios

It almost goes without saying that this sketch of agricultural development in western Europe over the past 250 years or so can depict only the broad frame. We need, to gain a better impression of the images within the frame, to look more closely at the national level where a variety of natural, social, economic and political conditions and actors, discourses and practices have interacted to influence the development of agriculture in specific cases. But how do we begin to compare variations in agricultural development across nation-states and regions? Taking the centrality

---

[2] That seasonality still features greatly in contemporary agriculture can be seen in the hundreds of thousands of seasonal harvest workers drawn from the fringes of Europe (or from outside it) to work mostly on big capitalist farms similar to those that flourished prior to the rise of family farming at the end of the nineteenth century (Koning, 1994).

of states as agents of change in industrial societies as our departure point, we will commence by posing four general questions:

Why do 'state elites', and their civil society allies and agents, choose to intervene in agriculture with a view to modernising it?

What do these elites understand by agricultural modernisation?

What sort of power resources do they believe to be necessary if their efforts are to be effective?

What implications follow for the farming population and farming from these modernising policies?

Our general answers to these four questions will introduce four hypothetical possibilities in the form of elementary ideal-typical scenarios of the aims and capacities of state elites intent on inducing 'agricultural modernisation'. As these ideal-typical scenarios consciously exaggerate and simplify complex historical realities, the pictures they provide us with may be empirically possible, but they are not empirical descriptions or explanations in any strict sense. The prime purpose of our ideal-typical scenarios is rather to provide interpretive benchmarks that can facilitate the reading and comparison of real world cases. Ever since Max Weber's pioneering methodological discussions, modelling *via* ideal-types, or scenario modelling, has been a potentially useful tool in the comparative historical and social sciences (Burger, 1987).

The four general answers are at one in assuming that strong states and their civil society allies and agents, for all their diversity, organisational complexity and irrespective of the bases of their strength, can be regarded as agents for certain purposes. They further converge in assuming that state elites are central to the capacity of states to exercise agency. It is these elites who we will take as finding the prospect of modernising agricultural change appealing and as coming to a clear-headed conception of what qualifies as modernising agricultural change. We will further assume that these state elites can come to a firm view of what sort of power resources the state needs to have at its disposal, and to deploy strategically and skilfully, if it is to have a realistic chance of converting visions of modernising agricultural change into concrete policy and farming realities.

By way of answer then to the question of *why* state elites might desire agricultural change of a modernising kind, we will suggest that our ideal-typical state elites have cogent political (as against economic) reasons for desiring such change in the first instance. Basically these political reasons reflect the view that a modernised agriculture is a necessary condition of having a stable modern state and a food-secure modern (*i.e.* industrial or post-industrial) society. Couched in functionalist language Eric Hobsbawm (1997: 214) comes at the problem somewhat differently by speaking not of reasons but of the functional contributions agriculture can make to the well-being of modern society. The three specific contributions he identifies cluster around modern agriculture's provision of food and raw materials, labour power reserves and capital.

Our second ideal-typical scenario suggests that state elites take agricultural modernisation to be at once transformative of the *status quo* and historically progressive[3]. For agricultural change to be transformative it must seek to change the existing system of production (whether in whole or in part) fundamentally as against superficially or cosmetically. Transformative agricultural change qualifies as historically progressive in so far as it is viewed as significantly improving on the past. A broad series of transformative and historically progressive changes may come to be regarded as an agricultural 'revolution'.

Our third ideal-typical scenario addresses the sort of power resources state elites perceive to be necessary if transformative change in agriculture is to be pursued effectively. General and specific forms of infrastructural power are relevant here[4]. General infrastructural power rests on two foundations: the state's capacity to build sufficient consent among its subjects or citizens to underpin its own political authority; and, secondly, its capacity to use coercion (as required) to deal with those who stand in its way. These general consensual and coercive power resources are drawn upon in fashioning the specific infrastructural power resources states rely on in stimulating, regulating and co-ordinating economic activities as well as in planning their future development. A strategically important specific power resource where agriculture is concerned – one resting on the consensual foundations of general infrastructural power – is the state's capacity to co-opt the co-operation of civil society actors and professional and private interests in corporatist partnership-type arrangements

Our fourth ideal-typical scenario suggests that 'subordination' to the values and production logic of manufacturing industry is a major consequence for the farming population and agriculture of the state's modernising efforts. We find

---

3 This suggestion is in keeping with TIPPS' (1973: 202) still valid observation that '...there is general agreement that whatever else it may be, modernisation is a type of social change which is both transformational in its impact and progressive in its effects'.

4 MANN (1988: 5) sees the state's infrastructural power is a matter of 'the capacity of the state to actually penetrate civil society, and to implement logistically political decisions throughout the realm'.

an early expression of this subordination thesis in the 1848 remark of Marx and Engels ([1848] 1997: 7) that the countryside would increasingly be subjected 'to the rule of the towns' within the emerging order of global capitalism. Similarly Eric Hobsbawm (1997: 206) observes how, in the 1848–1875 period, 'what a growing part of agriculture all over the world had in common was subjection to the industrial world economy'. The subordination in question here became more pronounced as the farmers' assigned role as cheap food providers developed, the numerical size of the farming population contracted with labour-saving mechanisation, urban employment opportunities opened up to rural migrants and the representatives of organised farming interests became progressively more co-opted by the state.

With these four elementary ideal-typical scenarios of the strong state intent on pursuing agricultural modernisation in hand, we are now ready to consider how modernising agricultural change has actually materialised in the cases the contributors of this volume examine.

IV. The Politics of Agricultural Modernisation

Eight of the contributions to this volume look at efforts to modernise agriculture in Spain, the Netherlands, Greece, Hungary, Germany, Switzerland, Belgium and England at various periods. Four others compare developments in two or more cases in such states as Austria, Portugal, France, Liechtenstein and Italy. The chapters contributed by Fritz Georg von Gravenitz and Niek Koning are distinctive for ranging in scope beyond Europe's geographical borders.

In reviewing these contributions in light of our discussion above, we will begin by asking whether political reasons were uppermost in attempts at initiating agricultural change? The crucial importance of the state for developments in agriculture is at the heart of most of our contributions, as is the centrality of perceived economic and political crisis conditions to the desire of state elites to intervene. Niek Koning argues, in his wide-ranging overview, that the state became significantly more important for agricultural development after the industrial revolution. In the second half of the nineteenth century in particular the transport revolution helped to bring about a fundamental change in agricultural markets that saw periodic chronic scarcity giving way to recurrent oversupply. The ensuing crisis of capitalist agriculture, greatly stimulated by prolonged crisis market conditions, was eventually resolved in favour of family farming. Thus began a long period where, on account of the perceived disadvantages of family farming in the field of scientific research and formal education, 'farm progress became increasingly dependent on government support' (Koning, below).

Crisis conditions threw up a variety of challenges that state elites found difficult to ignore. The rural unrest generated by the agricultural crisis of the late nineteenth century, for instance, induced states in Central Europe not only to re-define their trading policies but, at least as importantly, to establish and support what Jonathan Harwood calls 'peasant-friendly' plant-breeding stations. State elites would also encourage the emergence of new scientifically educated professions such as the agronomists. Daniel Flückiger describes for Switzerland how agronomists influenced agricultural practices directly by taking up farming and teaching upon graduation, and indirectly by staffing the emerging agricultural associations and finding employment in large numbers in the state administration. Another set of newly created professionals became crucial to the food control regulatory arrangements appearing in parts of Spain in the three decades preceding World War I. The Spanish state legislated, according to Gloria Sanz Lafuente, but the law's implementation depended – just as in many other countries – on different arms of the state deploying an increasing number of scientifically trained professionals.

Economic as well as political crisis conditions, Anton Schuurman argues, induced the Dutch state elites to intervene extensively in the 1930s in order to re-shape the agricultural institutional matrix in ways approximating to the sort of extensive state intervention characteristic of the period of 'organised capitalism'. Ernst Langthaler describes how Austria and England alike, in spite of having highly contrasting political regimes, intervened heavily in their agricultural sectors in response to the crisis conditions presented by World War II. Contrary to the common interpretation of the crisis-ridden interwar period as one that saw 'the end of globalisation' (James, 2001), Fritz Georg von Graevenitz conceives of the emerging international sugar regime as moving from non-binding non-governmental accords to binding, governmental pacts, thus providing the foundation for the post-1945 international sugar-trading regime. Effectual internationalism proved, in this account, to be a successful strategy of complementing national economic policies under conditions of severe crisis.

A further manifestation of crisis conditions, in the form of fascism's defeat in World War II, spurred autocratic regimes in Portugal and Spain in the 1950/60s to pursue the contrasting strategies of agricultural modernisation described by Dulce Freire and Daniel Lanero. And the post-war Soviet conquest of Hungary was followed by a crisis for the new autocratic Hungarian state in relation to agriculture in the 1950/60s that ultimately was resolved, according to Zsuzsanna Varga, in ways that paved the way for the 'Hungarian agricultural miracle'.

In what sense (if any) can the contemplated change discussed by our contributors be viewed as 'revolutionary' in the modernising sense of being transformative and historically progressive? Only two of our authors use the term 'agricultural revolution' explicitly. Niek Koning refers, when using the term, to 'a period of agricultural growth and adjustment of farm techniques'. In spite of its vagueness Ernst Langthaler opts to retain the concept and suggests that agricultural change to qualify as revolutionary must be relatively rapid in its speed, fundamental in its extent and progressive in its direction. Socrates Petmezas reminds us that Greece – along other southern European societies – missed the first agricultural revolution altogether. And it becomes clear from Paul Brassley's figures for the United Kingdom that the increase in agricultural output in the two decades after World War II was 'more than twice the annual rate of increase in the classic agricultural revolution period of the eighteenth and nineteenth centuries' (Brassley, below).

Whether distinctive conceptions of transformative agricultural change emerged in the decades following World War II is a question a number of the contributors to this volume pose. Central to the post-war modernisation that appealed to the Iberian technocratic and economic elites was a desire to exploit the economic opportunities the post-war economic recovery was opening up at home and abroad. Much the same was true of Greece, where breaking the ecological and institutional deadlock Socrates Petmezas identifies depended not alone on a strong state at home but equally heavily on the capacities of the industrial states of western Europe to absorb a large part of the rural population of the South (whose foreign work experience paradoxically often commenced in the agricultural sectors of the industrialised host states).

If the post-war push for agricultural change in southern Europe is explicitly presented as 'modernising', it arguably merits the description 'revolutionary' by virtue of aspiring to be transformative, comprehensive and historically progressive. Certainly the modernisation effort in these southern European cases was motivated by a desire to boost the productivity of land and labour significantly

Others, of course, had earlier travelled along a similar road to that taken in southern Europe in the post-1945 period. For example, the version of modernisation embraced within the Dutch 'agricultural institutional matrix' from its inception in the late nineteenth century was built heavily upon a normative preference for using science, technology and improved farming practices to make agriculture progressively more efficient, productive and competitive in export markets.

Broadly speaking we find state elites, even when the circumstances they faced differed widely, converging in the view that making agriculture more competitive was historically progressive by virtue of being good for the wider society and for the farming population as well. How to proceed, of course, was another matter and here the political differences between autocratic and liberal democratic states counted for much. Among the cases considered in this book post-war Hungary is distinctive in opting for collectivised agriculture as the best means of expanding land and labour productivity.

What specific power resources were deployed by state elites in pursuit of agricultural modernisation? Where the state's more general infrastructural power resources to achieve their goals are concerned, the authors agree that much more use was made of consent than of coercion. The most obvious exception here is post-conquest Hungary where for long the state's overriding concern was to use coercion to advance a model of modernisation that entailed re-organising agriculture structurally and ideologically along Soviet-style lines with a view to rendering it better able to supply the food needs of an industrial society.

In all other cases state elites attempted to secure the consent of farmers by appealing to their individual and collective self-interest and to their sense of duty as citizen producers in industrial or industrialising societies. Typically agricultural change was presented officially as good for *all* farmers at *all* times. The broad promise was to increase the incomes and working conditions of the farming population and to emancipate the semi-subsistence peasant from what was often depicted as ignorance, drudgery and poverty. But the pretension to be modern at almost all costs was by no means unique to governing and administrative elites. Quite often, as Daniel Flückiger demonstrates for interwar Switzerland, individuals, sections or even generations of farmers shared this view to a great extent.

In the third quarter of the twentieth century, when the productivist phase reached its pinnacle in agriculture, a political consensus came to be built around the official view, heavily influenced by agri-business, that only an ever-diminishing number of farmers had any realistic chance of staying on the land in the longer term. Indeed, shrinking the agricultural labour force came to be officially accepted as a necessary condition of agricultural progress – even if this was seldom as sharply formulated as in Spain where Rafael Cavestany, Minister for Agriculture from 1951 to 1957, encapsulated his modernising project in the motto 'fewer farmers, more agriculture'. A similar determination, though usually less directly communicated, was practiced in Holland, Switzerland and other already highly industrialised societies.

In the second phase of the productivist era, when significantly post-productivist ideas were already on the rise, Flavio Ruffini and his colleagues argue that off-farm employment creation became central for rural development in the Alpine regions. A question closely related to the growing prominence of 'rural development' policies throughout Europe, but one not pursued by any of the contributions in this volume, is where the 'best' balance between agricultural decline and increasing off-farm economic activity lies? This question is of crucial relevance since it is still agriculture, through its cultivation of the soil, its use of plants and animals and its maintenance of the landscape, that provides much of the basic infrastructure for rural tourism.

We have already seen how professionalisation, and the deployment of a whole set of new professions (such as the agronomists), became a critically important power resource for state elites intent on changing agriculture *via* building consent for modernisation. Nowhere is this clearer than in the sphere of agricultural education and advice. It was through agricultural education and extension that state elites not only elaborated their conception of commercialised agriculture but provided the detailed instructions as to how the various practices of peasant farming had to be adjusted to the needs of a rapidly growing agri-business, operating on the basis of adding monetary value through the transformation of fresh food into storable commodities.

There were occasions when state functionaries became adept at deploying themselves and their professional approach skilfully and to telling effect in building consent with the farming population. Agronomists as agents of modernisation in Swiss agriculture, according to Flückiger, appealed particularly successful to young farmers. For many of them to be unambiguously 'modern' became a significant marker of identity that allowed them to distance themselves not only from their parents and their way of farming but even from their teachers who often had a background in practical farming.

How effective were the modernising efforts in agriculture? And what considerations were decisive in accounting for effectiveness and ineffectiveness? To have the consent and support of at least the main farming organisations was something that state elites typically considered to be crucial. Such consent and support had to be won, nurtured and maintained, though it is obvious that in some instances (such as in the Netherlands, England and Switzerland for much of the twentieth century) modernising state elites were simultaneously pushing an open door as well as fighting against opposition when dealing with organised farming interests. The *quid pro quo* for the representatives of farming interests intent on working closely with the state was that corporatist relations with ministries and other state agencies promised to confer on organised farmers a significant influence over agricultural policy-making. Again and again opposition has taken the form of farmer protests and demonstrations against state authorities as well as their main representatives, as Schuurman's account of the Netherlands illustrates.

If the uses of consent were many and varied, the Hungarian case shows that there were definite limits in the longer term to the use of coercion as a power resource in the pursuit of agricultural modernisation. As long as the Hungarian state depended on dispossessed peasants to work the land, the latter, as Zsuzsanna Varga shows, retained a crucial source of collective power, something that became painfully obvious to the ruling elites as production and productivity slumped on the collective farms. Eventually, in response to this, the central political authorities came to accept at least informally that major concessions to the farming population were unavoidable.

What post-war Hungary illustrates, therefore, is how state-imposed agricultural change could become popularly contested in a way that undermined effectiveness. Under very different circumstances traditional agrarian elites in post-war Portugal, intent on preserving the *status quo*, possessed sufficient power to veto much of the state's modernising efforts[5]. In effect, the Portuguese dictatorship's autonomy was compromised by its dependence on the political support of the traditional elites. That different arms of the state sometimes take different stances and even work at cross-purposes also comes out in Serge Schmitz's discussion of the fate of the Belgian dogcarts. At one point members of the Agricultural Commission were able to veto (albeit only temporarily) a Senate proposal to ban the dogcarts, thus frustrating the covert agenda of suppressing peasant agriculture by invoking a discourse centred on real or imagined animal welfare.

This brings us to the crucial question of whether 'subordination' for the farming population and agriculture was the main consequence of the modernising efforts of European states and their elites? In most of the contributions to this volume we find the desired change in agriculture being presented as good for farmers, especially by promising them easier working lives. Yet even when official modernising efforts argued explicitly in terms of improving the welfare of the farming community, political elites hardly ever lost sight of the primacy of agriculture's role in serving the economic needs of the wider industrial society. On this fundamental question there was no substantial difference between democratic and authoritarian regimes. Matters seem to be more complicated when we encounter that popular strand of political rhetoric that presents agriculture and the peasant population as not only having a special political and cultural value but as being the very backbone of the national economy. Yet when it came to a choice between a policy of modernising agriculture or keeping the peasants on the land, the same pro-modernisation choice was made regardless of the ideological orientation of the ruling elite and their allies within the farming population. It was therefore hardly surprising that the period of post-war accelerated agricultural modernisation was marked by an unprecedented 'drift from agriculture'. Such out-migration came to be considered as one of the necessary

---

5 It has been contended that these elites had little incentive to opt for modernisation as long as they benefited under 'the long-standing policy of remunerative prices and low wages imposed by the Salazar regime' (Clar, 2008: 184).

conditions for the rapid conversion of agriculture from a low into a very high labour productivity sector. The question therefore arises: were those who packed their bags and left winners or losers? Given that they had no future to speak of by remaining on the land, those who left were, in Socrates Petmezas's judgement, better off even if leaving was 'an extremely painful experience for many emigrants'.

A real impoverishment of the agricultural world occurred as the knowledge-based authority or power to specify how food was to be produced and processed for human and animal consumption was either lost or greatly diminished. In the course of the twentieth century such power of specification was lost to urban-based food safety regulators, agronomists, and the food processing and retail giants. How technological modernisation tipped the balance of knowledge-based power (*qua* authority) away from farmers and the farming population and towards non-farming technical specialists comes across keenly in Paul Brassley's account of the introduction of pig artificial insemination in post-war England where the authority of scientists came to eclipse that of the hitherto authoritative breeders and breed societies.

Can we therefore say that the family farmers who became cheap food producers by virtue of their generally weak market position became accomplices in their own subordination, incapable of taking an independent and critical stance? It would appear that for a long time in the last century organised farmers were perceived (and often saw themselves) faring relatively well as consenting partners with the state in pursuing managed modernisation under corporatist arrangements. Most European states – post-independence Ireland is a notable exception for much of the twentieth century (Moser & Varley, 2012) – did move to incorporate their agricultural sectors, but in conformity with industrial norms and aspirations[6]. Under the prevailing corporatist arrangements the boundaries between the state and organised farmers became somewhat blurred. While it is true that state agencies often acted in defence of agricultural interests, it is equally important to remember that the representatives of organised farmers more often than not assumed the role of implementing the policy of industrial states in the countryside.

Writing of very different circumstances, some of our contributors suggest that farmers (whether or not formally organised) did have a substantial capacity for collective agency that was powerful enough to sometimes win important concessions within a strengthening modernisation process. Jonathan Harwood interprets

6 In his classic account Philipp Schmitter suggests that central to the ideal-typical corporatist system of interest representation is the way certain organized interests are selectively 'recognized or licensed (if not created) by the state and granted a deliberate representational monopoly within their respective categories in exchange for observing certain controls on their selection of leaders and articulation of demands and supports' (Schmitter, 1974: 93).

the Bavarian plant-breeding stations of the 1890s, designed to meet the needs of peasant farmers, as a concession to the farmer mobilisation provoked by the central governments decision to reduce tariff protection in 1891. Another notable case is post-war Hungary where state elites ruthlessly used coercive power to push through collectivisation only to find that they were still powerless to make collectivised agriculture more productive than the agriculture practiced by family farmers.

V. Conclusions

At the beginning of his contribution Niek Koning writes that the 'relationship between agriculture and the state is as fascinating as it is complex'. This statement captures a great deal while still leaving us with the unavoidable challenge of coming to grips with the complexity of the relationship, at least in some of its many manifestations. While broadly engaging with this challenge this book concurs with the conclusions of earlier volumes in this COST series which stress the specific importance of states, and their civil society allies, to the process of agricultural modernisation over recent centuries (Vivier 2008; Pinilla 2009; Landsteiner & Langthaler, 2010). To say this does not preclude accepting that technological knowledge, commercial incentives and favourable institutional conditions (Olsson & Svensson, 2011) were also necessary to the transformation of European peasantries into a diverse class of commercial farmers who, simultaneously, often regarded themselves and behaved like peasant-minded individuals. But this still incomplete transformation would have been substantially different were it not for the world-changing possibilities opened up, particularly after the Second World War, by access to the vast mineral resources of the lithosphere. This important facet of the history of agricultural modernisation has not been treated by any of the contributors to this volume. It therefore deserves more attention from historians and social scientists in the near future.

The gradual, but eventually substantial, enlargement of the mineral resource base of twentieth century European agriculture is the starting point for our 'integration through subordination' thesis. It provides the crucial material dimension of the context in which states and their civil society allies have sought to integrate their agricultural sectors into the industrial world. Industrial states intervened in agriculture not only to head off crisis conditions but, just as importantly, to define the substance and direction of what was to pass for 'progress' under modern conditions. Herein lies the nub of the 'integration through subordination' thesis. It does not mean that farming was a victim of industrialisation, but rather that the normative ideal of industrial production and its associated practices became centred on transforming agriculture in industrial society into something that would fundamentally rely on the same resource base – and therefore function in the same way and share the same values – as manufacturing industry since the time of the thermo-industrial revolution. Integrating industrialising agriculture *via* subordination therefore meant displacing peasant farming, not agriculture as such. In this process the farming population lost much of its capacity to represent itself as predicted by Karl Marx. This loss, however, was not primarily because of its own inadequacies but rather because the experience and language of peasant agriculture was neither appreciated nor understood by those powerful elites who looked to industrial modernity for their normative standards (Scott, 1998).

The impressive success of the dominant model of agricultural modernisation is perhaps best illustrated by the unprecedented increase in food production as well as in the spectacular productivity gains achieved mainly in the third quarter of the twentieth century. On the debit side of the equation we see the huge loss of social, animal and plant biodiversity in the countryside as well as the dwindling knowledge in the farming community of how biotic resources can be used in a sustainable way. Modernised agriculture, in other words, presents us with something rather different than simply industrialised agriculture. This conclusion may be disappointing to the technocratic worldview intent on creating a modern agriculture in the image of manufacturing industry, but it gives the historian and social scientist much to ponder and explore further.

Bibliography

Aldenhoff-Hübinger, Rita (2002), *Agrarpolitik und Protektionismus. Deutschland und Frankreich im Vergleich 1879-1914*, Göttingen, Vandenhoeck & Ruprecht.

Auderset, Juri, Bächi, Beat & Moser, Peter (2012), 'Die agrarisch-industrielle Wissensgesellschaft. Akteure, Diskurse, Praktiken', in Beat Brodbeck, Martina Ineichen & Thomas Schibli (eds), *Geschichte im virtuellen Archiv. Das Archiv für Agrargeschichte als Zentrum der Geschichtsschreibung zur ländlichen Gesellschaft*, Studien und Quellen zur Agrargeschichte, Band III, Baden, Hier und Jetzt, p. 21-38.

Bairoch, Paul (1989), 'Les Trois révolutions agricoles du monde développé. Rendements et productivité de 1800 à 1985', *Annales. Économies, Sociétés, Civilisations*, 44, p. 317-353.

Baumann, Werner (1993), *Bauernstand und Bürgerblock. Ernst Laur und der Schweizerische Bauernverband 1897-1918*, Zürich, Orell Füssli.

Behrisch, Lars (2008), 'Agrarian Statistics in Late *Ancien Régime* France and Germany', in N. Vivier (ed.), *The State and Rural Societies: Policy and Education in Europe 1750-2000*, Turnhout, Brepols, Rural History in Europe, 4, p. 35-55.

Boserup, Ester (1965), *The Conditions of Agricultural Growth: the Economics of Agrarian Change under Population Pressure*, London, George Allen & Unwin.

BRASSLEY, Paul (2008), 'Agricultural Education, training and Advice in the UK', in N. VIVIER (ed.), *The State and Rural Societies: Policy and Education in Europe 1750-2000*, Turnhout, Brepols, Rural History in Europe, 4, p. 259-278.

BURGER, Thomas (1987), *Max Weber's Theory of Concept Formation: History, Laws and Ideal Types,* Durham, Duke University Press.

CLAR, Ernesto (2008), 'Farm Policy under the Salazar and Franco Dictatorships in Portugal and Spain: Towards an Authoritarian Model of Intervention in Agriculture?', in Nadine VIVIER (ed.), *The State and Rural Societies: Policy and Education in Europe 1750-2000*, Turnhout, Brepols, Rural History in Europe, 4, p. 177-194.

CLAR, Ernesto & PINILLA, Vicente (2009), 'The Contribution of Agriculture to Spanish Economic Development, 1870–1973', in Pedro LAINS & Vicente PINILLA (eds), *Agriculture and Economic Development in Europe since 1870*, London, Routledge, p. 311-332.

COLLINS, Edward J. T. (ed.) (2000), *The Agrarian History of England and Wales, Vol. VII 1850-1914*, Cambridge, Cambridge University Press.

CROTTY, Raymond D. (2001), *When Histories Collide: the Development and Impact of Individualistic Capitalism*, Walnut Creek, CA, AltaMira Press.

DAVID, Thomas, MACH, André & STRAUMANN, Tobias (2008), 'The Questioning of Selective Protectionism in Switzerland. Cartel Law Reform and Corporate Governance Changes in the 1990s', in Margrit MÜLLER *et al.* (eds), *Pathbreakers: Small European Countries Responding to Globalisation and Deglobalisation*, Bern, Peter Lang, p. 449-476.

FEDERICO, Giovanni (2005), *Feeding the World: an Economic History of Agriculture, 1800-2000,* Princeton, Princeton University Press.

FEDERICO, Giovanni (2012), 'Natura non fecit saltus: the 1930s as the Discontinuity in the History of European Agriculture', in Paul BRASSLEY, Yves SEGERS & Leen VAN MOLLE (eds), *War, Agriculture, and Food: Rural Europe from the 1930s to the 1950s,* New York/ London, Routledge, p. 15-32.

FITZGERALD, Deborah (2003), *Every Farm a Factory: the Industrial Ideal in American Agriculture*, New Haven/London, Yale University Press.

GALLEGO MARTINEZ, Domingo & PINILLA, Vicente (1996), 'Del librecambio matizado al proteccionismo selectivo: el comercio exterior de productos agrarios y alimentos en Espana entre 1849 y 1935', in *Revista de Historia Economica*, XIV, 1996, p. 371-420.

GEORGESCU-ROEGEN, Nicholas (1960), 'Economic Theory and Agrarian Economics', in Nicholas GEORGESCU–ROEGEN (1976), *Energy and Economic Myths: Institutional and Analytical Economic Essays,* New York et al., Pergamon Press, p. 103-145.

GYSEL, Zacharias (1854), *Der Schaffhauser Bauer, wie er sein sollte, und wie er nicht ist, wie er ist, und wie er nicht sein sollte. Eine Geschichte aus dem Leben, den Bauern des Kantons Schaffhausen zur Beherzigung und Kurzeweil erzählt von einem Klettgäuer*, Schaffhausen, Brodtmann.

HALBEISEN, Patrick, MÜLLER, Margrit & VEYRASSAT, Béatrice (eds) (2012), *Wirtschaftsgeschichte der Schweiz im 20. Jahrhundert*, Basel, Schwabe.

HOBSBAWM, Eric [1962] (1977), *The Age of Revolution 1789-1848,* London, Weidenfeld & Nicolson.

HOBSBAWM, Eric [1975] (1997), *The Age of Capital 1848-1875,* London, Weidenfeld & Nicolson.

HOFREITHER, Markus (2009), 'Origins and Development of the Common Agricultural Policy', in Michael GEHLER, (ed.), *Vom gemeinsamen Markt zur europäischen Unionsbildung. 50 Jahre Römische Verträge 1957-2007,* 5, Wien, Köln, Weimar, Böhlau, p. 333-348.

ILBERY, Brian & BOWLER, Ian (1998), 'From Agricultural Productivism to Post-productivism', in Brian ILBERY (ed.), *The Geography of Rural Change*, London, Longman, p. 57-84.

JAMES, Harold (2001), *The End of Globalization: Lessons from the Great Depression*, Cambridge, Mass., Harvard University Press.

KLEINEWEFERS, Henner (1972), 'Wirtschaftspolitische Konzeption und Umweltproblematik: das Beispiel der Agrarpolitik', *Schweizerische Zeitschrift für Volkswirtschaft und Statistik*, 108/3, p. 283-328.

KONING, Niek (1994), *The Failure of Agrarian Capitalism: Agrarian Politics in the United Kingdom, Germany, the Netherlands and the USA 1946-1919,* London, Routledge.

KOSELLECK, Reinhard (1979), '"Erfahrungsraum" und "Erwartungshorizont" – zwei historische Kategorien', in Ders., *Vergangene Zukunft. Zur Semantik geschichtlicher Zeiten*, Frankfurt/Main, p. 349-375.

KRAUSMANN, Fridolin (2006), 'Vom Kreislauf zum Durchfluss. Österreichs Agrarmodernisierung als sozialökologischer Transformationsprozess', in Andreas DIX & Ernst LANGTHALER (eds), *Grüne Revolutionen: Agrarsysteme und Umwelt im 19. und 20. Jahrhundert* (Jahrbuch für Geschichte des ländlichen Raumes 3), Innsbruck, Studien-Verlag, p. 17-45.

LANDSTEINER, Erich & LANGTHALER, Ernst (eds.) (2010), *Agrosystems and Labour Relations in European Rural Societies*, Turnhout, Brepols, Rural History in Europe, 3.

LAUR, Ernst (1907), *Landwirtschaftliche Buchhaltung für bäuerliche Verhältnisse*, Aarau, Emil Wirz.

LAUR, Ernst (1911), *Grundlagen und Methoden der Bewertung, Buchhaltung und Kalkulation in der Landwirtschaft*, Berlin, Paul Parey.

MANN, Michael (1988), *States, War, and Capitalism: Studies in Political Sociology*, Oxford, Blackwell.

MARX, Karl & ENGELS, Friedrich [1848] (1997), *The Communist Manifesto,* Oxford, Oxford University Press.

MOKYR, Joel (2002), *The Gifts of Athena: Historical Origins of the Knowledge Economy*, Princeton, Princeton University Press.

MOSER, Peter (1994), *Der Stand der Bauern*, Frauenfeld, Huber.

MOSER, Peter (2012), 'Die Agrarproduktion: Ernährungssicherung als Service public', in Patrick HALBEISEN, Margrit MÜLLER & Béatrice VEYRASSAT (eds), *Wirtschaftsgeschichte der Schweiz im 20. Jahrhundert*, Basel, Schwabe, p. 568-631.

MOSER, Peter & VARLEY, Tony (2012), 'Corporatism, Agricultural Modernization, and War in Ireland and Switzerland, 1935-1955', in Paul BRASSLEY, Yves SEGERS & Leen VAN MOLLE (eds), *War, Agriculture, and Food: Rural Europe from the 1930s to the 1950s*, New York/London, Routledge, p. 137-155.

OLSSON, Mats & SVENSSON, Patrick (eds) (2011), *Growth and Stagnation in European Historical Agriculture*, Turnhout, Brepols, Rural History in Europe, 6.

PATEL, Kiran Klaus (2009), *Fertile Ground for Europe? The History of European Integration and the Common Agricultural Policy since 1945*, Baden-Baden, Nomos.

PFISTER, Christian (1995a), *Im Strom der Modernisierung. Bevölkerung, Wirtschaft und Umwelt im Kanton Bern 1700-1914*, Bern, Haupt.

PFISTER, Christian (1995b), *Das 1950er Syndrom. Der Weg in die Konsumgesellschaft*, Bern, Haupt.

PINILLA, Vicente (ed.) (2009), *Markets and Agricultural Change in Europe from the 13th to the 20th Century*, Turnhout, Brepols, Rural History in Europe, 2.

SCHMITTER, PHILIPPE C. (1974) 'Still the Century of Corporatism?', *Review of Politics*, 36, 1, p. 85-131.

SCOTT, James (1998), *Seeing Like a State: How Certain Schemes to Improve the Human Condition have Failed*, New Haven/London, Yale University Press.

SIEFERLE, Rolf Peter *et al.* (2006), *Das Ende der Fläche. Zum gesellschaftlichen Stoffwechsel der Industrialisierung*, Köln, Böhlau.

SOMBART, Werner (1902), *Der moderne Kapitalismus*, Leipzig, Duncker und Humblot.

SPOERER, Mark. (2010), '"Fortress Europe" in Long-term Perspective: Agricultural Protection in the European Community, 1957-2003', *Journal of European Integration History*, 16, 2, p. 143-162.

STUBER, Martin *et al.* (eds) (2009), *Kartoffeln, Klee und kluge Köpfe. Die Ökonomische und Gemeinnützige Gesellschaft des Kantons Bern OGG (1759-2009)*, Bern, Haupt.

TAVERNIER, Yves (ed.) (1972), *L'Univers politique des paysans dans la France contemporaine*, Paris, Presses de la fondation nationale des sciences politiques.

THIRSK, Joan (ed.) (1987), *England's Agricultural Regions and Agrarian History, 1500-1750*, London, Macmillan.

TIPPS, Dean C. (1973), 'Modernization Theory and the Comparative Study of Societies: a Critical Perspective', *Comparative Studies in Society and History*, 15/2, p. 199-226.

UEKÖTTER, Frank (2010), *Die Wahrheit ist auf dem Feld: eine Wissensgeschichte der deutschen Landwirtschaft*, Göttingen, Vandenhoeck & Ruprecht.

VAN MOLLE, Leen (2008), 'A State for the Peasants or the Peasants for the State? The Two Faces of Belgian Agricultural Policy, 1830-1914', in Nadine VIVIER (ed.) (2008), *The State and Rural Societies: Policy and Education in Europe 1750-2000*, Turnhout, Brepols, Rural History in Europe, 4, p. 159-176.

VAN DER PLOEG, Jan Douwe (2003), *The Virtual Farmer: Past, Present, and Future of the Dutch Peasantry*, Assen, Royal van Gorcum.

VIVIER, Nadine (ed.) (2008), *The State and Rural Societies: Policy and Education in Europe 1750-2000*, Turnhout, Brepols, Rural History in Europe, 4.

VOGT, Gunter (2000), *Entstehung und Entwicklung des ökologischen Landbaus im deutschsprachigen Raum*, Bad Dürkheim, Stiftung Ökologie und Landbau.

WEBER, Eugen Joseph (1976), *Peasants into Frenchmen: the Modernization of Rural France 1870-1914*, Stanford, Stanford University Press.

WRIGLEY, Edward Anthony (2010), *Energy and the English Industrial Revolution*, Cambridge, Cambridge University Press.

2. The evolution of farm policies: a long-term global perspective

Niek KONING

I. Introduction

The relationship between agriculture and the state is as fascinating as it is complex. Scientific interpretations are inevitably related to political struggles centred on land reform, free trade or protection, sustainability and so on and so forth. The picture is complicated by the enormous variability in situations. Past priorities may not apply today, nor may present priorities be appropriate tomorrow. Besides, there are strong differences between regions. The issues in Europe are not the same as those in Africa or in Latin America.

This chapter presents a nutshell interpretation of the evolution of farm policies through different periods and in different parts of the world. I start in the pre-industrial era, when population growth still caused agricultural prices to rise. This effect could provide sufficient incentives for farm progress through private initiative – until the slowly expanding capabilities of a society were exhausted and continued population growth led to a Malthusian crisis (Section 2). I then discuss how, in the latter part of the nineteenth century, the Industrial Revolution brought about a regime change in international agricultural markets as chronic scarcity gave way to recurrent oversupply. Under this new condition, farm progress became increasingly dependent on government support, especially of family farms (Section 3). I survey the policy responses to this change in different regions and the consequences this had for development (Sections 3 and 4). I then focus on the multilateral regulation of international agricultural markets (Section 5). This started in the 1930s-1940s as an attempt at regulation through managed trade. However, this approach was thwarted because the US and the EU acted as free riders. The reforms that started with the Uruguay Round of GATT negotiations (1986-1993) have allowed both blocks to whitewash this behaviour. This excursion is followed by a broader discussion of recent farm policy reforms (Section 6). Next, the attention is shifted to the future. I discuss the possibility that increased competition between food, feed and fuel might cause a return to scarcity in international agricultural markets (Section 7). I conclude with a brief section on policies that would make possible global food security and sustainable pro-poor growth.

II. Farm progress in the pre-industrial era

Through the ages, agriculture's importance for welfare, food security and the treasury has provoked government intervention. However, the nature of this intervention has undergone profound change and may change again in the future. In pre-industrial agrarian societies, in Europe and elsewhere, rulers taxed rather than supported their farmers. Sometimes they initiated tenure reforms, and sometimes large infrastructural, reclamation or irrigation projects. But for the rest they did little to encourage farm progress. The 'agricultural revolutions'[1] that these societies underwent – from slash-and-burn systems to sophisticated rotations with zero fallow – were mainly the result of private activities by farmers and landowners.

These agricultural revolutions had various drivers and conditions – population growth, environmental change, the previous exploration of technical options – and involved changes in social structures that stimulated or hampered further development. However, an important condition that allowed agricultural revolutions in pre-industrial market economies to occur was Malthus' law that, given the dearth of fertilizer and the costliness of transport, population growth raised agricultural prices and cheapened farm labour. This stimulated farmers to adjust their techniques and intensify their production (Boserup, 1965). It created rural markets for non-farm activities and enhanced market exchange, all of which accelerated the diffusion of innovations (Boserup, 1981). Larger-than-family farms played an important role in this process. Economies of scale were limited and often counterbalanced by advantages of family farms in labour relations (Allen, 1988). However, advantages in information, risk tolerance and market transactions favoured larger farms in the introduction of innovations. Indeed, innovations were normally first adopted by larger farmers (Grigg, 1982: 158) – even in those cases where they were soon taken up by small farmers (see Thoen, 1990 for medieval Flanders). As a consequence, larger farmers usually had a technical lead. Rising output prices enhanced this lead by stimulating investment while falling wages made it difficult for small farms to compensate for this by increased self-exploitation.

In this way, agricultural revolutions were possible even when government policies were not especially 'enabling' (see Hoffman, 1996 for early modern France). To be sure, because knowledge infrastructure was little developed and high risks enhanced the weight of tradition, collective learning was a slow process. Sooner or later, an agricultural revolution exhausted the cognitive and institutional capabilities of a

[1] The term 'agricultural revolution' has often been used in a more specific sense, *e.g.* to refer to the beginnings of farming, or the farm innovations in eighteenth century Britain. In this chapter, it is used as a broader concept that refers to a period of agricultural growth and adjustment of farm techniques.

society. Then food prices skyrocketed, squeezing the demand for non-farm products, making unemployed artisans swell the ranks of small farmers, and inducing the latter to over-exploit their plots in an effort to minimize their dependence on food markets (Meuvret, 1946). Society was pushed into a Malthusian spiral of soil degradation, food insecurity and disruption that ended in demographic stagnation or collapse. This drove food prices down and raised wages, temporarily reversing the process of intensification – until a new demographic upswing initiated a new cycle (also Wood, 1998).

III. Global industrialization and the shift to supportive farm policies

The Industrial Revolution did not immediately break this Malthusian regime. As earlier commercial revolutions had done, its textile factories and horse-drawn transport stimulated the demand of farm products rather than their supply, reinforcing the effect of population growth on agricultural markets. Between 1775 and 1875, agricultural prices remained high and, outside Britain and North America, farm wages low. Larger farms kept the lead in farm progress, and innovations remained a product of practical entrepreneurship in and around farming. In this period, agriculture became the focus of the ascending economic liberal model. Common lands were divided, property rights commercialized, government bodies privatized and tollhouses abolished. The repeal of the British Corn Laws (1846) heralded a worldwide liberalization of agricultural trade. In Britain, it stabilized agricultural prices without entailing a price decline. It did not hamper the Victorian 'high farming' revolution – the beginning of external input agriculture but still in a framework of large farms (Koning, 1994 and referenced literature).

All this changed when a new phase of the Industrial Revolution broke the Malthusian connection between population and prices (Schultz, 1945). From the latter part of the nineteenth century, cheap transport and industrial fertilizer boosted global food supply. Electricity, industrial chemistry, and internal combustion caused minerals to replace farm-produced materials. And in several countries, industrial competition in agricultural labour markets increased sharply. This led to falls in agricultural prices and rises in farm wages that caused a squeeze on farm profits (see Figures 2.1 and 2.2). The new developments did not boost the economies of size in agriculture, so that there was no corrective shake-out of small farms. While low profits eroded the technical lead of the large farms, rising wages reinforced the advantage that small farms derived from cheap family labour. As a consequence, large farms declined and the share of family farms in agricultural production increased[2]. Objective barriers

[2] Koning (1994) and referenced literature. See also case studies like Winstanley (1996) for Lancashire, Garrabou *et al*. (2001) for Catalonia, or Gastón Aguas and Lana Berasain (2007) for Navarra.

like 'efficiency wages' in other sectors and social-psychological adaptation to these barriers limited the mobility of family farmers (Akerlof, Yellen, 1986; Haagsma, Koning, 2002). Rather than leaving a sector with low earnings, they tightened their belts and seized upon the new technical and market opportunities to defend their incomes by raising production – initiating a treadmill by which their incomes were permanently depressed (Cochrane, 1959). In a free market, a balance between the growth in supply and that in demand was often only achieved when this treadmill squeezed its own fuel supply by reducing farm profits and thereby investment. It led to crowding and slowdown of farm progress, rather than to the efficient equilibrium of the standard economic model.

These developments induced profound policy changes in western countries. Liberal land reform gave way to redistributive measures that streamlined the shift from large farms to family farms. Government-sponsored experimental stations and farm schools, and extension agents took over the role of large farmers and landowners as initiators of innovation. One government after the other intervened to moderate the fall in agricultural prices. In the twentieth century, a new agricultural revolution followed that was based on high-yielding seeds, agri-chemicals and mechanization. Family farms rather than large farms were at the core of this revolution. The evolution of techniques and institutions adapted itself to family farms. Agricultural machinery was down-scaled[3] scale-sensitive activities were separated from primary agriculture or outsourced to contractors (Olmstead, 1975); and agricultural co-operatives emerged as an interface that reduced transaction costs in the relations between family farmers and those who bought their produce and sold them inputs (Farr, 2007; Garrido, 2007). The new model of agricultural development was only possible because government intervention overcame the risk-aversion of family farmers and their disadvantages in fields like information and consolidation, and mitigated the profit squeeze so that the frugality of family farmers left some margins for investment.

[3] See examples in Koning (1994: 27, 193).

Figure 2.1. Real wheat prices (5-year moving average) and farm wages, England and Wales

Source. Own calculations based on data in MITCHELL (1975: 191-195, 736; 1990: 737-741, 756-757).

Figure 2.2. Real wheat prices (5-year moving average) and farm wages, United States

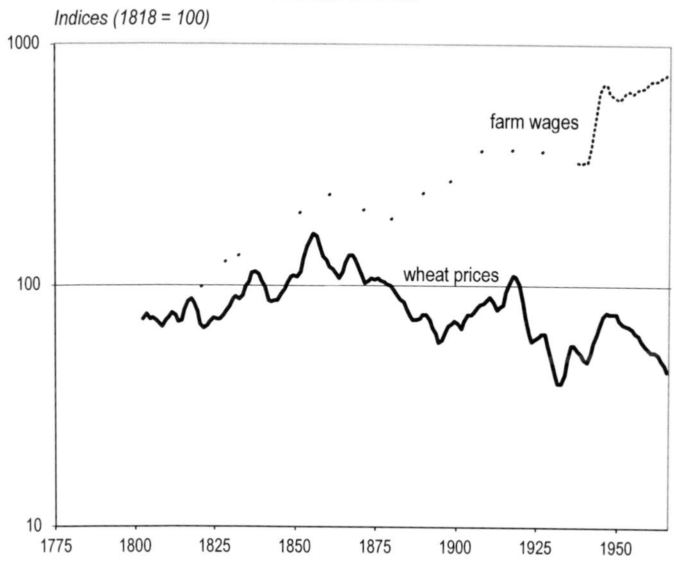

Source. Own calculations based on data in MITCHELL (1993: 129-130; 696-698) and US Bureau of the Census (1976: 207-209).

The introduction of supportive farm policies followed different pathways in various countries. In the late nineteenth century, most West European countries resorted to protection to shield their farmers from the fall in international agricultural prices. Other western countries still maintained free trade in farm products in this period. Most of them had a special advantage in agriculture. Rather than resorting to protection, they introduced policies to accelerate innovations. They weathered the price fall, and when international prices recovered after 1900, dynamic agricultural development resumed. A second fall of international agricultural prices around 1930 made all western countries resort to protection. By then, all of them had also engaged in government support of farm research, education and infrastructural programs, so that in these respects there was a policy convergence. After World War II, government support of agriculture further developed, aided by the general shift to social-Keynesian policies with which western countries responded to the traumatic experience of the 1930s and the communist challenge (Fahey, 2002; Koning, 1991). Denmark and the United States briefly tried to return to free market policies in the 1950s, but these experiments were short-lived and the outcomes disappointing (Koning, 1986 and referenced literature). After 1984, New Zealand abandoned protection. Its production costs in dairy farming are only half those in other prominent dairy countries like the US, Denmark or the Netherlands. Moreover, the country benefited from the simultaneous introduction of a milk quota system in the EU, which reduced the latter's share in the world dairy market from one-half to one-third.

Recent farm policy discussions have generated a new interest in the history of agricultural protection. Tracy (1989) and, more cautiously, Federico (2005a) have suggested that agricultural protection had political rather than economic causes. Conversely, O'Rourke (1997) shows that the policy choices that European countries made in the late nineteenth century were related to differential impacts of falling prices on their farm profits and real wages. Tracy was convinced that many more European countries could have adjusted their agriculture in a free market in the late nineteenth century by shifting to livestock products, which were more price-elastic than staples like grains. In reality, however, international markets for livestock products were soon oversupplied by a few well-situated countries like Denmark and the Netherlands (Bairoch, 1976).

Several authors have highlighted negative effects of agricultural protection in Germany (Gerschenkron, 1966; Puhle, 1986; Tracy, 1989). Indeed, in a static analysis, agricultural protection caused deadweight losses and reduced the buying power of non-farm groups. But a dynamic approach shows up matters in a different light. Productivity growth in German agriculture was among the highest in Europe (Figure 2.3). Agricultural protection probably accelerated overall growth, allocational distortions being offset by an increase in effective demand (Bairoch, 1976)[4]. In addition, this policy most likely raised real wages as agriculture was relatively labour-intensive and had a large share of total employment[5]. Overall, the domestic effects of agricultural protection seem to have been quite favourable (see also Webb, 1978). To be sure, protection alone did not enable farm progress. In France and Italy, where tenure relations gave little security to small farmers, and agricultural research and education lagged far behind those in Germany, productivity growth in farming was sluggish in spite of protection (Figure 2.3)[6].

Figure 2.3. The growth of agricultural productivity per head and per hectare in eight countries of Western Europe, 1870-1910

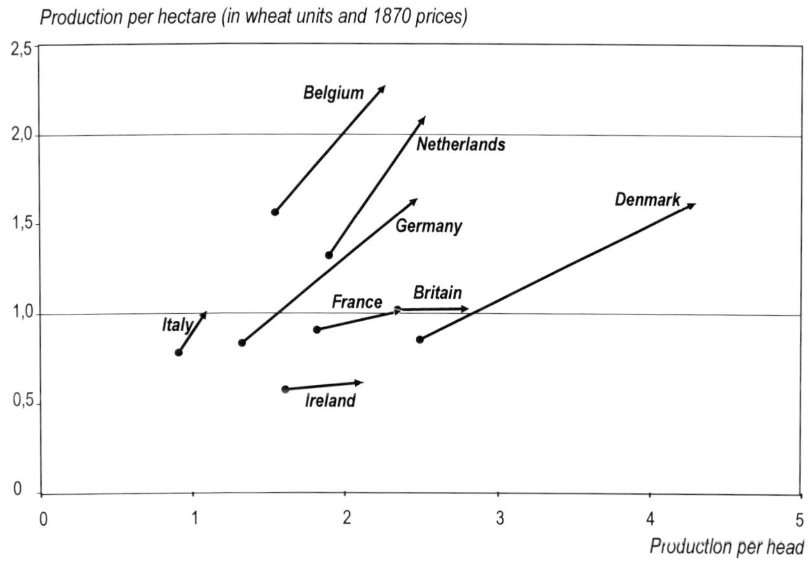

Source. VAN ZANDEN (1991).

[4] See also evidence in O'ROURKE (2000), based on a sample of 10 countries, which suggests that protection had a positive effect on economic growth. (O'Rourke finds Bairoch's explanation implausible, surmising instead that protection may have facilitated the reallocation of labour in some countries where industry was more strongly protected than agriculture, but this is not demonstrated by his analysis.)

[5] See also the quantitative analysis in O' ROURKE (1997) indicating that in France agricultural protection may have raised real wages by 3.5 to 4.5 per cent. In any case, the negative effect on the costs of living of working-class households has been strongly overrated. Something similar is true for the supposed harm to small livestock farmers. Livestock production was not less protected than grain and grain tariffs did not drive up feed costs (KONING, 1994 and referenced literature).

[6] Nevertheless, the protection of agriculture seems to have sustained the demand for industrial products (BAIROCH, 1976). This was important, because French and Italian industries were more dependent on the domestic market than their British counterparts, which, as the 'first workshop of the world', had gained an established position in the world market.

The free market adjustment thesis is also contradicted by the British experience. At the onset of the price fall around 1880, Britain possessed the most technically advanced agriculture in the world. However, industrial competition for labour had raised farm wages and the country no longer had a comparative advantage in agriculture. Nevertheless, agricultural free trade was maintained until 1930. According to standard economic theory, free market adjustment might have involved a strong reduction, or even total elimination of agriculture, but if a farm sector managed to survive to some extent it would see a recovery of profits and productivity growth. In reality, farm profits remained low and productivity stagnated throughout this period. This was not due to a technological ceiling, but to widespread disinvestment. As a consequence, British agriculture fell far behind the European productivity frontier (Koning, 1994 and referenced literature; see also Figure 2.3)[7].

The last case for the free market adjustment thesis is the adjustment of agriculture in New Zealand after it abandoned protection in 1984. This has been hailed as a success because it was followed by an increase in productivity growth (Sandrey, Reynolds, 1990). However, this increase was limited to horticulture and may have been due to pre-liberalization investments (Philpott, 1994). In the livestock sector, productivity growth remained unaltered in spite of the massive release of marginal resources (Philpott, 1994; Lawrence, Diewert, 1999).

The regime change in agricultural markets and the turn to government intervention was the theme of my book *The Failure of Agrarian Capitalism* (Koning, 1994). My idea was that the evolution of western farm policies could be analyzed by combining, on the one hand, a reconstruction of modern capitalism as a sequence of several politico-economic configurations and, on the other hand, a distinction between two main phases of the industrial revolution that impacted differently on the agricultural sector. I remain convinced that this provides a useful framework for understanding the complex evolution of agriculture and agricultural policy (also Garrabou, 1997; Moser, 2000; Baumann, Moser, 1999). However, the approach should be further elaborated. On the one hand, the evolution in the first phase of modern industrialization should be recognized as the last instance of the more general pre-industrial dynamics that I have dealt with in the preceding section. On the other hand, the analysis of the regime change in agricultural markets should be broadened to other regions of the world,

---

7 Both TRACY (1989) and FEDERICO (2005a) are silent about this episode, which is nevertheless an anomaly in their interpretation. Federico also denies the existence of an agricultural oversupply problem during the second wave of agricultural protection around 1930 (FEDERICO, 2005b). However, his methodology only shows that markets still balanced supply and demand at some price. The problem, of course, is that this price was too low to prevent a serious increase in the gap between farm and non-farm earnings.

which were confronted with it but responded in different ways. This is the subject of the next section.

IV. Diverging responses in the developing world

Even if western farm policies had an economic rationale, economic-liberal authors are right when they state that political causes played a critical role. In fact, the latter go much deeper than the political market mechanisms (low organization costs of landed minorities) on which they focus. European societies had gone through a millennia-old history of agricultural intensification, socio-economic differentiation and state formation. Their institutions had travelled with European emigrants to new settlement areas. Where temperate conditions precluded tropical export crops inducing a self-centred development, they underwent a parallel evolution as in old Europe. The ensuing configuration engendered a class-oriented articulation of rural interests, political classes with an eye for longer-term national interests and middle classes whose interests could not be ignored. When the regime change in agricultural markets undermined the traditional mode of agricultural development, it was these conditions that caused political markets to generate pressures for supportive farm policies.

The importance of these conditions becomes clearer when one compares the evolution of farm policies in western countries with what happened in other parts of the world. Like Western Europe, many East- and South Asian countries had been old core areas of social development. When the situation in agricultural markets changed, a comparable mix of rural unrest, middle-class interests and *raison d'état* led to 'developmental states' (Öniş, 1991), which also introduced supportive farm policies. In the Japanese Empire before World War II, agricultural protection and public investment in irrigation, research and infrastructure entailed important productivity increases in agriculture. Rural incomes contributed to industrialization as a demand factor (Francks *et al.*, 1999; Ohkawa, Rosovsky, 1961; Ohkawa, Shinohara, 1979). This pattern, which was also aided by land reform, continued in Japan, South Korea and Taiwan in the post-war decades. In Asian colonies of European countries, farmers were not protected. Here one could observe phenomena of stagnation and 'involution' that reminds one of pre-industrial Malthusian crises (Myrdal, 1968; Geertz, 1963). However, while the latter had occurred because an agricultural revolution had been exhausted, here an agricultural revolution was nipped in the bud. After independence, several governments introduced supportive farm policies (see *e.g.* overview in Dorward *et al.*, 2004: 88-89). Together with the high-yielding varieties from international research, these led to the Green Revolution that became an engine of industrialization[8].

Like agricultural protection in Germany before World War I, that in East Asia has been blamed for freezing farming structures, for retarding growth and for harming poor consumers (Anderson *et al.*, 1986; Beghin *et al.*, 2003; Diao *et al.*, 2002). These contentions are backed by standard equilibrium models, but the 'welfare losses' indicated by such models say nothing about how farm productivity, poverty or GDP would have evolved over time had farmers not been protected. The fact is that, in countries like South Korea and Taiwan, the increase in agricultural protection after 1970 was followed by new increases in farm output and incomes. This may well have caused the continuation of agriculture's contribution to the domestic demand pull for industrial growth, even if the relative importance of this contribution declined (Francks *et al.*, 1999; Park & Johnston, 1995).

In Latin America after 1492, European markets for tropical crops induced the rise of plantations that used coercive labour systems to prevent workers from setting themselves up as independent peasants. It created a social divide between planter elites and rightless workers, whose low living standards hampered the development of consumer goods industries and reinforced the export dependence of the plantations. When international agricultural prices declined, this 'disarticulated' structure (De Janvry, 1981) made the agrarian elites stick to open trade policies to secure their exports and use their socio-political dominance to shift the burden to the rural poor. In the end, they evicted many workers to pave the way for cost-cutting mechanization. It allowed a development of a kind, but the ensuing growth was limited by low land productivity, by social tensions that raised transaction costs and by poverty-constrained domestic markets (Alesina, Rodrik, 1994; Johnston, Kilby, 1975). Land reform and the conflicting trade policy interests of large and small farmers emerged as vital issues in this setting.

In Sub-Saharan Africa, iron-armed warriors and endemic diseases from which Europeans had no immunity postponed the colonial scramble until the late nineteenth century, so that it coincided with the decline in international agricultural prices. It limited the establishment of European-owned farms and plantations and caused a decline of larger indigenous farms. It made African farming even more a smallholder agriculture than it already was (Bundy, 1972; Huijzendveld, 1997; Munro, 1976). As in Asia, colonial governments failed to protect indigenous farmers. Relative abundance of land for some time provided an outlet for population growth, but this

[8] In China and Vietnam, accelerated agricultural growth became possible through the relaxation of policies that taxed agriculture for the sake of forced industrialization.

safety valve was gradually closed. Higher post-war prices induced new investment by smallholder farmers (Munro, 1976). In the 1960s, per capita incomes in Sub-Saharan Africa were higher than in Southern Asia, but continued population growth and new declines in the agricultural terms of trade led to a return of the vicious cycle of poverty and soil degradation, especially after 1980 (Koning, Smaling, 2005 and referenced literature). Unlike in Asia, national independence brought no turn to more supportive farm policies. Sub-Saharan Africa's hand cultivator societies were less differentiated, had property rights in people rather than material assets, and had more fluid and personalist socio-political relations (Bayart, 1989; Goody, 1976). This was not conducive to Asian-type developmental states. Interests organized themselves in clientelist factions rather than in class-based structures. Politicians saw themselves obliged to remunerate large numbers of supporters with public sector jobs, while farmers were too weakly organized to prevent them from footing the bill (Bates, 1981; Djurfeldt *et al.*, 2005). Rural malaise caused a flight out of agriculture. However, manufacturing and modern services were also sluggish. As a consequence, the rural exodus led to a proliferation of marginal activities and increased jostling for public sector jobs. The whole situation exacerbated the clientelism and internal aggression at all levels of African societies. It fuelled redistributive conflicts between factions that eroded social capital and the quality of governance (Ikelegbe, 2001; Patterson, 1998) and could all too easily end in violent conflict (Peters, 2006; Woods, 2003).

The stagnation in Africa and the successes in Asia have rekindled the debate on the relation between farm policies and development. Half a century ago, Johnston and Mellor (1961) pointed to the importance of agriculture as a starter engine of modern economic growth. In their view, agricultural growth was needed to provide opportunities for upstream and downstream activities as well as the savings, labour, and wage goods that successful industrialization required. Subsequent research has highlighted the importance of agricultural growth for avoiding pronounced income inequality and for stimulating the domestic demand for the products of emerging non-farm sectors (Adelman *et al.*, 1989; Delgado *et al.*, 1998; Hazell, Roell, 1983). The ascent of the liberal globalization agenda after 1980, however, made popular the idea that this home market effect was becoming less important because globalization would allow export demand to substitute for domestic demand. This was not borne out by experience. Apparently, new industries needed the domestic market as a training school before they were able to compete internationally. Besides, it has been suggested that agricultural growth has vital external effects on social capital and skills (Koning, 2002; Timmer, 1995). In any case, there is today a renewed recognition of the importance of supportive farm policies for pro-poor growth (World Bank, 2007). History suggests that such policies should also include protection in many cases.

V. Multilateral regulation of agricultural trade?

When many European countries turned to protection in the late nineteenth century, they did so because international prices declined. Unless combined with supply management, however, protection itself will further depress international prices by raising domestic production, exacerbating the oversupply at the global level. In the 1930s, when falling demand caused agricultural surpluses in several countries, and major exporting countries also introduced protection, supply control and managed trade became an important issue. The Monetary and Economic Conference of the League of Nations, the US Department of Agriculture in the New Deal Period, John Maynard Keynes in his blueprint for the post-World War II economic order, and various UN bodies all advocated international commodity controls in one form or another (Chimni, 1987; Gordon-Ashworth, 1984; Henningson, 1981; Keynes, 1943). All these authorities influenced the policy debate in the first post-war years, and led to a special position of agriculture in the General Agreement on Tariffs and Trade (GATT). In spite of its free trade philosophy, the GATT allowed countries to protect their farmers provided they controlled their production and exports, as well as to engage (if needed) in international commodity agreements to stabilize and support the international prices of primary commodities.

Nevertheless, a balanced multilateral system of managed trade did not emerge. OECD countries were hesitant to cooperate with international agreements that would support the prices of tropical products that they imported. Although they formally endorsed UNCTAD's Integrated Programme for Commodities (1974) that sought to increase the number of commodity controls, they thwarted its implementation and let the few existing control agreements collapse in the 1980s. Many economists see this as proof that free rider and rent-seeking problems make commodity controls inherently unviable (Bohman *et al.* 1996; Herrmann, 1986). In reality, the resistance of importing countries, and GATT rules that gave them a near-veto power may have been decisive (Chimni, 1987; Koning *et al.,* 2004; Maizels, 1992). Meanwhile, a coupling of protection with production and export controls as envisaged by the GATT was thwarted by the aggressive kind of protection pursued by the US and the EU. In the 1980s, competitive dumping caused a trade conflict between these two powers. In 1993, a compromise between them led to the WTO Agreement on Agriculture, prescribing reductions in price support measures but exempting direct payments under certain conditions. In the years that followed, both powers shifted increasingly from price policies to direct payments. Other countries could not follow this approach because of the high government cost involved, so that the obligatory reduction in price supports entailed a reduction in the support to their farmers. The Agreement on Agriculture gave developing countries more room for maintaining price supports. However, the US, the EU, the World Bank and the IMF have pressured them (in bilateral trade negotiations and negotiations on financial support) not to use this room. This whole policy evolution was surrounded by a discourse that depicted this evolution as 'trade liberalization', harking back to the situation that prevailed in the mid-nineteenth century. Model studies that show 'welfare benefits' but ignore the dynamics of agricultural markets and developing countries play an important role in this discourse (see FAO 2006 for a critique of these studies). Meanwhile, the income support given to OECD farmers has hardly decreased. Direct payments allow the US and the EU to continue exporting large volumes for prices below their own cost of production (Ritchie *et al.*, 2003) in a way that no longer violates multilateral trade rules, while other countries are obliged to reduce their tariff defences.

VI. Changes and continuity

The reform of agricultural trade policies was part of a broader movement of farm policy liberalization. In addition to reductions in price supports, this also involved a dismantling of parastatals in developing countries, and a general shift from public to private agricultural research that was coupled to a strengthening of intellectual property rights (Pardey *et al.*, 2006; Pingali, 2007). These policy changes coincided with changes in the private sector. Until the 1980s, agri-food markets were mainly marked by chain differentiation, price competition and an increasing standardization of products. However, this pattern has been altering ever since under the influence of forces such as increased concentration in agro-industries and retailing, the financial needs of highly capitalized farms, improved logistics, demand saturation and ever stricter food safety requirements. These forces stimulate product differentiation and a new vertical coordination within agro-industrial chains that involves farm management being more interfered with by buyers and sellers. Product differentiation also creates niches for new artisanal (including 'organic' or 'fair trade') products, but that has remained a side effect.

Set against these changes is a vital continuity (Coclanis, 2003): world agriculture is still marked by overproduction and a squeeze on farm profits. Claims by liberal economists that the problem of low farm earnings has ceased to exist (Gardner, 1992; Hill, 1996; OECD, 2002) are based on data on total incomes of farm households that include income supports and do not measure the gap in factor returns between farming and non-farming. Although this gap tends to diminish as farmer communities become more strongly culturally integrated into urbanizing societies (Haagsma, Koning, 2005), the level of per capita GDP at which this occurs has increased rather than decreased over recent decades (Timmer, 2007).

New agro-industrial developments interact with the secular overproduction dynamics of agricultural markets. The increased market power of traders and processors causes farm-gate prices and final-consumer prices to diverge in various cases (Gouin, 2007). Together with the loss of room for manoeuvre that quasi-integration involves for farmers, it reinforces the treadmill mechanism that generates overproduction. In its turn, overproduction weakens the market power of farmers against traders and processors, so that the two mechanisms become mutually reinforcing.

Rather than the alleged superior power of agrarian pressure groups that is blamed by liberal economists (World Bank, 2007), the continued squeeze on farm profits explains why, in spite of the liberal rhetoric, developed countries continue to support farm incomes. It also explains why the new chain coordination takes the form of a quasi-integration of agricultural production units, the ownership of which is left with self-employed farmers. Real integration would allow lower production and transaction costs, but would saddle traders and processors with the low profitability of primary agriculture.

Although the recent changes have not ended the secular overproduction dynamics of agricultural markets to which public and private actors have to respond, they are having important effects. Demands of chain integrators are creating new thresholds for farm producers. This is even true for organic or fair trade chains, where high certification costs can form a barrier to the participation of resource-poor farmers.

Liberal reforms have different effects in the developing world. Several successful Asian countries have become less dependent on farming and can moderate agricultural protection without endangering their economies (Dawe, 2007). In Latin America, liberal-economic macro-policies and repression of popular opposition paved the way for new export-led growth based on large farms. In some cases, this generated new employment. Horticulture in Chile is a case in point. (It should be noted, though, that even under the Pinochet regime Chile subsidized various farm inputs and stabilized the prices of major staple crops.) In other cases, however, liberalization created few additional job opportunities. In many places, low prices have driven many small farmers to cultivating illegal crops like coca, or turned them into new slum dwellers or illegal immigrants in the United States. Besides, the wild capitalism of the *latifundio* sector and the desperation of marginalized rural workers cause a scramble for fragile natural resources, leading to large-scale deforestation and depletion.

In Sub-Saharan Africa, the debt crisis allowed western donors to impose trade liberalization and reductions in government spending. The latter were first realized by cuts in public services, then by reductions in public sector wages and only in the last instance by public sector retrenchment. Farmers suffered from the competition of cheap imports and the neglect of roads and other public services. This only aggravated the agricultural crisis and the ensuing poverty trap in which the region was caught. It made access to, and control over, the public sector even more important for ambitious individuals, which aggravated the predatory tendencies in regional politics. Indeed, the 'bad governance' bogey with which international donors are blaming the failure of their liberal prescriptions is partly of their own making.

Finally, the shift from public to private agricultural research is narrowing the room for agricultural development in less-favoured areas (Pingali, 2007). Unlike the Green Revolution, the new Gene Revolution is led by corporations whose research agenda is guided by effective demand. It leads, for instance, to massive investment in herbicide tolerance but to underinvestment in the drought resistance that is vital for millions of poor farmers in less-favoured dryland areas. Developing countries cannot easily compensate for this through national research policies. While the germplasm of Green Revolution institutes like IRRI and CYMMIT was freely available to national researchers, the germplasm produced by private corporations is only available to those willing to pay the price.

VII. Long-term global food availability: continued abundance or new scarcity?

The more socially exclusive nature of farm progress and the failure to arrive at a balanced multilateral regulation of agricultural policies also involve risks for global food security in the future. Between now and mid-century, the world population is expected to grow from 6.5 to 9 billion, and demand for animal products to double (Steinfeld *et al.*, 2006), while the use of biomass for non-food purposes – including functionalized chemicals – will strongly increase. As a consequence, the global demand for farm-produced biomass can easily triple. The fact that, in the twentieth century, supply has been overabundant does not guarantee that this new increase in demand can be met effortlessly. The once plentiful space for reclaiming new fertile lands, tapping water reserves for irrigation, and boosting yields through agro-chemicals and varieties that invest more of their assimilates in seeds or tubers is gradually being depleted. At the global level, the biophysical potential for farm production is still adequate (Penning de Vries *et al.*, 1995; Wolf *et al.*, 2003), even allowing for claims on land and water for urbanization and biodiversity conservation (Koning *et al.*, 2008). However, rising prices of fossil fuels will increase food production costs and strengthen the demand for biofuels and biochemicals, which will compete for biomass with food production. Moreover, farmers are utility maximizers, not output maximizers. Production will run up against an economic ceiling long before the biophysical potential has been realized, especially since three quarters of the world's biophysical reserve capacity for food production lies in Latin America, Sub-Saharan Africa and former Soviet Union countries, where weak infrastructures, less favourable price relations and deep social inequalities make it rational for farmers to stick to less productive techniques (*ibid.*).

As a consequence, an adequate increase in global supply will partly depend on new technologies. Unlike current ecological techniques, which reduce emissions while minimising production losses, new farm technologies must aim at reducing emissions while increasing land productivity. In addition, new non-farm biomass production (marine systems, algae, etc.) should be developed, as well as bio-refinement technologies that allow whole plants rather than just seeds or tubers to be used in making valuable products. Investment in such technologies, and in the human capital that is needed to employ them, involves long gestation periods. To avoid unnecessary scarcity, such investments should be made in time. However, along with myopic expectations and financially constrained farmers, low current prices restrict the size of these investments (Ezekiel, 1938). If, after some time, it were to become more difficult for the global supply of food to keep up with demand, this could lead to soaring food prices, wreaking havoc in net food-importing poor countries (Koning *et al.*, 2008). Such cobweb ('pig cycle') effects might be exacerbated if government support for agriculture were to be strongly reduced in a final phase of international overabundance. In this sense, the present dismantling of price supports, the continuance of disguised dumping by developed countries, the phasing out of fertilizer subsidies in developing countries, and the worldwide reductions in support for farm research might pose serious threats. Besides longer-term cobweb effects, dismantling of price policies will also entail increases in short-term price volatility, which will likewise affect investment (Boussard *et al.*, 2006). No allowances have been made for such effects in the studies of long-term global food security that some established institutions have made. However, the recent spike in international food prices should be taken as a warning.

VIII. Which agricultural policies will enable global food security and sustainable pro-poor growth?

Three main lessons can be drawn from our survey. Firstly, under the evolutionary regime that has prevailed since the late nineteenth century, national and multilateral government intervention (including price and income supports) has become indispensable for a balanced development of the global agri-food economy. Secondly, different parts of the developing world are subject to different dynamic patterns, so that food security and sustainable pro-poor growth in these parts may require different types of intervention. Thirdly, it is not certain that the regime of abundant food supply at the global level will continue in the coming decades, so that responsible policies will have to reckon with the possibility of increased scarcity. These considerations lead to a number of conclusions for agricultural policies whose aim is to enable global food security and sustainable pro-poor growth.

Achieving these aims requires first of all the encouragement of smallholder-based agricultural growth in developing countries. This calls for public investment in infrastructures and research for small farmers, public coordination in establishing agro-industrial chains and supportive price policies. Besides, the weak and the poor should be supported, with priorities depending on regional configurations (land reform in Latin America, local justice for young farmers and women in Africa and social safety nets for farm labourers in Asia).

Other measures may help to prevent or mitigate any return to scarcity in international food markets. These include public investment in research for sustainable increases in yields, bio-refinement of food, effective meat substitutes and new non-farm production systems for biomass and renewable energy. Besides, it would be wise to stop subsidizing bio-energy and to discourage feedlot beef and using land that is suitable for cropping as pasture. Moreover, social safety nets should be introduced globally to prevent a retardation of the decline in demographic fertility in poor countries from enhancing the growth in world population.

Stimulating smallholder-based agricultural growth in developing countries and timely investment in increasing the carrying capacity for food production in the world at large requires the stabilization of international agricultural prices. Rather than the pseudo-liberalization of agricultural trade that is currently going on within the WTO and through bilateral agreements between rich and poor countries, this requires a multilateral system of managed trade. As regards tropical export crops, this means the establishment of international production controls based on export and production controls. With other crops, it means imposing disciplines like tradable export quotas and minimum import quotas on high- and upper-middle-income countries so as to keep world market prices above a desirable minimum level. Low-income countries should be exempted to create room for them to increase their production and exports. To prevent the demand of the affluent for animal products and non-foods to outcompete the demand for food for the poor, an international tax on the use of biomass for new non-foods could be introduced that is levied when international prices of food staples exceed a pre-agreed ceiling.

Bibliography

ADELMAN, Irma, BOURNIAUX, Jean-Marc & WAELBROECK, Jean (1989), 'Agricultural Development-led Industrialisation in a Global Perspective', in Jeffrey C. WILLIAMSON & Vadiraj R. PANCHAMUKHI (eds), *The Balance between Industry and Agriculture in Economic Development*, vol. 2, sector proportions, London, MacMillan Press, p. 320-339.

AKERLOF, George A. & YELLEN, Janet L. (1986), *Efficiency Wage Models of the Labor Market*, Cambridge, Cambridge University Press.

ALESINA, Alberto & RODRIK, Dani (1994), 'Distributive Politics and Economic Growth', *Quarterly Journal of Economics*, 109, p. 465-490.

ALLEN, Robert C. (1988), 'The Growth of Labor Productivity in Early Modern English Agriculture', *Explorations in Economic History,* 25, p. 117-146.

ANDERSON, Kym, HAYAMI, Yujiro & GEORGE, Aurelia (1986), *The Political Economy of Agricultural Protection : East Asia in International Perspective*, Sydney etc., Allen & Unwin.

BAIROCH, Paul (1976), *Commerce extérieur et développement économique de l'Europe au XIXe siècle,* Paris, Mouton.

BATES, Robert H. (1981), *Markets and States in Tropical Africa: the Political Basis of Agricultural Policies,* Berkeley, Berkeley University Press.

BAUMANN, Werner & MOSER, Peter (1999), *Bauern im Industriestaat. Agrarpolitische Konzeptionen und bäuerliche Bewegungen in der Schweiz 1918-1968*, Zürich, Orell Füssli.

BAYART, Jean-François (1989), *L'État en Afrique : la politique du ventre,* Paris, Fayard.

BEGHIN, John C., BUREAU, Jean-Christophe & PARK, Sung Joon (2003), 'Food Security and Agricultural Protection in South Korea', *American Journal of Agricultural Economics*, 85, p. 618-632.

BOHMAN, Mary, JARVIS, Lovell & BARICHELLO, Richard (1996), 'Rent Seeking and International Commodity Agreements: the Case of Coffee', *Economic Development and Cultural Change*, 44, p. 379-404.

BOSERUP, Ester (1965), *The Conditions of Agricultural Growth: the Economics of Agrarian Change under Population Pressure* , London, Allen & Unwin.

BOUSSARD, Jean-Marc, GÉRARD, Françoise, PIKETTY, Marie-Gabrielle, AYOUZ, Mourad & VOITURIEZ, Tancrede (2006), 'Endogenous Risk and Long Run Effects of Liberalization in a Global Analysis Framework', *Economic Modelling,* 23, p. 457-475.

BUNDY, Colin (1972), 'The Emergence and Decline of a South African Peasantry', *African Affairs,* 71, p. 369-388.

CHIMNI, B.S. (1987), *International Commodity Agreements: a Legal Study,* London etc., Croom Helm.

COCHRANE, Willard Wesley (1959), *Farm Prices - Myth and Reality,* Minneapolis, University of Minnesota Press.

COCLANIS, Peter A. (2003), 'Back to the Future: the Globalization of Agriculture in Historical Context', *SAIS Review,* 23, p. 71-84.

DAWE, David (2007), 'The Practical Experience with Agricultural Trade Liberalization in Asia', in Niek KONING & Per PINSTRUP-ANDERSEN (eds), *Agricultural Trade Liberalization and the Least Developed Countries,* Dordrecht, Springer, p. 175-195.

DE JANVRY, Alain (1981), *The Agrarian Question and Reformism in Latin America,* Baltimore, John Hopkins University Press.

DELGADO, Cristopher L., HOPKINS, Jane & KELLY, Valery A. (1998), *Agricultural Growth Linkages in Sub-Saharan Africa*, IFPRI Research Report 107, Washington, DC, IFPRI.

DIAO, Xinshen, DYCK, John, SKULLY, David, SOMWARU, Agapi & LEE, Chinkook (2002), *Structural Change and Agricultural Protection: Costs of Korean Agricultural Policy, 1975 and 1990*, USDA Agricultural Economic Report number 809, Washington DC, USDA.

DJURFRLDT, Göran, HOLMÉN, Hans, JIRSTÖM, Magnus & LARSSON, Rolf (2005), *The African Food Crisis: Lessons from the Asian Green Revolution*, London, CABI.

DORWARD, Andrew, KYDD, Jonathan, MORRISON, Jamie & UREY, Ian (2004), 'A Policy Agenda for Pro-poor Agricultural Growth', *World Development,* 32, p. 73-89.

EZEKIEL, Mordecai (1938), 'The Cobweb Theorem', *Quarterly Journal of Economics,* 53, p. 225-280.

FAHEY, Tony (2002), 'The Family Economy in the Development of Welfare Regimes: a Case Study', *European Sociological Review,* 18, p. 51-64.

FAO (2006), *Towards Appropriate Agricultural Trade Policy for Low-income Developing Countries,* FAO Trade Policy Technical Notes no. 14, Rome, FAO.

FARR, Ian (2007), '"Farmers" Cooperatives in Bavaria, 1880-1914: "State-help" and "Self-help" in Imperial Germany', *Rural History*, 18, p. 163-182.

FEDERICO, Giovanni (2005a), *Feeding the World. An Economic History of Agriculture, 1800-2000*, Princeton, Princeton University Press.

FEDERICO, Giovanni (2005b), 'Not Guilty? Agriculture in the 1920s and the Great Depression', *Journal of Economic History,* 65, p. 949-976.

FRANCKS, Penolope, BOESTEL, Johanna & KIM CHOO, Hyop (1999), *Agriculture and Economic Development in East Asia: from Growth to Protectionism in Japan, Korea and Taiwan*, London-New York, Routledge.

GARDNER, Bruce L. (1992), 'Changing Perspectives on the Farm Problem', *Journal of Economic Literature,* 30, p. 62-101.

GARRABOU, Ramon (1997), 'Políticas agrarias y desarrallo de la agricultura Española contemporánea: unos apuntos', *Papeles de Economía Española,* 73, p. 141-148.

GARRABOU, Ramon, PLANAS, Jordi & SAGUER, Enric (2001), 'Sharecropping and the Management of Large Rural Estates in Catalonia, 1850-1950', *Journal of Peasant Studies*, p. 89-108.

GARRIDO, Samuel (2007), 'Why did Most Cooperatives Fail? Spanish Agricultural Cooperation in the Early Twentieth Century', *Rural History,* 18, p. 183-200.

GASTÓN AGUAS, José Manuel & LANA BERASAIN, José Miguel (2007), '¿Condenado al absentismo? La administración del conde de Zaldívar en Cortes de Navarra (1859-1935)', in Ricardo ROBLEDO & Santiago LÓPEZ (eds), *¿Interés particular, bienestar público? Grandes patrimonios y reformas agrarias*, Zaragoza, Prensas Universitarias de Zaragoza, p. 223-255.

GEERTZ, Clifford (1963), *Agricultural Involution: the Processes of Ecological Change in Indonesia,* Berkeley, Los Angeles, University of California Press.

GERSCHENKRON, Alexander (1966), *Bread and Democracy in Germany,* New York, H. Fertig.

GOODY, Jack (1976), *Production and Reproduction: a Comparative Study of the Domestic Domain,* Cambridge, New York, Cambridge University Press.

GORDON-ASHWORTH, Fiona (1984), *International Commodity Control: a Contemporary History and Appraisal,* London, New York, Croom Helm & St. Martin's Press.

GOUIN, Daniel-Mercier (2007), 'La Gestion de l'offre dans le secteur laitier : un système de régulation toujours pertinent', *in* Jean-Marc BOUSSARD & Hélène DELORME (eds), *La Régulation des marchés agricoles internationaux: un enjeu décisif pour le développement,* Paris, L'Harmattan, p. 163-188.

GRIGG, David B. (1982), *The Dynamics of Agricultural Change: the Historical Experience,* London, Hutchinson.

HAAGSMA, Rein & KONING, Niek (2002). 'Endogenous Mobility-reducing Norms', *Journal of Economic Behavior & Organization*, 49, p. 523-547.

HAAGSMA, Rein & KONING Niek (2005), 'Endogenous Norms and Preferences and the Farm Income Problem', *European Review of Agricultural Economics*, 32, p. 25-49.

HAZELL, Peter B. R. & RÖELL, Alisa (1983), *Rural Growth Linkages: Household Expenditure Patterns in Malaysia and Nigeria*, IFPRI Report no. 41, Washington, D.C., IFPRI.

HENNINGSON, Berton jr. (1981), *United States Agricultural Trade and Development Policy during World War II: the Role of the Office of Foreign Agricultural Relations,* Ph.D., University of Arkansas.

HERRMANN, Roland (1986), 'Free Riders and the Redistributive Effects of International Commodity Agreements: the Case of Coffee', *Journal of Policy Modelling,* 8, p. 597-621.

HILL, Berkeley (1996), 'Monitoring Incomes of Agricultural Households within the EU's Information System – New Needs and New Methods', *European Review of Agricultural Economics,* 23, p. 27-48.

Hoffman, Philip T. (1996), *Growth in a Traditional Society: the French countryside 1450-1815,* Princeton, Princeton University Press.

Huijzendveld, Franciscus Dominicus (1997), *'Die Ostafrikanische Schweiz': plantages, planters en plattelandsontwikkeling in West-Usambara, Oost-Afrika, c. 1870-1930*, Amsterdam, N.W. Posthumus Reeks.

Ikelegbe, Augustine (2001), 'The Perverse Manifestation of Civil Society: Evidence from Nigeria', *Journal of Modern African Studies,* 39, p. 1-24.

Johnston, Bruce F. & Kilby, Peter (1975), *Agriculture and Structural Transformation. Economic Strategies in Late-developing Countries*, New York, Oxford University Press.

Johnston, Bruce F. & Mellor, John W. (1961), 'The Role of Agriculture in Economic Development', *American Economic Review*, 51, p. 566-593.

Keynes, John Maynard (1943), 'The International Regulation of Primary Products', reprinted in Donald E. Moggridge (ed.), *Collected Writings of John Maynard Keynes*, London, MacMillan & Cambridge University Press, p. 65-92.

Koning, Niek (1986), 'Beperkingen van landbouwinkomenspolitiek in kapitalistische industrielanden', *Tijdschrift voor Sociaalwetenschappelijk Onderzoek van de Landbouw,* 1, p. 113-133.

Koning, Niek (1991), 'Agricultural Prices Support: Real History and the Liberal View', *Tijdschrift voor Sociaalwetenschappelijk Onderzoek van de Landbouw*, 6, p. 44-65.

Koning, Niek (1994), *The Failure of Agrarian Capitalism: Agrarian Politics in the United Kingdom, Germany, the Netherlands and the USA, 1846-1919*, London, New York, Routledge.

Koning, Niek (2002), *Should Africa Protect its Farmers to Revitalise its Economy?*, IIED Gatekeeper Series no. 105, London, IIED.

Koning, Niek, Calo, Muriel & Jongeneel, Roel (2004), *Fair Trade in Tropical Crops is Possible: International Commodity Agreements Revisited*, North-South discussion paper no. 3, Wageningen University & Research Center, North-South Center.

Koning, Niek & Smaling, Eric (2005), 'Environmental Crisis or "Lie of the Land"? The Debate on Soil Degradation in Africa', *Land Use Policy,* 22, p. 3-11.

Koning, Niek *et al.* (2008), *Long-term Global Availability of Food: Continued Abundance or New Scarcity?,* NJAS Wageningen Journal of Life Sciences, 55.

Lawrence, Denis & Diewert, Erwin (1999), *Measuring New Zealand's Productivity. Report Prepared for the Department of Labour, Wellington*, Reserve Bank of New Zealand and the Treasury.

Maizels, Alfred (1992), *Commodities in Crisis: the Commodity Crisis of the 1980s and the Political Economy of International Commodity Policies,* Oxford, Clarendon Press.

Meuvret, Jean (1946), 'Les Crises de subsistances et la démographie de la France d'Ancien Régime', *Population,* 2, p. 643-650.

Moser, Peter (2000), 'Eine "Sache des ganzen Volkes"? Überlegungen zum Prozess der Vergesellschaftung der bäuerlichen Landwirtschaft in der Industriegesellschaft', *Traverse 2000/1,* p. 64-79.

Munro, John Forbes (1976), *Africa and the International Economy 1800-1960. An Introduction to the Modern Economic History of Africa South of the Sahara,* London, Totowa, Rowman & Littlefield.

Myrdal, Gunnar (1968), *Asian Drama: an Inquiry into the Poverty of Nations,* New York, Pantheon.

OECD (2002), *Farm Household Income: Issues and Policy Responses,* Paris, OECD Publishing.

Ohkawa, Kazushi & Rosovsky, Henry (1961), 'The Indigenous Components in the Modern Japanese Economy', *Economic Development and Cultural Change*, 9, p. 476-501.

Ohkawa, Kazushi & Shinohara, Miyohei (1979), *Patterns of Japanese Economic Development: a Quantitative Appraisal*, New Haven, Yale University Press.

Olmstead, Alan (1975), 'The Mechanization of Reaping and Mowing in American Agriculture, 1833-1870', *Journal of Economic History*, 35, p. 327-352.

Öniş, Ziya (1991), 'The Logic of the Developmental State', *Comparative Politics,* 24, p. 109-126.

O'Rourke, Kevin H. (1997), 'The European Grain Invasion, 1870-1913', *Journal of Economic History*, 57, p. 775-801.

O'Rourke, Kevin H. (2000), 'Tariffs and growth in the late 19th century', *Economic Journal,* 110, p. 456-483.

Pardey, Philip G., Beintema, Nienke, Dehmer, Steven & Wood, Stanley (2006), *Agricultural Research: a Growing Global Divide?,* IFPRI, Washington D.C., IFPRI.

Park, Albert & Johnston, Bruce (1995), 'Rural Development and Dynamic Externalities in Taiwan's Structural Transformation', *Economic Development and Cultural Change*, 44, p. 181-208.

Patterson, Amy S. (1998), 'A Reappraisal of Democracy in Civil Society: Evidence from Rural Senegal', *Journal of Modern African Studies,* 36, p. 423-441.

Penning de Vries, Frits, van Keulen, Herman & Rabbinge, Rudy (1995), 'Natural Resources and Limits of Food Production in 2040', in Johann Bouma, Arie Kuyvenhoven, Bert Bouman, Jan Luyten & Hubert Zandstra (eds), *Eco-regional Approaches for Sustainable Land Use and Food Production*, Dordrecht, Kluwer.

Peters, Krijn (2006), *Footpaths to Reintegration: Armed Conflict, Youth and the Rural Crisis in Sierra Leone*, PhD thesis, Wageningen University.

Philpott, Bryan (1994), *Productivity Growth by Type of Farming: 1972-1993,* Research Project on Economic Planning paper 259, Wellington, Victoria University.

PINGALI, Prabhu (2007), *Will the Gene Revolution Reach the Poor? Lessons from the Green Revolution,* Mansholt lecture, Wageningen January 26th, 2007.

PUHLE, Hans-Jürgen (1986), 'Lords and Peasants in the Kaiserreich', *in* Robert G. MOELLER (ed.), *Peasants and Lords in Modern Germany: Recent Studies in Agricultural History*, Boston, Allen & Unwin, p. 81-109.

RITCHIE, Mark, MURPHY, Sophia & LAKE, Mary Beth (2003), *United States Dumping on World Agricultural Markets*, Minneapolis, Institute for Agriculture and Trade Policy.

SANDREY, Ron & REYNOLDS, Russell (1990), *Farming without Subsidies: New Zealand's Recent Experience*, Wellington, MAF Policy Services.

SCHULTZ, Theodore W. (1945), *Agriculture in an Unstable Society,* New York-London, McGraw-Hill.

STEINFELD, Henning, GERBER, Pierre, WASSENAAR, Tom, CASTEL, Vincent, ROSALES, Mauricio & DE HAAN, Cees (2006), *Livestock's Long Shadow: Environmental Issues and Options*, Rome, FAO.

THOEN, Erik (1990), 'Technique agricole, cultures nouvelles et économie rurale en Flandre au bas Moyen Âge', *Flaran,* 12, p. 52-67.

TIMMER, C. Peter (1995), 'Getting Agriculture Moving: Do Markets Provide the Right Signals?', *Food Policy,* 20, p. 455-472.

TIMMER, C. Peter (2007), *A World without Agriculture: the Structural Transformation in Historical Perspective,* Wendt Lecture, May 8, 2007, American Enterprise Institute, Washington, DC.

TRACY, Michael (1989), *Government and Agriculture in Western Europe 1880-1988,* New York, Harvester Wheatsheaf.

WEBB, Steven B. (1978), *The Economic Effects on Tariff Protection in Imperial Germany,* 1879 to 1914, unpublished PhD thesis University of Chigaco.

WINSTANLEY, Michael (1996), 'Industrialization and the Small Farm: Family and Household Economy in Nineteenth-century Lancashire', *Past & Present,* 152, p. 157-195.

WOLF, Joost, BINDRABAN, Prem, LUIJTEN, Jan & VLEESHOUWERS, Leo (2003), 'Exploratory Study on the Land Area Required for Global Food Supply and the Potential Global Production of Bioenergy', *Agricultural Systems,* 76, p. 841-861.

WOOD, James W. (1998), 'A Theory of Preindustrial Population Dynamics', *Current Anthropology,* 39, p. 99-135.

WOODS, Dwayne (2003), 'The Tragedy of the Cocoa Pod: Rent-seeking, Land and Ethnic Conflict in Ivory Coast', *Journal of Modern African Studies*, 41, p. 641-655.

WORLD BANK (2007), *World Development Report 2008: Agriculture for Development, yellow cover draft*, Washington D.C., The World Bank.

3. Agricultural policy and the Dutch agricultural institutional matrix during the transition from organized to disorganized capitalism[1]

Anton SCHUURMAN

I. Introduction

Agricultural policy cannot be studied in isolation. No one has better demonstrated this than Michael Tracy in his magisterial *Government and Agriculture in Western Europe, 1880-1988*. Not only did this book do historical science a huge service by showing that contemporary European agricultural policy cannot be understood without placing it in its historical context, but it also made some important suggestions that can help explain the pattern of the development of agricultural policy over time. In particular, Tracy emphasises the explanatory importance of the general attitude in a country towards economic policy and more specifically to trade, the degree of public willingness to support agriculture and the relative importance of agriculture in the economy and in politics.

Of course Tracy is less concerned with constructing a general interpretive and explanatory framework than he is with identifying the developmental tendencies at work in Western Europe (especially in Britain, France, Germany and Denmark) from the 1880s to the end of the 1970s. And here a major feature of Tracy's approach is to view economic change as being born out of crisis conditions – much of his discussion focuses on the effects of the Great Depressions of the 1880s and 1930s.

While Tracy gives prominence to the major downturns of the 1880s and 1930s *The End of Organized Capitalism* by Scott Lash and John Urry carries the sort of historical and comparative approach found in Tracy a step further and in a more ambitious direction – its concerns range across economy, society and culture outside as well as inside Europe. A critical difference between Tracy and Lash and Urry is that the latter commence their account with the upswing in the economic cycle of the 1890s that marked what proved to be a critical 'expansive phase of the Kondratieff cycle' (Lash & Urry, 1987: 8). It was this upswing that would inspire the transition

1 I thank Peter Moser and Tony Varley for all their dedicated support in improving the paper. I also thank the anonymous referees for their insightful comments on a previous draft. The paper stems from the NWO-research project Democratization and Modernization in the Dutch Countryside, 1840-1920 whose participants include Paul Brusse, Wim Coster, Piet van Cruyningen, Pim Kooij, Ronald Rommes, Anton Schuurman and Remco Visschers.

from the era of liberal capitalism (whose end German writers have associated with the downward phase of the Kondratieff long wave that commenced in the mid-1870s) to what Lash and Urry term 'organized capitalism'.

The period of organized capitalism, which would be distinctive not alone for its economic characteristics but for new patterns of state intervention and 'modernist' culture, was set to endure till the late 1960s when it began to disintegrate and to be replaced by the 'disorganized' capitalism associated with a dynamic new wave of globalisation. What we have then in Lash and Urry is the bold suggestion that the massive and varied changes that occurred over the course of the twentieth century can be described and understood in terms of the transition from 'organized' to 'disorganized' capitalism. At the same time Lash and Urry want us to be sensitive to historical difference and complexity. They are constantly aware, in focusing on the transition from organized to disorganized capitalism in the USA, Germany, Britain, Sweden and France, that the patterns of observable change do not take shape everywhere at the same time or speed.

My task in this paper will be to use Tracy and Lash and Urry to provide a context for discussing agricultural policy in the Netherlands over the course of the twentieth century. My main goal is not so much to give a complete and detailed description of the construction and contention of Dutch agricultural policy in the twentieth century as to analyse it in the light of Tracy's and Lash and Urry's conceptual frameworks. The hope is that this will facilitate international comparisons using the Dutch case, as well as connecting the study of agricultural policy more generally with the developmental tendencies of the contemporary world. The specific question I will pose is: how well does the distinction between 'organized' and 'disorganized' capitalism illuminate the patterns of change to be observed in Dutch agricultural policy? Pursuing this question will involve describing how Dutch agriculture became organized and then showing how it reacted as the transition to disorganized capitalism began in earnest. So as to proceed, we will first need to say some additional words about Lash and Urry's account of the dynamics of organized and disorganized capitalism.

II. Organized and disorganized capitalism

Where the economy is concerned, organized capitalism came to revolve around the dominance of large-scale extractive and manufacturing industries located in big industrial cities. The emphasis on standardised mass-produced commodities left relatively little choice for consumers. State regulation of the economy sought to overcome problems created by the market and state-controlled welfare provision put in an appearance (though here the differences between cases like Sweden and the USA are very striking).

The period of disorganized capitalism commencing in the late 1960s and early 1970s was characterized by a decline in the relative importance of extractive and manufacturing industries and by a notable increase in the importance of smaller scale service industries and manufacturing industries geared to serving the consumer tastes of an increasingly sophisticated and culturally diverse rising middle class. Conditions of rapid and sweeping globalisation released competitive pressures that constantly urged the need for flexibility and cost cutting. The state's attempts at economic regulation became more restricted and less effective; and there occurred a general decline of faith in large-scale planning (Daniels *et al.*, 2008: 78). Another notable development was that rising costs (partly associated with bureaucratic inefficiencies), demands for lower taxation, and growing social inequality come to challenge the idea of a centralized welfare state. In essence the transition from organized to disorganized capitalism would involve a shift from a seller's to a buyer's market. And as the power of market forces expanded with accelerating globalisation, the power of nation-states tended to become more restricted.

III. The pre-twentieth century development of the Dutch agricultural institutional matrix

The term 'agricultural institutional matrix' will be used to designate the complex of actors and institutions that embodied, guided, directed and organized the agricultural sector in the Netherlands. Contained within this matrix we find market forces, science, state bodies (including government) and private enterprise, all of which are linked together in complex ways. It is far too simple, in my view, to discuss agricultural policy purely in terms of relationships between the national government and the agricultural sector. The term agricultural institutional matrix permits us to capture the complexity resulting from the dense sets of relationships that can exist between a multitude of state and non-state actors and forces. Thus the early Dutch agricultural institutional matrix was the result of developments in the nineteenth century that saw state policy becoming embedded in the activities of an expanding agricultural civil society. During the nineteenth century this agricultural civil society had constructed a national economic sector whose aims would be pursued in a market environment by co-operatively organized farmers committed to using scientific knowledge. It came to be accepted that the state should safeguard this sector from adverse conditions, help to educate the agricultural population and generally lay down the 'rules of the game' (Schuurman, 2008).

At the beginning of the twentieth century the Dutch agricultural institutional matrix comprised a set of distinguishable organisations. The most recent and important addition to the matrix was the department of agriculture (founded in 1898) that would become the directorate of agriculture in 1906. This was not yet a full-blown ministry of agriculture, but for the first time there was a central state agency possessing considerable autonomy and that could coordinate the sector on the political level. Its partner was the Royal Dutch Agricultural Committee, which had been founded in 1884. Represented on this committee were the regional and local agricultural organizations and societies that had been established since the 1840s and that still remained the most important bodies in Dutch agriculture.

These farmers' societies had been responsible for agriculture's greater visibility in the public domain. Their members engaged in public discussion by writing in newspapers and journals; and the societies themselves published magazines in which new agricultural methods and implements were discussed and disseminated. They further organized exhibitions in order to spread new knowledge and tried to raise the standard of agricultural practice through competition. Since 1876 the Netherlands had its own national agricultural college in Wageningen and this quickly became the leading national institute for agricultural education and research. Alongside it several agricultural schools and experiment stations were founded. These experiment stations became the foundation of a research programme aimed at the scientific improvement of agriculture. At the same time, the farmers' organisations and the Dutch government were expanding the extension services. The Wageningen graduates started to populate the experiment stations, the directorate of agriculture, the professional organizations of the farmers' societies and associations. They could regularly be heard at agricultural meetings and their writings appeared in agricultural magazines and government publications.

Between them these actors, at the forefront of the emerging late nineteenth century agricultural institutional matrix, created a common vision for Dutch agriculture that remained dominant and valid until the last quarter of the twentieth century. At its core was the desirability of technological progress to raise productivity and to encourage competitiveness. Just around the turn of the century the rise of the Christian (mostly Roman Catholic) farmers' unions did change the character of the organizational landscape somewhat, but it did not fundamentally alter the direction of the Dutch agricultural institutional matrix. As elsewhere, there was a certain limited (albeit significant) co-operation among the higher echelons of what in reality was a religiously and ideologically segmented society.

The confessional farmers' unions were in agreement with the liberal agricultural societies that state intervention had to be minimised, and that technological progress to raise productivity and promote competitiveness in agriculture was to be encouraged. However, what did change was the role of farmers' unions at the local level. These became much more involved in the everyday lives of ordinary farmers than had been the case with the nineteenth century farmers' organizations. The confessional farmers' unions wanted to organize the entire countryside. They were not content therefore to limit themselves to the organisation of the men; they also established women's and youth organisations. Their model was one of managed modernisation. The farmers' unions were also much more strongly connected with the cooperative movement from around 1900 than had been the case with the earlier farmers' associations (Smits, 1996).

Dutch agricultural policy (as with economic policy in general) did nevertheless retain its liberal character. Reflecting its position on the West-European delta, the Netherlands was an open economy; and Dutch agriculture was heavily market and export orientated. The national state, as a vital element of the agricultural institutional matrix, came to be tasked with helping to enhance agriculture's competitiveness, performance and productivity and with safeguarding the quality of its products (Smits, 2008).

IV. Organized Capitalism, c. 1930- c. 1970

IV.1. The Great Depression

The prolonged depression of the 1930s occasioned a dramatic U-turn in the agricultural policy of the Netherlands. With agriculture and the countryside severely hit by the depression, the Dutch state became deeply involved in the agricultural economy. An extreme situation called for extraordinary measures and the emergency conditions were set to continue as war with Germany became imminent. The state's newly expanded role found expression in the way the directorate of agriculture became a ministry in its own right in 1935 (Trienekens, 1985; van Kamp, 2005).

The measures taken by the government can be read off from the laws that were enacted. In 1931 the Wheat Law came into effect, followed in 1932 by the Crisis Dairy Law and the Crisis Pig Law. In 1934, these laws were brought within a single framework: the Agricultural Crisis Law. It was this legislation that provided the state with far-reaching competences to settle questions relating to the production, quality, price, as well as to the processing and marketing of most agricultural products. In 1937, in response to the way the small farmers on the sandy soils were severely hit by the economic downturn of the 1930s, a special governmental Small Farmers Service was created.

All these measures proved costly and came to consume a major part of the national budget. Between 1933 and 1936 it is estimated that the government spent almost 200 million guilders annually, thus providing the agricultural sector with a total income of somewhere between 400 to 500 guilders out of a national budget of around 800 million guilders (Bieleman, 2008: 314).

The U-turn that this increased state intervention represents can be explained by using Tracy's framework. In the 1930s a large part of the population still worked in agriculture (20 per cent of the labour force). Agriculture was strongly organized at the local, regional and national levels. The different segments of the institutional matrix were well connected, not least by many personal ties. People knew each other from Wageningen, from their region of origin and from the organizations they belonged to. Their careers often involved occupations that straddled several segments of the agricultural institutional matrix (Krips-van der Laan, 1985). Furthermore, the agricultural organizations had good contacts with political parties. During the inter-war period the confessional parties had a majority in parliament and were always part of the government (Griffiths & de Vries, 1987). The general climate was favourable towards government intervention (Scott, 1998). National policies were strongly *en vogue*.

The implications of all this were that the 1930s marked the end of the period of modern globalization that had started in the nineteenth century (Held *et al.*, 1999). Not only did the general climate favour state intervention but also politicians and economists were agreed as to the wisdom of the new approach. The crisis of the 1930s, and the response it evoked, would determine the mindset of politicians till the closing decades of the twentieth century.

IV.2. Modernisation as a positive buzzword

After the war the government returned to the political challenge of making Dutch agriculture ever more competitive. Convinced that the population would continue to grow swiftly with the linked danger of growing unemployment, it initiated an active emigration policy. At the same time it tried to modernize the economy. Industrialization was to be the keyword in striving for this goal. So as to move in the right direction and prevent an upsurge of social problems, the modernization and industrialization policies were accompanied by a strong emphasis on traditional and religious family values. It is therefore not surprising that the Ministry of Social Care (*Maatschappelijk Werk*) was founded in 1952 (Schuyt & Taverne, 2000).

Where agriculture was concerned much attention was paid to promoting rationalization, specialization, intensification, mechanization and increasing the scale of farming operations. The institutional matrix became formalized, professionalized and for the most part paid for by the taxpayers' money. As consumers these taxpayers could avail of cheaper food products in return.

Another notable development was that the business world got an official place at the table. After 1954 commodity boards and sectoral boards were founded; and these were given powers to regulate economic relations between the production chains. The sectoral boards were given responsibility for wage control, quality control and structural policy. A new Agricultural Board, which replaced the Agricultural Council that had been an advisory council since 1945, was tasked with regulating and stimulating activities in the sector, improving the quality of production, animal health, advising the government and representing and promoting the interests of the sector. Farmers' and workers' organizations were represented. To fund its activities, the Board could raise taxes autonomously. In practice, it was through intensive consultation rather than through exercising its capacity to regulate that the Agricultural Board exerted most influence on official government policy (Krajenbrink, 2005).

The policy adopted towards agriculture was completely in line with the general policy of stimulating the economy more broadly and of introducing comprehensive welfare state measures to reconcile the interests of employers and employees, thereby catering to the needs of those who for whatever reason found themselves outside the labour market. Modernity, built on technocratic knowledge, saw engineers, scientists, social scientists and economists being recruited to aid the state in striving for progress (van Vegchel, 1995).

Because agriculture was significantly better organized than other sectors (thanks to its long history of organization and strong and overlapping binary identities – agriculture versus other sectors, the outer-regions against the Randstad, the countryside against the city and religion against secular beliefs) – state agricultural policy tended to be the most successful. All this is evidenced in the strong rise in productivity, farming incomes and in the prominence of the Netherlands in competitive export markets. The farmers became used to a process of permanent change and innovation. They came to expect a situation where they were being permanently retrained (Douw & Post, 2000).

Of course, a drawback with this system was that the agricultural institutional matrix had settled on a general policy and direction that gave the individual farmer little freedom to deviate. When a farmer asked for advice he received it from experts all of whom had been trained in the same beliefs and technologies. When he asked for money from a bank, or wanted to get a subsidy from the state, he was subjected to the same decision-making criteria – based on the same conceptions of efficiency, rationalization and rising productivity (van der Ploeg, 2003). All this meant that Dutch agriculture became rather uniform and standardized, as did the landscape and even the biodiversity (Saris, 2007: 303-308).

As long as there were rising incomes, prosperity and financial compensations for misfortunes, there was almost no opposition to the *status quo*. Moreover, in the 1950s and 1960s the pillarization (vertical segmentation according to religious denomination and secular ideologies) of Dutch society was still strong. There were still strong feelings of solidarity and compliance with the ideas of the leaders. What opposition there was to the dominant policy of scaling-up and rationalization found its most visible expression in the Farmers' Party (*Boerenpartij*). Founded in the 1960s, this party had its best electoral result in 1967 when it won seven of the 150 seats in the national parliament. As well as defending the interests of small farmers, the Farmers' Party attracted some support from elements that were dissatisfied with the modernization process more generally (Nooij, 1969).

What impact did the appearance of the European Economic Community and the Common Agricultural Policy (CAP) have on Dutch agriculture? A major positive consequence was that a significant set of new market opportunities became available (Meester, 2008; Burger, 1993). It was also the case, of course, that the CAP worked to undermine the regulatory character of the boards, even if their framework was not altered and they were still used both for the implementation of agriculture policies and for consultation purposes.

The fact that the CAP left structural policy to the national governments can also be viewed as positive for Dutch agriculture. The government agencies could continue doing what they did best: creating a good environment for an increasingly more productive agriculture. In the 1950s and 1960s national agricultural policy mostly involved land consolidation, water management, regional planning, agricultural and social extension and farm household extension (Karel, 2005; van den Bergh, 2004). The countryside was more or less completely re-arranged for more productive agricultural use, a process that had commenced during the depressed 1930s. Not only were many city parks created in the 1930s, but the newly formed government service for land and water (1936) got access to a labour supply that allowed it to undertake projects that otherwise would have been impossible to pursue and complete. In 1939, for instance, 72,000 labourers were deployed of which 57,000 were engaged in land consolidation work (Buiter & Korsten, 2006: 34).

Two other events influenced the practice of the Dutch land consolidation: the inundation of Walcheren at the end of the Second World War and the flooding of large parts of Zeeland and South-Holland in 1953. The experience with the reconstruction of these regions was applied elsewhere. Although some attention was paid to landscape and nature, modernising agricultural interests (and, more specifically, those elements eager to implement the industrial model in the agricultural sector) were clearly predominant.

IV.3. Disorganized capitalism: new contestations and new solutions

At the beginning of the 1970s there could still be considerable optimism about the established order of organized capitalism. The prospects for society, the economy and agriculture looked good. Certainly Sicco Mansholt, the Dutch Commissioner of European Agriculture (1958-1972), was intent on continuing with the apparently successful policies from the recent past that thought in terms of ever more rationalization, intensification, specialization, mechanization and up-scaling. But, as we now know, the period of organized capitalism was beginning to come to an end as globalisation accelerated, the Keynesian and interventionist nation-state began to be eclipsed and Fordism and the welfare state began to slip into a deepening crisis. A new economic policy loomed into view that emphasised the control of inflation, financial deregulation, tax reform, competition policy, privatization measures, labour market reforms and the reduction of subsidies (Woodward, 1999). 'All that is solid about organized capitalism, class, industry, cities, collectivity, nation-state, even the word', write Lash and Urry (echoing the famous words of *The Communist Manifesto*), 'melts into air' (Lash & Urry, 1987: 312-313). As time went by agriculture would have to adapt itself to radically changing conditions; and, in turn, the agricultural institutional matrix would have to be reconstructed, redefined and indeed reinvented.

The reaction of the Dutch agricultural institutional matrix to the new conditions can be divided into two periods: one covering the 1970s and 1980s and a second the 1990s. In the 1970s the matrix was strongly on the defensive: it wanted to keep things as they were and continued to apply old solutions. A different approach is evident in the second period when the matrix was obliged to take serious account of the new realities.

The new context within which Dutch agriculture had to operate was characterized by features such as:

- an increasing number of farmers feeling threatened by the accelerating pace of change;
- the appearance and practice of alternative styles of farming;

- the countryside becoming a contested terrain;
- agriculture beginning to produce surpluses;
- agriculture causing environmental problems;
- consumers became more powerful;
- CAP still important within the EC, but no longer its fundamental engine;
- state policy embracing supply-side economics (van Dijk *et al.*, 1999: 65-86; Kooij, 1999; Schuurman, 2007; van der Ploeg & Wiskerke, 2005).

In a sense the fundamental problem was that agriculture was being punished for its own success – the sector was producing too much (Meester, Strijker, 1985). Yet despite overproduction being a recognised problem, a solution was at first sought in the 'high modern' response of organized capitalism that involved scaling-up and the pursuit of even greater efficiencies. As the surpluses and environmental problems grew in size and significance organic farming, already stimulated by new social trends that saw middle class prosperity and 'new social movement' consciousness create consumers that were less homogeneous in their tastes, began to challenge mainstream agriculture. Between them more well-to-do health and environmentally conscious consumers and niche farmers were demanding and producing new products for a different sort of market. Of course while these consumers asked for better quality products, they still wanted low prices. Many farmers felt weary of all the changes. They (and their families) wanted to participate in the consumer society with higher incomes, improved working conditions and even days off.

Another major development was that the increasing mobility made possible by rising car ownership and the process of suburbanization was changing the countryside. Space was needed for dwellings, recreation and nature preservation. Under the impact of occupational and settlement changes farmers were becoming an ever-shrinking minority in the countryside. They began to lose their newly won power in the town councils and in the governance of the rural communities – although at a slower rate than the drop in their numbers would have predicted (Munters, 1989). At the end of the twentieth century even the district water boards were no longer farmer strongholds (van Meurs, 2002). As new groups started to play a role in the countryside, and began to make claims on its use that were not always compatible with those of agriculture, the countryside was increasingly faced with satisfying the demands of nature lovers and recreation seekers as well as of efficient agriculture (Kooij, 1999). Horses that had replaced the tractor in the 1950s were beginning to return as hobby horses in the 1990s (van Cruyningen, 2006).

What these developments signified was that the old taken-for-granted assumption that the countryside's and agriculture's sole purpose was to produce sufficient food for the Dutch population (and for export markets), and to offer a constantly decreasing number of farmers a good income, was coming under intense pressure (van der Molen *et al.*, 1978). The agricultural institutional matrix itself began to show cracks. In 1974 there was a demonstration in the Galgewaard football stadium in Utrecht that became famous for the reason that the farmers were protesting not only against the Dutch government but also against the leaders of the three most important farmers' organizations (who had then held office for two decades). The Galgewaard farmers were protesting against the constant pressure to scale up their businesses, the pressure of work and especially against the unacceptably low prices they had to take for their agricultural produce (Krajenbrink, 2005: 248-255). This protest differed from that of the Farmers' party in the 1960s in that the Galgewaard farmers belonged to the target group of Dutch agricultural policy. In the face of serious discontent the farmers' traditional leaders may have retreated somewhat from their positions, but agricultural policy did not change in its essentials. It was in these circumstances that new farmers' organizations were established and became visible in the 1990s and in the early years of the new century (van Broekhuizen *et al.*, 1997; van der Ploeg, 2003).

What became evident to many was that the farmers had grown too dependent on state initiatives, rules and regulations. And with slowed economic growth the differences between small and large farmers, arable farmers and intensive livestock farmers and the North and the South were thrown into focus. Each of these interests called for its own customised policies. It became harder to find solutions for problems from 'the commanding heights' (Yergin & Stanislaw, 2002). The difficulties experienced within the agricultural matrix were aggravated by the way other social groups, whose interests were often incompatible with those of agriculture, were now demanding that their claims on the use of the countryside be taken seriously. Matters were not helped by the way it was becoming progressively clearer that agriculture had severe environmental problems. Intensive pig and poultry farming, in particular, had given rise to a surplus of manure that could no longer be spread on the fields (Frouws, 1993).

As the period of organized capitalism gave way to disorganized capitalism, 'agriculture' and the 'rural' came to be decoupled as two distinguishably separate dimensions of the new countryside. The one-dimensional framing of agriculture as a national economic sector geared to producing food, which had existed since the 1840s, was coming to an end. Of course agriculture and the countryside remained as food producers, but other powerful interests – especially the retail giants – and such international institutions as the World Trade Organization (WTO) were placing a range of new demands on producers (Ilbery, 2005). Increasing production could no longer be the only goal; this had to be pursued together with achieving higher quality standards and sustainable production methods. In this way the influence of consumers and the general public would grow at the expense of agricultural interests.

Agriculture continues to play an important role in the care of the landscape, but again it is not the only player (Schuurman, 2003). It still has a large influence on regional development, but this is now a shared responsibility (Wiskerke, 2007). It is still the most important component of the European Union budget, but increasingly its dominant position here is being ceded to other policy priorities (Sapir, 2004).

Because the position of agriculture had become more complex, agricultural policy had to broaden its scope as well. The various changes we have been describing would inspire the appearance of different kinds of organizations and a different kind of institutional matrix. Of course, the flexible, diverse and customised solutions that now found favour dovetailed nicely with the spirit of supply-side economics, the emphasis on a demand-driven approach, the expanded role of the market and constant innovation (Ministerie van LNV, 2005)[2].

Since the 1990s many changes have been made to the Dutch agricultural institutional matrix. Organizations within the matrix have been repositioned, redefined and merged; some have even disappeared. Many of these organizations went back to the end of the nineteenth and the beginning of the twentieth centuries and had been much expanded, professionalized and brought within the domain of the state since the 1950s. Some departments, such as the Agricultural Extension Service and the Agricultural Research institutes (which had grown out of the experiment stations), were privatized. The Agricultural Research institutes merged in 1998 with the Agricultural University to form Wageningen University and Research Centre. The Government Service for Land and Water Use became the Service for the Rural Areas. When the Agricultural Board ceased to function in 1998 most of its tasks pertaining to public law were transferred to the commodity boards. It was succeeded by a private organization, 'LTO Netherlands', as a partner for consultation and an advocate for agricultural interests. The mostly liberal farmers' societies merged with the confessional farmers' unions into regional farmers' organizations. What also occurred was that the Ministry came to acknowledge that Dutch agriculture must include more than just primary producers. Since the mid-1980s there has in consequence been a bi-annual consultation between top Ministry personnel and important private enterprises such as Numico, Unilever, Ahold, Heineken, CSM, the Rabobank, Friesland Coberco, Cebeco, Arcadis, Cargill and AVEBE (Bekke & de Vries, 2001: 57).

[2] For an inspiring vision on the future of the Dutch agricultural sector, see Zachariasse (2000).

The change from agriculture's old one-dimensional character finds expression in the name of the Dutch Ministry for Agriculture. Up to 1959 it was called the Ministry for Agriculture, Fisheries and Food Supply. Then for 30 years it was known as the Ministry of Agriculture and Fisheries. This name changed in 1989 to the Ministry of Agriculture, Nature Conservation and Fisheries and changed again in 2003 to the Ministry of Agriculture, Nature and Food Quality. These names are more then just new labels. They reflect the acceptance of the changing place of agriculture and the reconfiguring of its position (and that of the Dutch agricultural institutional matrix) in order to survive in the new economic and political climate.

IV.4. The agricultural Sonderweg

When we look again at Tracy's explanatory framework we can see how the critical circumstances in which the agricultural institutional matrix has had to confront its problems has undergone profound change. For one thing the general attitude towards economic policy had altered; increasingly the consensus now is that the market rather than the state should play the major role. No longer was the general public in favour of agricultural support – Dutch farmers were often perceived as polluters and as animal unfriendly. Nor was the strategic argument deemed valid any longer; at a European level society was now self-supporting and self-sufficient in food. What had also changed was that industry and services had now to trade in a more highly competitive European and world market. The Dutch government was no longer willing or able to intervene to save important Dutch industries (like the aircraft manufacturer Fokker) from bankruptcy or foreign takeover.

Dramatic agricultural changes were in train as the number of farms and farmers decreased continuously to very low numbers (a few per cent of the working population). Furthermore, the relative power of agriculture had declined even if the economic importance of agriculture (as part of agri business) remained high and agriculture continued to retain its importance among Dutch exports. With farmers and their families decreasing so much in number, they no longer constituted a decisive strand of the electorate. Seen against this background it is hardly strange that support for agriculture would have dropped after the 1970s. What indeed is remarkable is that farmers and agriculture, compared with other groups in society and other sectors of the economy, managed to fend off major change until the 1990s (with the exceptions of the 'Interim-law' of 1984 to control the growth of pig farming and the super levy introduced in the same year to control the increase in milk production).

Although we can date the beginnings of disorganized capitalism to around the late 1960s/early 1970s, we have seen how it was only since the 1980s that the trend of political change – less government and more market – became really clear. The first cabinet of Lubbers (1982-1986) commenced a policy of reducing government spending and promoting privatization. Yet state intervention in agriculture remained very strong. This Sonderweg of agriculture can mainly be explained by the manner agriculture was well organized and agricultural policy became the European Union's responsibility and therefore difficult to change (Siemes, 2008; van Dijk *et al.*, 1999).

The peculiar position of agriculture also expressed itself in organizational terms. Around 1965, in a clear shift of policy, the Roman Catholic bishops of the Netherlands encouraged the Catholic organizations to become more open to working with non-Catholic groups (Kennedy, 1995: 95; Smits, 1996: 50). This cleared the way for the merging of socialist and Catholic labour unions (1976) and for the amalgamation of Protestant and Catholic political parties (1980). Significantly, however, the farmers' unions and societies did not merge until the middle of the 1990s.

But, as has been said, this Sonderweg of agriculture was far from successful. The crisis could not be resolved in the old way; in fact, it only grew deeper (Frouws, 1993; Siemes, 2008). Moreover, agriculture was severely hit by animal diseases in 1996 (BSE), in 1997 (classic swine fever), in 2001 (foot-and-mouth disease) and again in 2003 (fowl plague) (Mourits *et al.*, 2008). It had therefore become clear in the 1990s that the underlying changed position of agriculture in society, culture and the economy had to be acknowledged if new solutions were to be found. As in the 1880s and 1890s, these new solutions had partly appeared in the preceding decades when the tendency for preserving the *status quo* was still dominant.

One of the main reasons for the continued strong role of the state in agricultural affairs in the 1970s and 1980s was certainly the CAP. As long as the CAP remained the most important EU policy area, and dominant states like France were major beneficiaries under it, it was very difficult to bring in major reform. The growing importance of environmental politics (since 1975), along with the Single European Act (1986) and the European Monetary Union (1999), would prove to be significant in weakening the resistance to reform. After 1989, as a result of the enlargement process, there was also the fear of an even costlier CAP. Further pressure for reform would come from the 1994 GATT Uruguay Round Agreement on Agriculture; and from the commencement in 2001 of negotiations (still not finalized) in the WTO Doha Development Agenda.

In spite of the current period of low or even diminishing prices on the world market, the future seems bright once again for the agricultural sector in the Netherlands. The economic 'rise of the rest' promises a rise in demand for agricultural products, not only for food products but also (as in the period before 1840) for raw materials and biofuel (Meester, 2008; Amsden, 2001). There will, however, be no return to the agricultural practices of organized capitalism. The ecological, social, political and cultural contexts in which agriculture is now practiced have simply changed too much for this to happen.

V. Conclusion

The development and repositioning of the Dutch agricultural institutional matrix provide a suggestive case of how agriculture is restructured in the broader transition of society. When Dutch society shifted from liberal to organized capitalism in the interwar period, so too did agriculture follow more-or-less immediately. But in the transition from organized capitalism to disorganized capitalism that began in the late 1960s agriculture lagged behind for quite some time. How can this Sonderweg be explained?

When the agricultural institutional matrix made its U-turn in the 1930s, Dutch agriculture was already highly developed. State, market, science and private enterprise were institutionally linked in ways that made Dutch agriculture a competitive, efficient and highly productive sector. Relatively speaking, agriculture was one of the most modern sectors of the economy. The agricultural leaders seized the opportunity offered by the crisis of the 1930s. Many new policies were introduced and state spending on agriculture grew swiftly.

After the Second World War Dutch agriculture regained its market orientation. State policy was directed primarily at providing a knowledge and technical infrastructure, securing good production conditions, supporting the reorganization of agriculture and catering to the needs of the farmers and farm labourers who were no longer able to earn a decent living in the sector. Although in certain circles agriculture and the countryside were viewed as traditional, there can be no doubt but that agriculture's economic organization was a very modern part of society.

Why were Dutch agricultural organizations relative to other organizations slow in adapting to the transition from organized to disorganized capitalism? During the 1950s and 1960s the Dutch agricultural institutional matrix had been strongly organized. It might even be said that agriculture qualifies as one of the most successful instances of organized capitalism in the Netherlands. For this reason alone it was inclined to continue in the direction that up to then had proved so successful; and being so superbly organized it was in a good position to resist change. By virtue of its strong organization and well connectedness with Dutch political parties and government, but mainly reflecting the way agricultural policy had been 'Europeanised', the Dutch agricultural institutional matrix changed very slowly. During the 1970s and 1980s

Dutch agriculture remained a sector where state intervention continued to be critically important.

The shift to more neoliberal economic policies (national and international) did not cause agricultural economic conditions to improve. And as the European agricultural policy finally began to change, the Dutch agricultural institutional matrix and agricultural policy came under increasing pressure to follow suit. Ever since agricultural policy in the Netherlands has become notably more flexible, less hierarchical and more diverse and directly linked to the market than was the case in the post-1930s period. Once again public-private arrangements have undergone change. The Dutch agricultural institutional matrix has repositioned and broadened itself to meet the challenges of the changed position of agriculture, of the new circumstances faced by the nation-state and of the changes that have made for accelerated globalisation.

Drawing inspiration from Tracy and Lash and Urry, this chapter has analysed the changes (and their timing) in state policy towards agriculture in the Netherlands since the late nineteenth century. Obviously Tracy's framework needs to be updated to take account of the major changes in recent decades. Nowadays strategic considerations are no longer restricted to food security in times of war; more and more they extend to guaranteeing food safety, conserving the landscape, preserving the rural sphere and advocating sustainable agriculture. In Lash and Urry's account of the 'end of organized capitalism', and in their key concepts of organized and disorganized capitalism, we have been provided with the bases of a more comprehensive and revealing interpretation of the dynamics of agricultural policy change (and change in the agricultural institutional matrix) in the Netherlands over the course of the twentieth century. My account has shown how the peculiarities of the Dutch case (especially what has transpired in the early decades of disorganized capitalism) demonstrate the absence of any perfect fit between the model and reality. In the most recent period, agriculture has faced many of the same problems as other sectors of the economy and society. For a relatively long time, in view of its specific position, it could resist these problems, but in the end it too had no choice but to accommodate itself to the new realities.

Bibliography

AMSDEN, Alice H. (2001), *The Rise of the Rest: Challenges to the West from Late-industrialization economies,* Oxford, Oxford University Press.

BEKKE, Hans & DE VRIES, Jouke (2001), *De Ontpoldering van de Nederlandse landbouw. het ministerie van Landbouw, Natuurbeheer en Visserij, 1994-2000,* [The Ministry of Agriculture, 1994-2000] Leuven/Apeldoorn, Garant.

BIELEMAN, Jan (2008), *Boeren in Nederland: geschiedenis van de landbouw, 1500-2000,* [History of Farming in the Netherlands, 1500-2000] Amsterdam, Boom.

BUITER, Hans & KORSTEN, Jan (2006), *Land in aanleg: de dienst landelijk gebied en de inrichting van het platteland,* [The Service for the Rural Area] Zutphen, Walburg Pers.

BURGER, Ary (1993), *Voor boerenvolk en vaderland: de vorming van het EEG-landbouwbeleid. [The Making of the CAP],* Documenten en onderzoek PSCW. Amsterdam, Het Spinhuis.

DANIELS, Peter, BRADSHAW, Michael, SHAW, Denis & SIDAWAY, James (eds) (2008), *An Introduction to Human Geography. Issues for the 21st Century,* Third Edition. Harlow, Pearson Education.

DOUW, Leo & POST, Jaap (eds) (2000), *Growing Strong. The Development of the Dutch Agricultural Sector. Background and Prospects,* The Hague, LEI.

FROUWS, Jaap (1993), *Mest en macht: een politiek-sociologische studie naar belangenbehartiging en beleidsvorming inzake de mest-problematiek in Nederland vanaf 1970. [Manure and Power],* Wageningen, Vakgroep Rurale Sociologie.

GRIFFITHS, Richard T. & DE VRIES, Marian (1987), 'The Agricultural Lobby', in Richard T. GRIFFITHS (ed.), *The Netherlands and the Gold Standard, 1931-1936. A Study in Policy Formation and Policy,* Amsterdam, NEHA, p. 47-62.

HELD, David, MCGREW, Anthony, GOLDBLATT, David & PERRATON, Jonathan (1999), *Global Transformations. Politics, Economics and Culture,* Cambridge, Polity Press.

ILBERY, Brian (2005), 'Changing Geographies of Global Food', in Peter DANIELS, Michael BRADSHAW, Denis SHAW & James SIDAWAY (eds), *An Introduction to Human Geography. Issues for the 21st Century,* Second Edition, Harlow, Pearson Education, p. 168-185.

KAREL, Erwin H. (2005), *De maakbare boer: streekverbetering als instrument van het Nederlandse landbouwbeleid, 1953-1970. [The Rural Area Development Programme, 1953-1970],* Historia Agriculturae, 37, Groningen/Wageningen, NAHI.

KENNEDY, James C. (1995), *Nieuw Babylon in aanbouw. Nederland in de jaren zestig. [The Netherlands in the Sixties],* Amsterdam, Boom.

KOOIJ, Pim (1999), *Mythen van de groene ruimte. [Myths of the green area],* Wageningen, Wageningen Universiteit.

KRAJENBRINK, Evert Jan (2005), *Het landbouwschap: 'zelfgedragen verantwoordelijkheid' in de land- en tuinbouw, 1945-2001, [The Agricultural Board, 1945-2001],* Den Haag, Eigen beheer.

KRIPS-VAN DER LAAN, Hilde M. F. (1985), *Praktijk als antwoord. S.L. Louwes en het landbouwcrisisbeleid. [S.L. Louwes and the Agricultural Policy in the Great Depression],* Historia Agriculturae, 16, Groningen, NAHI.

LASH, Scott & URRY, John (1987), *The End of Organized Capitalism*, Cambridge, Polity Press.

MEESTER, Gerrit (2008), 'Toekomst van het EU-beleid voor landbouw, voedsel en groen' [Future of the CAP], in Huib SILVIS, Arie OSKAM & Gerrit MEESTER (eds), *EU-beleid voor landbouw, voedsel en groen. Van politiek naar praktijk*, Second edition, Wageningen, Wageningen Academic Publishers, p. 387-415.

MEESTER, Gerrit & STRIJKER, Dirk (1985), *Het Europese landbouwbeleid voorbij de scheidslijn van zelfvoorziening*, Voorstudie 46. [The CAP beyond Self-sufficiency] Den Haag, WRR.

MINISTERIE VAN LNV, (2005), *Kiezen voor landbouw: een visie op de toekomst voor de Nederlandse agrarische sector,* [Choices for Agriculture], Den Haag, Ministerie van Landbouw, Natuur en Voedselkwaliteit.

MOURITS, Monique *et al.* (2008), 'Veterinair en fytosanitair beleid' ['Veterinary policy'], in Huib SILVIS, Arie OSKAM & Gerrit MEESTER (eds), *EU-beleid voor landbouw, voedsel en groen: van politiek naar praktijk*, Second edition. Wageningen, Wageningen Academic Publishers, p. 175-190.

MUNTERS, Quirinus J. (1989), *De stille revolutie op het agrarische platteland: boeren en openbaar bestuur 1917-1986 [Farmers and Public Governance, 1918-1986],* Assen, Van Gorcum.

NOOIJ, Adrianus T.J. (1969), *De boerenpartij. Desoriëntatie en radikalisme onder de boeren. [The Farmers' Party],* Utrecht, Eigen beheer.

VAN DER PLOEG, Jan Douwe VAN DER (2003), *The Virtual Farmer: Past, Present, and Future of the Dutch Peasantry*, Assen, Van Gorcum.

PLOEG, Jan Douwe VAN DER & WISKERKE, Han (eds) (2005), *Het landbouwpolitieke gebeuren: liber amicorum voor Jaap Frouws, [Aspects of Agrarian Policy],* Wageningen, Circle for Rural European Studies.

SAPIR, André (2004), *An Agenda for a Growing Europe, the Sapir report*, Oxford, OUP.

SARIS, Frank (2007), *Een eeuw vogels beschermen. [A Century of Bird Protection],* Zeist, KNNV Uitgeverij.

SCHUURMAN, Anton (2003), 'Vertel, Muze, vertel. Geschiedenis, ruimte en cultureel erfgoed' ['History, Space and Cultural Heritage'], in Michiel A. W. GERDING (ed.), *Belvedere en de geschiedenis van de groene ruimte*, Historia Agriculturae, 33, Wageningen/Groningen, NAHI, p. 9-49.

SCHUURMAN, Anton (2007), 'From Citizen to Citizen-consumer. The Development of the Welfare State and the Consumer Society', in Hester MOERBEEK, Anke NIEHOF & Johan VAN OPHEM (eds), *Changing Families and their Lifestyles. Liber amicorum voor Kees de Hoog*, Mansholt publication series, Volume 5, Wageningen, Wageningen Academic Publishers, p. 213-228.

SCHUURMAN, Anton (2008), *The Construction of Dutch Agriculture Inc: from Liberal Capitalism to Organized Capitalism,* Paper for Rural History Session at the ESSHC 2008 in Lisbon, Wageningen (Manuscript).

SCHUYT, Kees & TAVERNE, Ed (2000), *1950. Welvaart in zwart-wit. [1950. Prosperity in Black and White],* Den Haag, Sdu Uitgevers.

SCOTT, James (1998), *Seeing like a State: How Certain Schemes to Improve the Human Condition have Failed,* New Haven/London, Yale University Press.

SIEMES, Hans (2008), *Braks. Carrière met een diploma handmelken. [Braks. A biography],* Zwolle, Waanders Uitgevers.

SILVIS, Huib, OSKAM, Arie & MEESTER, Gerrit (eds) (2008), *EU-beleid voor landbouw, voedsel en groen: van politiek naar praktijk, [EU-Policy for Agriculture, Food and the Green Area],* Second edition. Wageningen, Wageningen Academic Publishers.

SMITS, Jan-Pieter (2008), 'Technological Change, Institutional Development and Economic Growth in Dutch Agriculture,1870-1939', in Pedro LAINS & Vicente PINILLA (eds), *Agriculture and Economic Development in Europe since 1870*, London, Routledge, p. 97-117.

SMITS, Mari (1996), *Boeren met beleid. Honderd jaar Katholieke Nederlandse Boeren- en Tuindersbond, 1896-1996 [Dutch Catholic Farmers- and Gardeners Union, 1896-1996],* Nijmegen, Valkhof Pers.

TRACY, Michael (1989), *Government and Agriculture in Western Europe 1880-1988,* New York, Harvester Wheatsheaf.

TRIENEKENS, Gerard (1985), *Tussen ons volk en de honger. Voedselvoorziening 1940-1945. [Food Supply, 1940-1945],* Utrecht, Matrijs.

VAN BROEKHUIZEN, Rudolf Erik, KLEP, Leo & OOSTINDIE, Henk (eds) (1997), *Renewing the Countryside: an Atlas with two Hundred Examples from the Dutch Rural Society,* Doetinchem, Misset.

VAN CRUYNINGEN, Piet (2006), 'Van trekpaard naar luxepaard. De mechanisatie van de landbouw' [The Mechanisation of Agriculture], in Dolly VERHOEVEN *et al.* (ed.), *Gelderland, 1900-2000,* Zwolle, Waanders, p. 345-350.

VAN DEN BERGH, Simon M. (2004), *Verdeeld land: de geschiedenis van de ruilverkaveling in Nederland vanuit een lokaal perspectief, 1890-1985. [Land Consolidation from a Local Perspective, 1890-1985],* Historia Agriculturae 35, Groningen/Wageningen, NAHI.

VAN DER MOLEN, H., VAN DER BRUGGE, G. J. ter, VAN DIEST, A. (1978), *Omstreden landbouw. Bijdragen over technische en ecologische aspecten van de Nederlandse landbouw in de naoorlogse tijd. [Contested Agriculture],* Aula-boeken 635, Utrecht/Antwerpen, Het Spectrum.

VAN DIJK, Gert, KLEP, Leo & MERKX, A. J. (1999), *De corrosie van een ijzeren driehoek. Over de omslag rond de landbouw. [On the Reversal of Agriculture],* Assen, Van Gorcum.

VAN KAMP, Hans (2005), *Dien Hoetink. 'Bij benadering'. Biografie van een landbouwjuriste in crisis- en oorlogstijd. [Dien Hoeting. A biography],* Historia Agriculturae, 36, Groningen/Wageningen, NAHI.

VAN MEURS, Rudie (2002), *De laatste dagen van een boerenrepubliek. Polderdistrict Betuwe, Vijfentwintig jaar werk en strijd. [Polder District Authority Betuwe],* Utrecht, Matrijs.

VAN VEGCHEL, Gerrit (1995), *De metamorfose van Emmen. Een sociaal-historische analyse van twintig kostbare jaren, 1945-1965. [The Metamorphosis of Emmen],* Amsterdam, Boom.

WISKERKE, Han (2007), *Robuuste regio's: dynamiek, samenhang en diversiteit in het metropolitane landschap. [Dynamics, Coherence and Diversity in the Metropolitan Landscape],* Wageningen, Wageningen UR.

WOODWARD, Nicholas (1999), 'The Search for Economic Stability. Western Europe since 1973', in Max-Stephan SCHULZE (ed.), *Western Europe. Economic and Social Change since 1945,* London/New York, Longman, p. 63-81.

YERGIN, Daniel & STANISLAW, Joseph (2002), *The Commanding Heights: the Battle for the World Economy,* New York, Simon and Schuster.

ZACHARIASSE, Vinus (2000), 'Towards a New Framework', in Leo DOUW & Jaap POST (eds), *Growing Strong: the Development of the Dutch Agricultural Sector. Background and Prospects,* The Hague, LEI, p. 157-179.

4. Health, local taxation and market competition: municipal laboratories and state food control in Spain, 1887-1914[1]

Gloria Sanz Lafuente

I. Introduction

Foodstuffs inspection (or food control) became established in Spain in the second half of the nineteenth century as issues relating to public health, market competitiveness and segmentation, the rising urban demand for food and local taxation assumed greater importance. As we will see, certain local actors in the cities were decisive in putting food control on to the political agenda before the central Spanish state began to take a serious interest in the subject. Generally speaking, food control research has formed part of the history of human and veterinary health, as well as of science history; it has advanced more recently on the basis of academic research conducted under the headings of social, economic and political history. Following early pioneering work dealing with the adulteration of food and food control at the local level (Ellerbrock, 1987; Teuteberg, 1994; Hietala, 1994), general accounts have been provided of food scares (Ferrières, 2002), conditions affecting the quality of food (Stanziani, 2005) as well as national attempts at food control via the introduction of regulatory codes (French & Phillips, 2000). Marc T. Law, for instance, has linked the regulation of food purity with dynamic market conditions and analyzed the influence of the US State Pure Food Regulation on the price commanded by dairy products (Law, 2003).

Over the last few years a number of influential studies have come to focus on cases of food control at the municipal level. Prominent exponents of this line of research include Lucie Paquy (Grenoble), Peter Scholliers (Brussels), Derek J. Oddy (Society of Public Analysts, London) and Alessandro Stanziani (Paris). All these authors have contributed towards expanding the analytical framework for studying food safety. As Peter Scholliers (2007: 77) observes, the recent literature has come to see food control as part of the history of hygiene and health as well as relating it to 'price and competition, protectionism, producers' interests, product classification, quality norms, and the relationship between state and private interests'. Evidence relating to how national food laws come to be applied at local level can be found in the substance of local food and health legislation, in local monitoring regimes and the

[1] This paper derives from the research projects: NISAL: SEJ2007-60845 and SALMON: HAR2010-20684-C02-01. Translation: Korey Wheatley.

employment of new professionals, in the politics of municipal decision-making, and in developments relating to trade and market competition. The emerging body of local food control research has thus come to form part of a history of food safety in Europe (Bruegel & Stanziani, 2004: 7-16; Sanz Lafuente, 2006a: 81-118) while dovetailing with research relating to the development of living standards (Sanz Lafuente, 2006b).

Against such a background this paper will consider some of the food control activities of the municipal chemical laboratories of Spain in the 1887-1914 period (Sánchez León, 2001: 201-206, Guillem-Llobat, 2008). Using data from cities of different size, I will concentrate my attention on the inspection of food other than fresh produce (meat, fruit and vegetables) in the municipal laboratories of San Sebastián and Madrid. Unfortunately, with some exceptions, data covering the same years for all the cities were unavailable. The chronology and legal constitution of these laboratories will be reviewed, as will the factors that came to underpin technical food inspection. The quantitative results emanating from the inspection process will also be considered. To contextualize the emerging municipal regulatory regimes of San Sebastián and Madrid, the origins of the demand for scientific analysis of foodstuffs will be studied. As much as these origins can be associated in the late nineteenth century with concerns about fair and honest trading, food safety regulation was being legitimated more by reference to pubic health considerations by the beginning of the twentieth century.

II. The legal constitution of the municipal laboratories in Spain, 1887-1914

Some rudimentary food regulation existed during the *Ancien Régime*, but the institutionalized process of inspection at the municipal level that initiated modern food control in Europe had to await the nineteenth century. Municipal institutions and civil society private initiatives – based on voluntary associations or professional organizations like the Public Analysts Society of London (1874) – contributed significantly to the appearance and development of national legislation and institutions relating to food safety. Brussels had created a municipal laboratory as early as 1856; Antwerp followed in 1875, Paris in 1878 and Ghent in 1887. As for the German case, Vera Hierholzer comments: 'Until the end of the 1870s, the legal basis for food monitoring in Germany was a patchwork of state and city borders, by-laws, and administrative and police decrees' (Hierholzer, 2007: 117). Compared to the situation elsewhere, there was a delay of some years in the creation of municipal laboratories in several of the Spanish cities. A chemical laboratory had existed in Barcelona since 1867 (Vera y López, 1885: 7) and an Institute for Certifying Foodstuffs, directed by Calvet and organized through the initiative of the Barcelona City Hall, began functioning in 1882 (Casas Duran, 1988: 53; Puerto-Sarmiento & Cobo Cobo, 1983: 152). Under the influence of the French model, other notable developments had seen the appearance of the Barcelona Municipal Microbiology Laboratory (1886) (Casas Duran, 1988: 26-27) (an institution more orientated towards clinical than foodstuffs analysis), the Madrid Laboratory (1877) and the laboratories of Valencia (1881), Vitoria, Bilbao and Zaragoza (1886) (Junta Provincial de Sanidad, 1886a: 13-15: 1886b), and San Sebastián (1887) (Puerto-Sarmiento & Cobo Cobo, 1983: 165; Etxaniz Makazaga, 1998: 280).

The slow growth of the Spanish population during the nineteenth century, despite high birth rates, was related to the high mortality rate (30‰), a rate considerably higher than that found among the countries of central and northern Europe. There is no doubt but that the underlying reason for the high infant mortality rate was the lack of adequately funded public health institutions. The first half of the twentieth century witnessed a drop in mortality rates and a convergence with the rates found in other European countries (Nicolau Nos, 2005: 95; Pérez Moreda, 1985). Considerable historical importance has to be accorded, in accounting for this convergence, to the food control initiatives of physicians and hygiene specialists, such as Pedro Felipe Monlau and Francisco Méndez Álvaro. Their writings were published in the first half of the nineteenth century and addressed such pressing sanitary problems as the containment of outbreaks of epidemic diseases, the provision of clean drinking water and the sanitation of homes, cemeteries, and garbage dumps. In his inaugural speech as a member of the Royal Academy of Medicine, Méndez Alvaro, as well as criticizing the then parlous condition of public health in Spain, alluded in 1853 to the lack of official action. Significantly, he identified the municipalities as the appropriate sites in which major policy change should occur. *Apropos* food production, Méndez Alvaro had noted that the breakdown of guild-based controls and industrial flexibility had led to the absence of proper procedures for regulating manufacturing and distribution activities within the municipalities.

Before the creation of the first municipal laboratories, chemistry, pharmacology, and medicine were academically organized around specialized journals (Granjel, 1983; Muñoz Calvo, 1994; Moro Aguado, 1986; Fresquet Ferrer, 1990; Campos, 2003). Prior to the *General Health Law* of 1855, translations had been published of French studies of the forgery and alteration of foodstuffs and of the ways of identifying forgery and adulteration. The work of J. Garnier and Ch. Horel (1846) was translated by Magín Bonet, and the dictionaries of foodstuff forgery by M. A. Chevalier (1855) and J. L. Souberain (1876) were translated by Ramón Puig and J. R. Gómez Pamo respectively. In the late 1870s Francisco Javier Agreda published his dictionary of foodstuff and beverage forgeries, a work that addressed the chemical means of recognizing adulterations.

The biographies of municipal laboratory directors help us to place them centrally within the process of the internationalization of public health. For example, César Chicote Riego (Alvarez Sierra, 1965), a doctor in pharmaceuticals and the director of the municipal laboratory of San Sebastián (1887-1898) and Madrid (in 1898) was instrumental in compiling food safety guidelines that led to the Royal Decree of 1908. Chicote, an important figure in the internationalization of European science in Spain, had attended courses given by Laureano Calderón y Arana, a professor of science and pharmacology who was exiled in Paris and who had been a member of the International Committee to Reform Chemical Nomenclature. The work of the two chemical laboratories that Chicote directed became the basis of publications in the Amsterdam *Revue Internationale des Falsifications*, the *Chemiker-Zeitung*, and the *Zeitschrift für Nahrungsmitteluntersuchung, Hygiene und Warenkunde*. Toward the end of the nineteenth century, under the influence of French and Belgian literature, Cesar Chicote and José Moreno Madrid (a science faculty member) authored works dealing with food adulteration and the ways of detecting it.

While in 1822 a Penal Code and urban municipality rulings had already explicitly recognized fraud involving food supplies, the legal constitution of local chemical laboratories was mainly a feature of the last two decades of the nineteenth century. With the implementation of the *Ley General de Sanidad* (General Health Law) of 1855 and its modification in 1866, the first local health services, such as the *Junta Local de Sanidad* (Local Board of Health), and the local meat inspection service of 1859 (Sanz Lafuente, 2006b: 525-40), were finally established. Under the *Ley Municipal* (Municipal Law) of 1845 and 1877, hygiene and the control of food supply became the responsibilities of the local authorities. The Royal Decree of 5 January 1887 provided legal support and recommended the creation of municipal chemical laboratories that would monitor, test and analyze food. Under the designation of Local Hygiene Laboratory or Local Chemical Laboratory, these bodies would be responsible for putting in place a regime of food control that would extend to water, beverages, condiments and additives. In practice, the activities of the chemical and laboratory inspectors encompassed a whole range of tasks that covered water quality, disinfection services and vaccine preparation, along with collaboration with slaughterhouse veterinary inspectors. The range of tasks could even extend to bacteriological research, as was the case from 1886 at the Barcelona Municipal Microbiology Laboratory under the direction of Jaume Ferrán i Clua (Casas Duran, 1988: 26). The duties of the technical laboratory personnel extended to the inspection of centres of food production, distribution, and consumption (such as taverns and restaurants) within the municipalities. Their main work activity, aside from collecting and analyzing samples, was taken up with preparing reports, keeping records, and issuing analysis certificates to producers and distributors. They were not, however, able to impose sanctions.

As important as was the 1887 law it did not of itself create the Spanish municipal chemical laboratories; it only recommended their establishment. The 1908 law, heavily influenced by Belgian legislation, showed a marked interest not merely in the repression of fraud but in its detection and prevention via a regime of inspection. Most significantly, it recognised the need to construct municipal laboratories in urban centres of more than 10,000 inhabitants. The precepts of the 1908 law were repeated in subsequent legislation until 1923 so as 'to enhance the zeal of authorities in enforcing already established provisions'[2]. Pharmacists, physicians and scientists, as new groups of qualified professionals, formed the core technical staff of these new municipal laboratories. Later on these professionals were joined by veterinarians who were put in charge of slaughterhouse inspections.

Table 4.1. Number of recorded food analyses in five Spanish cities

Date	*Town*	*Analyses per 10,000 inhabit.*
1887-1899 (average)	San Sebastián	125.8
1903	Bilbao	18.6
1901/1904/1909/1912 (average)	Madrid	72.2
1907	Barcelona*	15.1
1907	Toledo	6.7

* Municipal Microbiology Laboratory.

Source. Municipality Archive: San Sebastián Laboratory yearbooks 1887-1899. Bilbao Laboratory yearbook 1903, Madrid Laboratory yearbooks 1901/1904/1909/1912. Toledo Laboratory yearbook Barcelona: Casas Durán, p. 54. 1900 Population Censuses.

The creation of the laboratories may have been promoted by the central state, but their appearance would depend critically on local initiatives and on local funding. The activities and organization of the laboratories were extremely diverse. Some of the early laboratories, as in Cádiz in 1901, were not at first explicitly concerned with food control: instead, their activities related mainly to disinfection. A comparison of the number of analyses carried out per 10,000 inhabitants points to the importance of specific local conditioning factors influencing the sort of priorities and laboratory

[2] Royal Decree of 26 of April 1920.

practices that came to be adopted. San Sebastián, a case that became an organizational model for others, had a greater number of food analyses per 10,000 inhabitants than the other cities.

Crisis conditions generated by recurring outbreaks of epidemic diseases during the first half of the century – as in San Sebastián in 1834, 1845, 1855 and 1885 – were also important in legitimating the need for municipal laboratories, as was the pressure coming from the 'new' health professions in the cities. The period between 1871 and 1900 saw the publication of 41 per cent of the local medical topographies (*topografías médicas*) produced in the 1801-1940 period (Urteaga, 1980: 23). Intensive international academic communication made its own independent contribution to the opening of the municipal laboratories in Spain. Linked to this was the birth of the 'social question' and the rise of a more 'scientific view' of hygiene problems that accompanied the 'bacteriological revolution' in Spain (Herold-Schmidt, 1999: 303-441). Nor were local political and economic considerations – linked to issues of local consumption taxes and competition between old and new products in a dynamic and segmenting market – irrelevant to the appearance of the new laboratories.

III. Food control conditioning factors

Before 1908 the detection of food adulteration in a range of products relied on a variety of criteria and methods (Oddy, 2007: 91). While laboratory work benefited from the technological advances of the time – the availability of microscopes, new alcohol meters and chemical agents for use in analysis – scientific working methods were still often poorly standardized and the laboratories had to contend with blurred and changing definitions of what constituted food purity. Article 15 of the 1888 San Sebastián Laboratory Regulation stated that the director was responsible for the 'choice of methods and procedures involved in analytical tasks [and for] assuming all the legal responsibility involved'; a similar statement was found in the Madrid Laboratory's regulatory code of 1902[3].

Although the purity of some products and their composition had already become the subject of legislation[4], the next major step was the publication in 1908 of standardized instructions regarding analysis and sampling work[5] and the compiling of the first technical instructions relating to foodstuffs, packaging, kitchen utensils and food containers (Royal Decree of 22 December 1908, extended and amended in

3 *Reglamento y tarifa del servicio del Laboratorio Químico Municipal de San Sebastián*, (1902).
4 For example: RO de 9 de junio de 1858 ... RO de 17 de Junio 1890... RO de 20 de enero de 1908.
5 Laboratorio Municipal de Higiene de Madrid (1906).

1920)[6]. These instructions represented not only the first comprehensive attempt at the standardization of laboratory work methods, but constituted the first attempt at defining the purity standards for a whole series of food products. As stipulated by the 1908 legislation:

> The lack of a precise definition of the main foodstuff products has enabled forgery, in view of the diversity of existing criteria of food acceptability, so in order to prevent any ambiguity, the definitions of what may be both practically and commercially considered a pure foodstuff product are established for each food item in the annexes to this Royal Decree, [...][7].

The attempt of the 1908 legislation to impose central standards of food purity and control can only be understood in the context of the limitations and ambiguities that previously had been rife. Chicote noted the difficulties with wine analysis in 1897 when he stated that his microscopic analysis gave him 'no idea of composition'. What he also found 'difficult to determine', in view of the different make-up and alcohol content of the vintages of different regions, was the amount of water in wine. The analysis of the nitric acid additive content in wine, whose effect was similar to that of gypsum, was also difficult to establish. Regarding the adding of banned artificial colorants, Chicote indicated that 'his research was extremely difficult and the problems involved could sometimes preclude accurate findings'. With regard to olive oil, adulterated by the addition of peanut, sesame, cotton, or rapeseed oil, Chicote felt obliged to say that: 'I must confess that there is no quick and thorough procedure' (Chicote y Riego, 1897: 194).

Both the building of a scientific consensus relating to the purity of foodstuffs and the establishing of appropriate legislation were processes driven by economic interests, scientific theories and social considerations (Salais *et al.*, 1998). The quest for a common regime of inspection was also given urgency by the diversity of purity norms subscribed to by the directors of the laboratories. Let us provide an illustrative example here. The addition of gypsum – potassium bisulphate – was a wine-production technique that had been in use for centuries, enabling producers to intensify colour and to accelerate and complete the fermentation process while reducing the risk of subsequent fermentation. Gypsum also acted as a preservative during transport in hot weather or when wine was manipulated and mixed with other products. The main health consequences of excessive gypsum in wine are digestive disorders. Whereas some of the legislation of the *Ancien Régime* had ruled on the use of gypsum, in 1888 there was still no technically standardized norm defining the maximum number of grams of potassium bisulphate that a litre of wine could contain

6 Royal Decree of 17 September, 1920.

7 Royal Decree of 22 December, 1908.

to qualify as 'good' in a laboratory analysis. This spawned a series of different local rulings. While in the laboratories of Madrid, Zaragoza, Bilbao and Valladolid some two grams per litre was considered the maximum authorized amount, as many as four grams per litre were permitted in Alicante. Only in 1892 did a Royal Decree finally establish a maximum of two grams per litre of gypsum, and this rule held good until the advent of the 1908 regulation.

The laboratory directors can be identified as an important force behind the standardization of the food purity rules. What proved especially significant here was the consensus the directors built in daily practice, through mutual communication and even through correspondence with laboratories further afield in Bordeaux and Paris. On such bases a generalized practice began to emerge among the first laboratories, with the notable exception of Alicante[8]. Building a consensus among the laboratory directors was achieved in a context where many private concerns were disposed to resist the regulatory regime emerging around the new municipal laboratories. Thus the director of the San Sebastián laboratory commented in 1888 that:

> It would be useless to list the thousands of drawbacks and inconveniences we have suffered. Some wineries sent us scurrilous anonymous notes, others threatened to contact their elected representatives to have us put out of business, and most of them made ridiculous objections, using the same delicacy they applied when contaminating their wines. (Stanziani, 2005: 333)[9].

What further gave impetus to the work of the municipal laboratories was the appearance of new chemical products such as the colouring additives created in the latter part of the century. The introduction of these additives had to overcome the resistance of traditionalists who swore by the old manufacturing methods. For instance, the toxic colorant (Naphthol S. Yellow) used in the making of sweets was decommissioned in 1905 in San Sebastián despite egg yolk being traditionally considered the 'natural ingredient' in its manufacture. Likewise, the natural colorant for pasta was considered to be saffron and not aniline[10]. The harmfulness of some of these traditional substances always emerged 'after the fact', allowing chemical food research to come to fairly firm conclusions. Compared with the traditional nature and acceptance by 'usage' of the standard additive, the mere novelty and artificiality of new products became the reason for their rejection on occasion. This of course is understandable in light of the way the toxicity of colorants such as bi-nitro naphthol was not clearly or widely known. At other times, conventional practices were changed virtually overnight. Within a few years, the use of copper as a green

8 Laboratorio Químico Municipal de San Sebastián (1888), p. 28.
9 Laboratorio Químico Municipal de San Sebastián (1888), p. 30.
10 Laboratorio Químico Micrográfico de San Sebastián (1907), p. 14.

colorant for vegetable preserves became widespread; its status in 1891 as a banned product changed to that of an authorized one in 1908 [11] – in maximum amounts of one milligram per kilo of product. In a context of the appearance of new food processing firms and increasing manufacturing sophistication, all developments mentioned in the chapter on 'colorants' in the 1908 decree, no concrete provision was made for usage restrictions.

The power that competitive market forces were able to exert, to the advantage of some products and to the disadvantage of others, would also contribute to the standardization of food purity and to changing conceptions of how this was to be defined (Stanziani, 2003). In 1899, in line with what was happening in other countries (Merki, 1994: 192-202), saccharine was classified as a 'medicinal product', thus forbidding its use in foodstuffs and beverages as a form of 'food adulteration'[12]. This made it a substance to be sought out by the laboratories and eliminated in sweets, confectionary and liquors[13]. Early in the century, other products possibly containing saccharine were included in the search. In 1906, the yearbook of the San Sebastián Municipal Laboratory indicated that 'the heavy fines imposed by city hall and the closure of two (lemonade and soda) factories due to their use of saccharine have contributed to the elimination of this type of adulteration'; similar opposition to the use of saccharine was found in the Royal Order on Food Adulteration published in the same year[14]. The monitoring of biscuits, confectionery, sodas, soft drinks, and syrups in the municipal laboratory of Madrid was related to the saccharine issue as well (Table 4.4). While the *Madrid Chemical Micrography Laboratory Bulletin* recognized that it was a 'harmless' product, it nevertheless justified its banning by citing both its reduced 'energy-giving potential' compared with sugar, and the 'major services provided by Medicine to Industry'[15]. Behind the attack on saccharine can be detected an explicit interest in defending the use of sugar as a food sweetener, especially when we find sugar's nutritional value being compared with that of saccharine. Eventually, in 1913, this standard and scientific convention was subject to modification by Royal Decree when all those punished earlier for their use of saccharine in food and beverages were granted a free pardon[16].

---

11 Royal Order of 9 December, 1909; Royal Order of 11 August, 1906.

12 Royal Order of 3 April, 1899; Law of 28 November, 1899.

13 Royal Order of 3 April, 1889 ; Royal Order of 17 September, 1909; Royal Decree of 19 November, 1916.

14 Laboratorio (1907), p. 13 ; Royal Order of 11 August, 1906.

15 Boletín del Laboratorio Químico-Micrográfico, November 1901.

16 Gaceta de Madrid, nº 37, 23 December, 1913.

IV. The demand for food analysis

Who were the prime movers in initiating the analysis of food within the municipalities? Spanish laboratories were partly self-financing as a result of the practice of charging analysis fees to private concerns. Essentially Spanish laboratory activity had two distinguishable sides: an official side that saw the laboratories processing samples provided by local and provincial municipal authorities as a public service, and a private side that involved individuals and corporations paying for laboratory services in accordance with a set fee structure. The revenue derived from this source helped fund the municipal inspection service. Such a funding model was not peculiar to Spain. Thus, in 1878, the services made available by the laboratory of Paris were being used by the police and prefecture as well as by private actors to whom a tariff list applied (Scholliers, 2007: 85; Stanziani, 2007: 110-111). In this respect, Paris differed from Brussels where the tests were offered *gratis* to the public. Frequent mention was made in the yearly reports of the Spanish laboratories to how important the charges paid by private users of laboratory services were to financing equipment and covering running expenses[17]. There was also an emphasis on measuring the precise level of the private demand for laboratory services within different municipalities. In 1888, private analysis fees ranged between 7.50 and 12.50 pts in San Sebastián, depending on the product and the type of analysis. Charges of this order made food analysis prohibitive for the bulk of consumers, bearing in mind that the average daily wage of urban workers ranged between 2.46 and 3.64 pts between 1856 and 1905, and that for most of the rural population wages were still very low (Maluquer de Motes, Llonch, 2005: 77). As of 1897 the analyses of bread, meat, fish, milk, wine, cider and spirits were provided gratis, but if a certificate was required then the set fee had to be paid[18].

The cost of a laboratory certificate and its usefulness for commercial publicity[19] – as a guarantee of quality – made certificates a valued resource in market competition and a basis of market segmentation for food producers and vendors. Certainly there was a strong commercial demand for certificates; in 1898, for instance, 85 per cent of all analyses in San Sebastián were accompanied by a request for a certificate. A comparison of the analyses requested by public and private clients over time points to the high level of private customer demand, accounting on average for 52.6 per cent of San Sebastián food analyses between 1888 and 1899. Yet from the beginning of the century the available data for Madrid, San Sebastián, and Bilbao show a fall-off in the level of private demand and a consolidation of the use public bodies were making

17 Laboratorio Químico Municipal de San Sebastián (1888: 27).

18 Ayuntamiento de San Sebastián (1897: 4 y ss.). (Art. 48 ordenanzas municipales).

19 Ayuntamiento de Madrid (1902: 67).

of the laboratories. The increased share of water analyses – 75 per cent of all Madrid analyses were of water in 1909 – further accentuated the drop in private demand, at least in the case of Madrid. Greater concern with the quality of water was the first sign of a transformation in the type of local health work being undertaken by the laboratories; and marked a change from their initial very high levels of dependence on private demand for analyses.

Table 4.2. Privately requested food analyses in four Spanish cities

Date	*% San Sebastián*	*% Madrid*	*% Bilbao*	*% Toledo*
1888	52.8			
1889	56.7			
1990	62.4			
1891	51.9			
1892	46.3			
1893	50.2			
1894	77.5			
1895	49.2			
1896	53.2			
1897	37.8			
1899	41.2			
1901		21.5		
1902		18.6		
1903		14.6	36.2	
1906	10.2	6.4		
1909		4.3		1.3
1913		13.3		

Source. San Sebastián: 1888-1899/1906 Laboratory Yearbooks. Madrid: 1904, 1909, 1912 Laboratory Yearbooks. *Bulletin of the Chemical Micrographic Laboratory 1901-1902*. Bilbao: 1903 Laboratory Yearbook. Toledo: 1910 Laboratory Yearbook.

Unfortunately the yearbook data do not provide further details on the nature of the demand for analysis originating among private food distribution and processing interests. The same is not true of the analysis requested by public sector customers, details of which are to be found in the yearbooks of laboratories. Although there is a certain amount of local diversity, analyses relating to food hygiene, requested by veterinary inspectors and chemists in the Madrid and San Sebastián laboratories and by the physicians of Toledo, represent the bulk of requests for analysis. Responsibility for inspection of food substances rested with the municipal laboratory, and its director could undertake inspections whenever he felt these were appropriate or when other competent authorities authorized an inspection.

Another prominent aspect of the work of the municipal laboratories was the manner local consumption taxes impinged on the demand for food analyses. The consumption tax on salt, spirits, liquors, alcohols, and certain foodstuffs and fuels had become an important element of the municipal tax structure and therefore an important source of tax revenue for local authorities (Martorell Linares, 1995; Vallejo Pousada, 1996). Not alone did the collection of this tax affect the urban population but it also had implications for the largely agricultural populations in the hinterlands of cities. The analysis of municipal laboratories became a crucial factor in establishing the market price for a product, especially those products whose price depended on determining degrees of alcohol content or purity accurately and precisely (Vera y López de 1885: 55).

Table 4.3. Public sector requested food analysis in three Spanish cities

Non-private food analysis requests	*San Sebastián*	*Toledo*	*Madrid*
Date	*1887-1898*	*1907/1909*	*1901/4/8/9/13*
Laboratory (chemical and veterinary inspectors)	70.06	20.4	51.3
Arbitration/Consumption	24.17	2.8	4.2
Deputyship, City Hall, Governor, Courts	5.70	12.3	18.8
Physicians community/private	0.03	57.3	25.7
Health Police	0.02	7.2	0.0
Other	0.02	0.0	0.0
	100	100	100

Source. See Table 4.2. Toledo: Laboratory Yearbook.

V. The character of laboratory analyses

A first step in tracing the working practices of the municipal laboratories is to study the sort of analysis being conducted and the sort of certification being granted to product-based substances. Bearing in mind the high percentage of privately requested analyses, the overall predominance of good quality reports (with the exception of 1896) was closely connected with the demand for certification for commercial purposes. Over time substances graded as 'good' dropped from 70 per cent to 40 per cent while those graded as showing the presence of non-toxic adulteration grew in number. A combination of the rise of industrial chemistry and an increasingly competitive urban market in Madrid led to an average percentage of 45 per cent of substances being classified as adulterated (either 'dangerous' or 'safe' frauds) between 1901 and 1912.

Figure 4.1. The tests performed by the San Sebastián laboratory showing dangerous and adulterated foodstuffs, 1891-1898

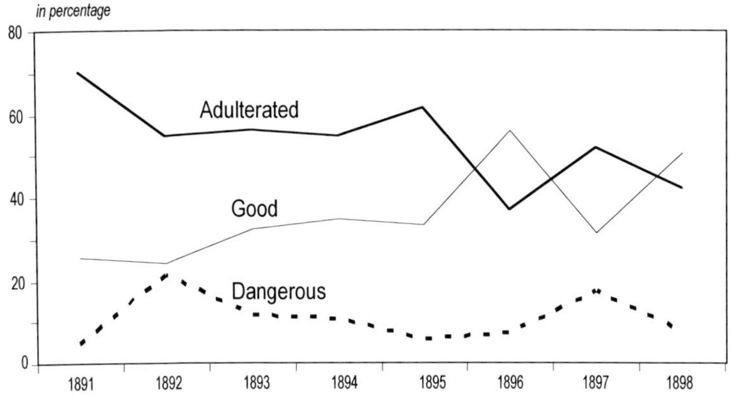

Source. San Sebastián 1891-1898 Laboratory Yearbooks.

Alcoholic beverages were an important source of taxation revenue for both the local and the central state in Spain. Even before the municipal laboratories had appeared, worries about adulteration had been influenced by local fiscal considerations (Pan Montojo, 1994; Sánchez Léon, 2001). Until the beginning of the twentieth century product-based inspection results in San Sebastián pointed to the importance of local cider consumption; in consequence, the work of the municipal laboratories can be said to have reflected concerns about fair and honest trading more than public health. Between 1888 and 1897 c. 78 per cent of product analyses in San Sebastián were carried out on four products: wine, cider, spirits and alcohol. The zeal in persecuting and excessively controlling *cizarra* – a new, non-fermented cider with a low alcohol content – which was considered a low-grade product and formed part of local consumption, was related to a tax system based on alcohol per centage (proof) content. The campaign against *cizarra* was not without controversy as it generated division and discord among the local elected councillors (Etxaniz Makazaga, 1998: 283-284). The logic behind disapproving of the product was based on the technical description of what was supposed to constitute a 'good' cider – a standard based on the cider produced in France which had a greater alcohol content. Local treasury interests and the convention of 'natural cider' came up against a social and cultural practice associated with the *cizarra* widely consumed in San Sebastián. In 1908, the technical regulations on cider stated: 'As watered-down cider is habitually drunk in some regions, its sale will be allowed, providing it is properly identified'[20].

Table 4.4. Chemical food analyses at the Municipal Laboratory of San Sebastián, 1888-1897 and Madrid, 1901/04/09/12

Madrid	%	*San Sebastián*	%
Alcohol, wine, spirits, and liquor	28.6	Alcohol, wine, spirits, and liquor	44.87
		Cider	34.50
Milk	24.0	Milk	10.86
Sodas, soft drinks, and syrups	6.8		
Bread, flour, yeast	2.4	Bread and flour	2.28
Chocolate and cocoa	2.2	Chocolate	1.02
		Sugar	0.82
Coffee	3.2	Coffee	0.81
Oil and vinegar	8.7	Oil and vinegar	1.31
Spices (paprika, saffron, etc.)	5.8	Spices	0.88
Noodles	2.5	Noodles	0.47
		Fats	0.39
Salt	2.7		
Cookies, confectionery	2.4		
Tea	2.1		
Preserves	1.6		
Sausages	1.2		
Others (15 products)	5.8	Others (12 products)	1.79
Total	100	Total	100

* The main percentage item of water was excluded in Madrid.

Source. San Sebastián. Laboratory Yearbook (1888-1897) Madrid. Laboratory Yearbook 1901/1904/1909/1912.

Results accumulated in Madrid at the start of the twentieth century show a greater diversity of analyzed products, although here again wines, liquors and spirits (with 28 per cent of the analyses) accounted for the single largest class of analyzed products (Saumell, Arnabal, Jordi, 2004). Products such as sodas, syrups, soft drinks and coffee were subjected to more analysis than yeast, flour and bread in Madrid. Where the range of analyzed foodstuffs is especially revealing is in the relative attention paid

20 Royal Decree of 23 December, 1908.

to ordinary foods (bread, flour, or fats) as against refined or luxury foods (chocolate, tea, coffee, liquor, etc). Generally speaking, the composition of the products analyzed does not correspond to the consumption patterns of the bulk of the population.

The commoditisation of alcohol-based products was the pretext for most of the food regulations, and the distinction between wine alcohols and industrial quality alcohols (starch-based) was a key element in defining laboratory practice. Determining the alcohol content of a beverage and its pureness became the guiding factor in the Spanish regime of analysis since some municipal taxes were levied proportionally to the pure alcohol content of products. Due to the importance of wine production and its central place in the tax structure, it is no mere coincidence that the alcohol content and purity of wine and related products accounted for the bulk of legislation. In keeping with business and protectionist criteria, the legislation would change substantially from the mid-nineteenth century. Whereas the imitation of natural wines using non-toxic substances had been authorized in 1860, it was in a context of competitiveness and alcohol overproduction that natural wines came to be defined in 1892, and the use of industrial alcohol in their preparation forbidden. The municipal laboratories were therefore exhorted to thoroughly check artificial wines[21]. Vinegar was considered to be 'natural' if it was made from wine, and thus other types of vinegar based on non-wine ingredients were excluded from the market until 1908. With a wine market afflicted with overproduction problems and undergoing significant technological advance, the link between the technology of viniculture and scientific production became much closer from the end of the nineteenth century (Pan Montojo, 2003). Thus, its strategic significance within the tax structure led to the concentration of analyses on wine products. The use of alcohols of different origin, and the introduction of products such as salicylic acid to delay fermentation for a certain period so as to avoid paying sampling duties, also led to controls on viniculture(Chicote y Riego, 1897: 227). Indeed, it was to control artificially coloured wines that a municipal laboratory was first proposed in Paris (Stanziani, 2007: 110).

The products analyzed in the San Sebastián Laboratory in the late nineteenth century may have fallen heavily within the category of alcohols, but the percentage of products considered to be 'adulterated' over the total range of analyzed products differed widely. The 'new products', containing more refined than ordinary foodstuffs, comprised a category showing a relatively high degree of adulteration. A consciously cautious approach can be detected (as Peter Atkins (1991: 317) mentions in the case of milk) in the handling of sophisticated products – the colorants in pasta or in preserves, for instance, or the first preservative agents added to sausages and

21 Royal Order of 23 February 1860; Royal Decree of 11 de March, 1892; Royal Order of 23 December, 1895.

cheese. The demand for luxury goods among certain sectors of the population, and who consumed products that most of the population could still barely afford, would also generate business for the municipal laboratories. Early in the twentieth century the highest percentages of detected adulteration in the Madrid analyses were found in such new products as condensed and powdered milk, biscuits and confectionery products, sodas and soft drinks, as well as in such refined consumer products as coffee, seltzer water, chocolate and cocoa. Other products, such as milk or ice, in high demand among urban consumers at the time, generated new and widespread health risks. The expansion of milk consumption in cities before World War I (Nicolau Nos & Pujol Andreu, 2004) occurred in a context where controls were being imposed on the widespread use of skimmed milk and the adulteration of milk with substances such as borax (Atkins, 1991: 317-39).

Table 4.5. Detected adulteration rates at the San Sebastián Municipal Laboratory, 1891-1897

Product	*% Adulterated*	*Product*	*% Adulterated*
Cider	81.58	Coffee	42.16
Pasta (noodles)	77.63	Spices	41.51
Cheese	75.00	Vinegar	41.14
Milk	72.79	Wine	38.96
Sausages	68.75	Spirits	34.69
Preserves	57.78	Fats	34.62
Cocoa	50.00	Ice Cream	31.71
Chicory	49.32	Bread	29.15
Seltzer Water	47.14	Edible Oil	28.44
Chocolate	46.75	Sugars	25.91
Tea	45.95	Salt	16.67
Beer	45.28	Flour	15.91
Liquor	44.00	Alcohol	15.32

Note. Excluded water.

Source. San Sebastián. Laboratory Yearbooks 1891-1898.

Table 4.6. Detected adulteration rates at the Madrid Municipal Laboratory, 1901/1904/1909/1912

Product	*% Adulterated*	*Product*	*% Adulterated*
Alcohol	100.0	Salt	55.0
Ice Cream	100.0	Beer	50.0
Condensed/Powdered Milk	100.0	Preserves	47.2
Seltzer Water	75.6	Fats, butter, margarine	43.9
Milk	74.3	Bread, flour, yeast	43.5
Cheese	74.1	Sausages	42.7
Pasta (noodles)	70.4	Spirits, wines, liquors	40.8
Chocolate and cocoa	68.1	Ham	32.8
Oil and vinegar	67.6	Vermouth	32.5
Ice	66.7	Spices (paprika, saffron, etc.)	30.0
Cookies, biscuits, confectionery	66.0	Honey	28.6
Sodas, soft drinks, syrups	65.4	Sugars	28.1
Coffee	62.1	Tea	25.0

Note. Excluded water.

Source. Madrid. Laboratory Yearbooks 1901/1904/1909/1912.

VI. Conclusions

The study of the municipal laboratories in Spain has commenced but the topic is by no means exhausted. There is every likelihood that new sources of local data and additional case studies will introduce new questions and open up new approaches in this research area in the future. The conclusions of my study indicate that a pre existing demand and the desire to raise local tax revenue were two important background features that contributed to the foundation of the Spanish municipal laboratories. Over time the persistence of adulteration as a local political issue reflected the importance of alcoholic beverages as a source of tax revenue as well as the attention the national state was prepared to pay to defining the purity of economically significant products such as wine. Other developments that gave impetus to the movement to establish the laboratories prior to the 1887 law were the memory of the old epidemic cycles, fresh experience of epidemics in the 1880s (especially in 1885), the dissemination of hygiene practices and international scientific debates, as well as the local demand for analyses from health professionals. The coincidence of the appearance of the laboratory movement and the new science of bacteriology proved especially important for the development of the law.

A number of specific conditioning factors came to influence the sort of analyses, based increasingly on shared standards, that came to constitute the focus of the working practices of the new laboratories appearing after 1887. Here we can mention the progressive (if uneven) urbanization of the municipalities and its impact on outbreaks of epidemic diseases linked to water quality, hygiene conditions and the prevalence of food adulteration. Considerations relating to public health – water analysis especially – the dependence of municipal tax revenue on beverage alcohol consumption and increasing market competition and segmentation in the urban areas all proved to be significant in influencing the sort of analyses that came to dominate in the new laboratories.

The work of the municipal laboratories came to focus on two aspects of the adulteration of food in particular. First, the laboratories set out to control the distribution and production of hazardous or harmful food products – dangerous or poisoned products – legitimating their activity here with reference to the preservation of public health. And secondly the laboratories sought to discover forged and altered food products – 'falsified goods' – which, although not necessarily health risks in themselves, did nonetheless constitute consumer fraud and an attack on fair and honest trading. The food products analyzed in the laboratories were classified as either 'good' or 'bad', with the latter including both products hazardous to health and products considered harmless in this regard. Criteria based on public health (hazardous products) as well as on fair and honest trading (products that were altered under commercial pressure so as to increase supply or extend shelf life) became the twin justifications for the establishment of chemical inspections in the new municipal laboratories.

Until 1908 the early municipal laboratories not only faced a lack of technical regulations dealing with the composition of different food products, but also lacked a set of standardized practices (endorsed by the laboratory directors) to analyze and to certify product purity. Fundamentally purity (and its synonym of quality) can be seen as a convention or norm that was built up and modified throughout the period as the possibilities of chemical analysis expanded in line with advances in scientific knowledge. Prior to 1908 the heads of laboratories, on the basis of communication among themselves and with their counterparts in laboratories in France, had begun to build a consensus as to how the daily practice of their workplaces should properly be organized. Underlying the drawing up of regulations and the adoption of conventions relating to food purity can be identified a certain desire to protect traditional products (such as sugar and full strength cider) against new competing ones. With the Royal Decree of 1908, published while Juan de la Cierva was a minister of the Maura government, the foundation was laid for the creation of an authoritative set of purity standards and food inspection procedures.

The municipal laboratories combined two sets of services: a public service provided *gratis* to local and provincial municipal authorities, and a private service offered to private individuals and corporations who paid for their test results according to a set fee structure. From the beginning of the twentieth century the available data for the Madrid, San Sebastián and Bilbao municipal laboratories show a drop in the business generated by private interests and a consolidation of the function of the laboratories in serving the needs of public bodies. This pattern can be related to the expanding professional role of the local medical/pharmaceutical staff of the state, one dimension of whose activity involved taking the initiative in subjecting foodstuffs to analysis within the municipalities. The importance of local consumption taxation was something else that had the effect of greatly increasing public sector demand for food purity analyses.

Until the beginning of the twentieth century the product-based inspection in San Sebastián shows the work of the municipal chemical laboratories to have been less directed by public health concerns than by concerns about fair and honest trading and the necessity to maximize municipal consumption tax revenue. Between 1888 and 1897 c. 78 per cent of product analyses carried out in San Sebastián were focused on the four products of wine, cider, spirits and alcohols. It is notable that the composition of the products analyzed does not proportionately correspond to the pattern of local consumption. What is striking is how much the activity of the municipal laboratories was so well integrated with the local tax collection regime as well as with commercial realities that reflected market competition between old and new food products in urban areas. The items showing the highest percentage of adulteration were dairy products and new 'products', a category that contained more refined than ordinary foodstuffs. Other features of the work of the municipal laboratories – such as testing water quality, disinfection services, vaccine preparation and collaboration with slaughterhouse veterinary inspectors – involved the laboratories becoming local public health agents in the struggle to combat the prevailing high human mortality rates.

Bibliography

ALVAREZ SIERRA, J. (1965), *El doctor don Cesar Chicote y el laboratorio municipal (texto impreso)*.

ATKINS, Peter J. (1991), 'Sophistication Detected: or the Adulteration of the Milk Supply, 1850-1914', *Social History*, 16, p. 317-339.

BRUEGEL, Martin & STANZIANI, Alessandro (2004), 'Pour une Histoire de la sécurité alimentaire', *Revue d'Histoire Moderne et Contemporaine,* 51, p. 7-16.

CAMPOS, Ricardo (2003), *Curar y gobernar: Monlau, Rubio, Giné: medicina y liberalismo en la España del siglo XIX*, Tres Cantos, Nivola.

CASAS DURAN, Frances (1988), *Història del Laboratori Municipal de Barcelona: introducció a un centenari (1887-1987),* Ed. Libres de la Frontera.

CHICOTE Y RIEGO, Cesar (1897), *Alimentos y bebidas. Investigación de sus alteraciones y falsificaciones*, Madrid, Tipográfica Fernando Fé.

ELLERBROCK, Karl-Peter (1987), 'Lebensmittelqualität vor dem Ersten Weltkrieg: Industrielle Produktion und staatliche Gesundheitspolitik', in Hans-Jürgen TEUTEBERG (ed.), *Durchbruch zum modernen Massenkonsum. Lebensmittelmärkte und Lebensmittelqualität im Städtewachstum des Industriezeitalters,* Münster, F. Coppenrath, p. 127-188.

ETXANIZ MAKAZAGA, José Manuel (1998), 'Del Laboratorio Químico Municipal al Instituto Municipal de Higiene', *Boletin de la R.S.B.A.P.,* LIV, p. 279-290.

FERRIÈRES, Madeleine (2002), *Histoire des peurs alimentaires. Du Moyen Âge à l'aube du XXe siècle,* Paris, Seuil.

FRENCH, Michael & PHILLIPS, Jim (2000), *Created not Poisoned? Food Regulation in the United Kingdom 1875-1938*, Manchester University Press.

FRESQUET FERRER, José Luis (1990), *Francisco Méndez Alvaro (1806-1883) y las ideas sanitarias del liberalismo moderado,* Madrid, Ministerio de Sanidad y Consumo.

GRANJEL, Mercedez (1983), *Pedro Felipe Monlau y la higiene española del siglo XIX,* Salamanca, Cátedra De Historia de la Medicina, Universidad de Salamanca.

GUILLEMI-LLOBAT, Ximo (2008), *El control de qualitat dels aliments. El cas valencià en el context internacional (1878-1936)*, Valencia, PUV.

HEROLD-SCHMIDT, Hedwig (1999), *Gesundheit und Parlamentarismus in Spanien. Die Politik der Cortes und die öffentliche Gesundheitsfürsorge in der Restaurationszeit (1876-1923),* Husum, Matthiesen Verlag.

HIERHOLZER, Vera (2007), 'The War Against Food Adulteration: Municipal Food Monitoring and Citizen Self-Help Associations in Germany', in Peter ATKINS, Peter LUMMEL & Derek J. ODDY (eds), *Food and the City in Europe since 1800*, Hampshire, Ashgate, p. 117-128.

HIETALA, Marjatta (1994), 'Hygiene and the Control of Food Finnish Towns at the Turn of the Century: A Case Study from Helsinki', *ICREFH,* II, p. 113-129.

JUNTA PROVINCIAL DE SANIDAD (1886a), *Reglamento de organización interior para el servicio del laboratorio químico municipal de la I. Villa de Bilbao,* Bilbao, 1886.

JUNTA PROVINCIAL DE SANIDAD (1886b), *Reglamento para el servicio del Laboratorio Químico Municipal de Vitoria,* Vitoria, 1886.

LABORATORIO MUNICIPAL DE HIGIENE DE MADRID (1906), *Instrucciones para el servicio de inspección de subsistencias*, Madrid, Imprenta Municipal.

LABORATORIO QUIMCO MICROGRAFICO DE SAN SEBASTIÁN (1888), *Memoria correspondiente al primer año económico de su instalación presentada en el mes de julio al Excmo. Ayuntamiento por su director Dr. César Chicote*, San Sebastián, La Voz de Guipúzcoa.

LABORATORIO QUIMCO MICROGRAFICO DE SAN SEBASTIÁN (1907), *Memoria resumen de los trabajos realizados durante el año 1906*, San Sebastián, 1907, Imprenta y Encuadernación de J. Baroja e Hijo.

LAW, Marc T. (2003), 'The Origins of State Pure Food Regulation', *Journal of Economic History,* 63, p. 1103-1130.

MALUQUER DE MOTES, Jordi & LLONCH, Montserrat (2005), 'Trabajo y relaciones laborales', in Albert CARRERAS & Xavier TAFUNELL (eds), *Estadísticas Históricas de España*, vol. II, Bilabo, Fundación BBVA, p. 77-154.

MARTORELL LINARES, Miguel Angel (1995), 'La reforma pendiente: La hacienda municipal en la crisis de la Restauración: el fracaso de la Ley de Supresión del Impuesto de Consumos', *Hacienda pública española,* nº 132, p. 143-152.

MERKI, Christoph M. (1994), 'Sugar Versus Saccharin: Sweetener Policy before World War I', in John BURNETT & Derek J. ODDY (eds), *The Origins and Development of Food Policies in Europe*, London / New York, Leicester University Press, p. 192-202.

MORO AGUADO, Jesús (1986), *Francisco Méndez Alvaro: la higiene española en el siglo XIX*, Valladolid, Consejería de Bienestar Social.

MUÑOZ CALVO, Sagrario (1994), *Historia de la farmacia en la España moderna y contemporánea,* Madrid, Síntesis.

NICOLAU NOS, Roser & PUJOL ANDREU, Josep (2004), 'Urbanización y consumo: La ingesta de proteínas animales en Barcelona durante los siglo XIX y XX', *Investigaciones de Historia Económica*, 3, p. 101-134.

NICOLAU NOS, Roser (2005), 'Población, salud y actividad', in Albert CARRERAS & Xavier TAFUNELL (eds), *Estadísticas Históricas de España*, evol. II, Bilabo, Fundación BBVA, p. 77-154.

NICOLAU NOS, Roser & PUJOL ANDREU, Josep (2004), 'Urbanización y consumo: La ingesta de proteínas animales en Barcelona durante los siglo XIX y XX', *Investigaciones de Historia Económica*, 3, p. 101-34.

ODDY, Derek J. (2007), 'Food Quality in London and the Rise of the Public Analyst, 1870-1939', *in* Peter ATKINS, Peter LUMMEL & Derek J. ODDY (eds), *Food and the City in Europe since 1800*, Hampshire, Ashgate, p. 91-104.

PAN MONTOJO, Juan (1994), *La bodega del mundo. La vid y el vino en España, 1800-1936*, Madrid, Alianza.

PAN MONTOJO, Juan (2003), 'Las industrias vinícolas españolas: desarrollo y diversificación productiva entre el siglo XVIII y 1960', in Carlos BARCIELA LÓPEZ &

Antonio Di Vittorio (eds), *Las industrias agroalimentarias en Italia y España durante el siglo xix y xx*, Alicante, Publicaciones de la Universidad de Alicante, p. 313-334.

Paquy, Lucie (2004), 'Santé publique, répression des fraudes et action municipale à la fin du xixe siècle: le laboratoire grenoblois d'analyses alimentaires', *Revue d'histoire moderne et contemporaine*, 51-3, juillet-septembre, p. 44-65.

Pérez Moreda, Vicente (1985), 'La modernización demográfica, 1800-1930. Su limitaciones y cronología', in Nicolás Sánchez Albornoz (ed.), *La modernización económica de España 1830-1930*, Madrid, Alianza, p. 35-56.

Puerto Sarmiento, Francisco Javier & Cobo Cobo, Josefa (1983), 'El Laboratorio Municipal de Madrid en el último tercio del siglo xix', *Dynamis*, Acta hispánica ad medicinae scientiarumque historiam illustrandam, nº 3, p. 149-172.

Salais, Robert, Chatel, Élisabeth & Rivaud-Danset, Dorothée (eds) (1998), *Institutions et Conventions. La réflexivité de l'action économique*, Paris, Éditions de l'EHESS.

Sánchez León, Pablo (2001), '...Con el nombre del vino. Alcoholes, fiscalidad y salud pública en el Madrid del siglo xix', *Hacienda Pública Española*, 158-3, p. 191-214.

Sanz Lafuente, Gloria (2006a), 'Perspectivas de historia de la seguridad alimentaria. Entre la ley y la práctica social de la inspección 1855-1923', *Revista Española de Estudios Agrosociales y Pesqueros*, 212, p. 81-118.

Sanz Lafuente, Gloria (2006b), 'Seguridad e higiene alimentaria en el mundo rural. La inspección municipal de carnes en la segunda mitad del siglo xix', in Jordi Bolos, Antonieta Jarne & Enric Vicedo, (eds), *Condicions de vida al món rural*, Lleida, Institut d'estudis Illerdencs, p. 525-540.

Saumell, Antoni, Arnabat, Ramon & Jordi Romeu (2004), *Estació de viticultura i enologia de Vilafranca del Penedès 1903-2003. Cent anys d'història*, Vilafranca del Penedès.

Scholliers, Peter (2007), 'Food Fraud and the Big City: Brussels' Responses to Food Anxieties in the Nineteenth Century', in Peter Atkins, Peter Lummel & Derek J. Oddy (eds), *Food and the City in Europe since 1800*, Hampshire, Ashgate, p. 77-90.

Stanziani, Alessandro (ed.) (2003), *La Qualité des produits en France, xviiie-xxe siècles*, Paris, Belin.

Stanziani Alessandro (2005), *Histoire de la qualité alimentaire (xixe-xxe siècles)*, Paris, Seuil.

Stanziani Alessandro (2007), 'Municipal Laboratories and the Analysis of Foodstuffs in France under the Third Republic: A Case Study of the Paris Municipal Laboratory, 1878-1907', in Peter Atkins, Peter Lummel & Derek J. Oddy (eds), *Food and the City in Europe since 1800*, Hampshire, Ashgate, pp. 105-116.

Teuteberg, Hans-Jürgen (1994), 'Food adulteration and the Beginnings of Uniform Food Legislation in Late Nineteenth-Century Germany', *ICREFH II*, p. 146-160.

Urteaga, Luis (1980), 'Miseria, miasmas y microbios. Las topografías médicas y el estudio del medio ambiente en el siglo', *Geocrítica*, 29, p. 5-50.

Vallejo Pousada, Rafael (1996), 'El impuesto de consumos y la resistencia antifiscal en la segunda mitad del siglo xix: un impuesto no exclusivamente urbano', *Revista de Historia Económica,* 2, p. 339-370.

Vera y Lopez, Vicente de (1885), *Laboratorios municipales de salubridad*, Madrid, F. Maroto e Hijos San Sebastián, La Voz de Guipúzcoa.

5. The modernisation of agriculture in Greece (c. 1920-1970): variation of a European Mediterranean Model?

Socrates D. PETMEZAS

I. Introduction

The aims of this chapter are to present the Greek case of agricultural modernization and to look at agricultural modernization in the Mediterranean area through the Greek mirror. The recent high growth rates in the northern Mediterranean area (from Portugal to Turkey) have made historians and economists more conscious of earlier positive dynamics in the Mediterranean economies and encouraged them to reveal virtues where previously they were seeing only shortcomings. Thus Giovanni Federico (2003: 99) has added a significant subtitle – *Successo o fallimento?* – to his chapter on agriculture in the recent *Storia Economica d'Italia*. It seems that the mood of economic historians is very much influenced by their contemporary experience and expectations.

It is difficult to go against this *Zeitgeist* but I will try to show, using the Greek case and referring to other northern Mediterranean countries, that Mediterranean agriculture was handicapped during most of the nineteenth and early twentieth centuries and that only recently has it been able to break out of the double trap of ecological and institutional deadlock. This emancipation was achieved via active state policy and public funding initiatives, but what proved really decisive were post-war technological innovations and – most importantly of all – the integration into Euro-Atlantic labour markets which had the effect of sponging out the surplus rural labour from the Mediterranean countryside, thereby clearing the way for the modernization of Greek agriculture.

While it is true that labour-intensive Mediterranean agriculture had been able, in spite of its evident retardation, to feed its expanding population, in many areas (as in Greece) this was done by taking an 'involutive' path that entailed stagnating labour productivity and falling crop yields, without securing the necessary income, or growing employment, for the rural population[1]. The out-migration of the excess underemployed rural population seems to be the single most important factor explaining productivity growth in the Greek case (and I believe in the Mediterranean

[1] For a discussion of the concept of 'involution', see GEERTZ (1963)

area in general) and this population movement was not structurally linked to the local (regional or national) process of industrialization and urbanization. Demo-economic developments in the encompassing 'global' economy better explain the 'pull' factor, while contextual crises in certain key agricultural sectors can be shown to have served as catalysts for setting the movement in motion.

II. The ecological and institutional deadlock in Greece *c.* 1910: a unique phenomenon?

During the second half of the nineteenth century the kingdom of Greece (covering the southern part of the present-day country) had a labour-intensive agriculture, producing typically Mediterranean products. The kingdom may have always ran a net surplus in agricultural trade, thanks to its expanding exports of labour-intensive plantations (mostly based on producing currants but also olive oil, wine and dried figs), but the country was an ever greater importer of wheat and other food products of extensive or tropical agriculture. On this basis Greek agriculture was able both to feed its population and to generate some of the urgently needed exchange to cover its structural (and growing) trade and current accounts deficits (Petmezas, 2009: 358-360).

This was a real achievement considering that Greece, like the other Mediterranean countries, had completely missed the so-called first agricultural revolution. The factor endowment, geo-morphology and climate usually associated with Mediterranean Europe made the adoption of the techniques and innovations of the first agricultural revolution expensive (*i.e.* economically unprofitable) and almost ecologically unsustainable, given the low and seasonally uneven rainfall and the paucity of the Mediterranean hydrological systems. Spanish historiography has advanced ample and adequate arguments on this score (Simpson, 1995: 265-266; Clar & Pinilla, 2009: 312-315), and I have just to add that the same arguments apply to the Greek case almost to the letter.

The growth in the volume of southern Greek agricultural production depended until c. 1880 on the ever increasing agricultural surface and, since then, on the change of the culture mix in favour of labour intensive export plantations, mainly currant vineyards. At the end of the nineteenth century, the continuous rise of the value of agricultural production and exports was exclusively dependent on the expansion of the international demand for currants, the main Greek export product. The chronic currants crisis that broke out in August 1893, as a result of the new prohibitive French tariffs on the import and use of non-edible currants, had the effect of completely destroying the delicate equilibrium between men and resources in the southern Greek countryside and of forcing a large part of its population to search for alternative employment and income sources. Those who stayed in agriculture reduced the fallow without adopting the use of expensive imported chemical fertilisers or better rotation systems. Impoverished small-owners were thus forced to adopt a predatory use of land resources that undermined the natural fertility of the soil. As a result, cereal agriculture entered an 'involutive' spiral with yields dropping significantly over the next three decades (Petmezas 2009: 359-360).

Southern Greek agriculture had thus fallen into an ecological deadlock. The Greek government, opposed to any intervention in the markets, was forced to engage with the problems of the rural world. It took more that ten years to find an appropriate solution of the currants crisis. This took the form of an imposition of monopsonistic control over the major export sector of Greek agriculture by a major private banking-industrial consortium. By such means did the currants economy survive as an appendix of large-scale industrial and financial concerns, but without recovering its position as the key sector of the Greek economy (Petmezas, 2009: 363).

Ecological deadlock in the southern provinces was combined with a major institutional deadlock in the northern lowlands. In these extensive, marshy and sparsely populated plains, large estates cultivated by sharecroppers were the dominant form of land tenure. Each share-cropping family had enough land to plough using its rudimentary implements and the landowners made sure that the land left fallow, just like the rest of the unused land, was in continuous strips that were rented out to the inversely transhumant shepherds who spent their winters on the plains. In the last decades of the nineteenth century, as commercialization and monetization made substantial advances, share-croppers preferred to put into commercial use their prerogatives of free usufruct in the communal resources and to limit as much as possible their involvement in cereal culture (whose product they shared with their landlords). For their part some of the landowners tried to place sharp limits on the free use of communal lands by sharecroppers and even to convert secure land tenure into short-term (and thus more insecure and more tightly controlled) leases. In Thessaly, now under Greek rule and possessing a parliamentary regime, such moves by large landowners (mostly influential Greek capitalists who had bought out the former Ottoman Muslim landowners) were met with strong opposition from the sharecroppers and their political supporters. Until the beginning of the twentieth century cereal production stagnated and then grew only because of real price incentives. The general international wheat price increase was strengthened locally by the imposition of substantial tariff protection and the reduction of transport costs that followed the completion of the Thessalian railway network (Petmezas, 2009: 360-361).

III. Looking at the Global Mediterranean Picture: the 'involutive' path of Mediterranean agriculture

Early in the twentieth century Greek agriculture, which by then was facing a double ecological and institutional deadlock, had a relatively satisfactory level of return to land (probably 80 per cent of the Italian and UK equivalents) but its labour productivity was still only a small fraction of the English (25 per cent) and even of the Italian (50 per cent) equivalents (Table 5.2). An alternative estimate points to a comparable difference *vis-à-vis* Spanish (80 per cent) and French (20 per cent) labour productivity in agriculture (Simpson, Carmona, 2003: 29). This performance was the result of half a century of relatively rapidly rising Total Factor Productivity (TFP) (see Table 5.1). Indeed, almost all the Mediterranean countries had shown a comparable development of their annual rate of Total Factor Productivity growth. Nearly half of the growth in agricultural output is explained by the rise of TFP – *i.e.* by the better use of available factors (and more efficient culture mix) – since technological change was rather limited.

Table 5.1. The Growth of Total Factor Productivity (TFP) of the Mediterranean agricultures

Portugal			*Italy*			*Spain*		*Greece*		
	TFP	*TFP/O*		*TFP*	*TFP/O*		*TFP*		*TFP*	*TFP/O*
1865-1802	0,73	52,00	1861-1881	0,89	0,49	1857-1890	1,14	1860-1887	1,73	0,47
			1881-1897	0,64	0,69	1890-1905	0,06	1887-1911	-0,22	-1,37
1902-1927	0,37	47,80	1897-1913	1,02	0,89	1905-1913	0,77	1911-1930	-0,87	-0,36
			1913-1920	-0,95	1,68	1913-1931	1,29	1911-1939	0,22	0,06
			1920-1925	3,31	1,03					
1927-1962	1,94	79,70	1925-1939	0,66	1,27			1911-1950	-0,57	-0,35
1962-1973	-0,79	-39,40						1950-1981	3,79	0,88
1973-1990	3,23	120,50								

Sources. For Portugal, see Lains (2009: 340); for Italy, see Federico (2009: 243); and for Greece, see Petmezas (2005). The estimates for Spain are based on my own calculations using Coll and Brigas (2000: 350).

This picture of rising TFP (which is consistent with other north-western European rates (Federico, 2003: 123) should not lead us to an optimistic conclusion. The turn of the century was characterised by crises which hit the key wine-producing and export sectors and which arose as well from overseas competition in corn and meat products. Overseas competition forced many European states to adopt protectionist policies in favour of cereal production and cereal producers. Protectionist tariffs in Spain and Italy favoured big landowners and cereal producers; they also put a brake on emigration (Fenoaltea, 1993; Simpson, 1995: 167) and possibly boosted agricultural incomes, but not labour productivity and economic efficiency. As a result the early twentieth century has been characterised either by short decelerations of TFP growth (as in the case of Spain, Portugal and Italy), or by a more abrupt fluctuation (as in the Greek case).

Table 5.2. Raw estimates of productivity levels in Southern Europe

c. 1910	*O/L (UK prices)*	*O/L (local prices)*	*O/Lm (UK prices)*	*O/Lm (local prices)*	*O/T (UK prices)*	*T/L*	*T/Lm*
France	58	57	86	85	143	41	60
Italy	31	30	47	45	165	19	28
Spain	31	24	32	25	54	58	59
Germany	56	58	97	100	168	33	58
UK	100	100	100	100	100	100	100
Greece	*22*	*25*	*33*	*34*	*74*	*29*	*44*
Ottoman Macedonian provinces (annexed in Greece in 1913)	14		22		46	32	

Source. See Petmezas (2000: 334; 2005: Table 7).

I think that with the exception of the advanced regions in Spain and Italy, most Mediterranean countries and regions faced the same problem of ecological deadlock. All these countries were experiencing population increase and shared a comparable geo-morphological and climatic environment as well as relative factor endowments (abundance of labour and the relative scarcity of land and especially of capital). They were all comparable in their levels of economic development and GDP composition (Kostelenos *et al.*, 2007: 195). The available agricultural technology (necessary to boost their stagnant labour productivity) was either too expensive to adopt by their under-capitalised farms or went against the efficient use of their endowed factors. In consequence one can say that, in the medium term, these countries experienced (and were locked into) stagnant labour productivity. Output growth to match the needs of a growing population depended upon the expansion of the agricultural surface or the adoption of a more 'abundant factor-intensive culture mix' that favoured their most abundant factor – labour. C. 1900 all the available agricultural land was already in use in southern Greece (Petmezas, 2009: Table 14.2), in Portugal (Lains, 2009: 339), in many parts of Italy (Galassi & Cohen, 1992: 147) and in Spain (Sanz, 1987: 241-244).

Thus Greece and the other Mediterranean countries have been able to feed their growing populations by using their abundant labour, and by (slowly) expanding the available agricultural surface, without introducing any significant (and costly) technological innovation. Of course, such a model of output increase was subject to ecological limits. Once marginal lands were put into cultivation, yields fell. The same happened if fallow was reduced in the absence of the use of costly fertilisers or the adoption of 'scientific complex rotations' (which meant advancing money to buy inputs and invest in irrigation). In the Mediterranean countries, as already pointed out, the adoption of such techniques was economically unrewarding. The alternative solution was the adoption of a culture mix that was oriented towards labour intensive and income-elastic products like wine, tobacco, olive oil, fresh and dried fruits. Such a choice entailed increasing exports and therefore a steadily growing international demand for such products in high income countries. This last condition was not always fulfilled (as the taste for Mediterranean products spread only slowly) and this led to many more-or-less serious crises that destabilised the social and economic equilibrium[2].

Relatively small countries, like Greece and Portugal, were specialized in just one or two labour intensive products (wine, currants), which made them more vulnerable than larger and more diversified countries (such as Italy, Spain, Turkey) to a major crisis in one market for agricultural products. These larger countries may have been less vulnerable, but they did face comparable structural constraints because international demand grew too slowly to absorb the rising volume of exports. Even if surplus export production did attract international demand, the danger of falling rates of return was present as labour productivity stagnated or fell. In other words agricultural expansion in labour intensive plantations was only a temporary relief, if the 'surplus' rural labour could not find alternative employment. To sum up, both the usual means of augmenting output (through expanding the agricultural surface or changing the culture-mix towards labour intensive commercial products) led to an ecologically 'involutive' path. The key variable for sustained growth was the increase of labour productivity or, more realistically, the redirecting of surplus labour towards alternative employment outlets.

The continuous specialisation in new labour-intensive products, such as oranges in late nineteenth century Spain, facilitated in particular by transport or marketing innovations, could defuse crises but it could not change the underlying adverse situation. The only way out was the adoption of the appropriate technological innovations as they became economically available. Spain, a country that faced

2 Inter-war competition in the market for Mediterranean agricultural produce is discussed by Pinilla & Ayuda (2006).

current accounts and labour-intensive-agriculture crises comparable to those of Greece and Portugal at the turn of the century, had been able to find a way out in its agricultural sector. If cereal production and the area of land in use stagnated in the southern hinterland (Andalusia and Old Castile), it seems that the seeded area expanded in the other provinces where, thanks to the introduction of fertilisers, yields were sustained and even grew (Sanz, 1987: 241-248). The early adoption of modern olive-oil presses was also a significant productive innovation (Simpson, 1995: 214-217). Organisational innovations were also possible and it is not surprising that it was only in the sector of speculative export products (wine, oranges etc.) that efficient cooperatives were founded and functioned autonomously (without depending on state support or guidance) in Greece and in Spain.

The so-called 'second agricultural revolution' of the late nineteenth century opened up new possibilities for the Mediterranean countries to respond to their ecological constraints by adopting selected seeds and animal breeds, using chemical fertilisers and specialised implements etc. Since food production in the semi-arid Mediterranean countries is more dependent on regulating the water availability and seasonality than on regulating temperature (as in water abundant western and northern Europe), investment in drainage and irrigation techniques was of critical importance in adding to the productive power of the land (yields). These innovations led to the intensification of the use of the scarce factor – land. Early in the twentieth century these innovations were the chief means of resolving the ecological deadlock. The main problem was that adopting such innovations simultaneously required buying inputs (seeds, fertilisers, implements) and investing (and therefore going into debt) in land reclamation and improvement. Only big landowners or well-off farmers could afford this option, something that was particularly difficult in the former Ottoman Balkans. Greece and the other independent Balkan countries became known for their free peasants with secure tenures but under-capitalised farms.

IV. Land reform in inter-war Greece: solution or problem?

The Greek case of land reform, with its profound and complete transformation of the land tenure system in the early inter-war period, stands out as notably different from the other Mediterranean countries. It is in this respect much closer to the eastern European countries which, as a means of legitimizing the new nation-states, experienced massive and unusually radical land reforms in the aftermath of the dissolution of the Austro-Hungarian, Russian and Ottoman empires. Ottoman Greece had been characterised by small-scale peasant agriculture in the lowlands where big landownership dominated. Usually the peasants were share-croppers paying rent to the landowners and paying taxes to the Ottoman state and to tax-farmers.

Like Serbia independent Greece had been transformed into a country dominated by small or medium landowners and cultivators in the nineteenth century. Large (usually cereal-producing) estates were exceptional and regionally limited to small enclaves. These estates were formerly owned by Muslim landlords and had been sold to rich and influential Christian capitalists right after political independence. The survival of Muslim-owned large estates in the Balkan countries was very much exceptional (as was the case in Greek Thessaly between 1881 and 1920), and understandably Muslim landowners were reluctant to invest in land reclamations and improvements as well as productive innovations. Where Christian capitalists and influential notables were able to acquire large estates (usually buying out their Muslim counterparts), they opted (as their predecessors had done) to adopt share-cropping and to try to restrict the secure tenure and other prerogatives of share-croppers. This led to embittered social relations but did not facilitate investment. Large estates cultivated by sharecroppers dominated the land tenure system in the lowlands of Kosovo, Macedonia, Epirus and Thrace, still under Ottoman rule; and, even though a considerable segment of the big landlords were now Christian, the majority of Christian peasants looked on the land tenure system as the embodiment of Muslim rule. At the turn of the century therefore the contending local national movements, sponsored by the neighbouring Balkan states, adopted the peasant demand for the expropriation of large estates in favour of the peasantry, as the best way to win over Christian peasants to their cause.

The result was that big landowners, especially those holding large cereal-producing estates covering thousands of hectares, had never been as numerous, socially influential and politically strong as in the other Mediterranean countries. During and after the Balkan Wars and the First World War, when Greece doubled in size upon annexing the provinces of Macedonia, Epirus and Thrace, the expropriation of large estates and their distribution to landless Christian cultivators was considered a necessary act to establish the legitimacy of the new regime and to win the hearts of the peasant population. The Greek defeat in the Asia Minor campaign and the ensuing massive Greek-Turkish population exchange totally reconfigured the ethnic composition of the respective countries. Probably a quarter of the Greek population c. 1924 was comprised of impoverished or/and dislocated refugees, most of whom were without steady employment or decent housing. Massive and radical land reform was an expedient that would reduce social and political tension in the northern provinces (with their small but important ethnic minorities) and in Thessaly. Land, which was formerly owned by large landowners or by Muslim and 'Bulgarian' peasants who were now forcibly or voluntarily out-migrating under the provisions of the Peace Treaties, was also distributed to incoming refugees. Seventy per cent of the rural population in the north were former landless share-croppers or settled refugees; in all, these received 75.4 per cent of the land then in agricultural use.

The Greek land reform of the 1920s had a controversial impact in the short run. It expropriated the large estates and distributed the land to the landless share-croppers and the incoming refugees without leading to a noticeable growth of output and productivity (Petmezas, 2009: 363-364). More than 145,000 refugee families were added to the 150,000 families of landless share-croppers and consequently the available land per family was reduced. Share-croppers ended up with less land than they had been renting beforehand. Under conditions of extreme financial stringency in the early 1920s, there were practically no funds available for accompanying financial support. The new small peasant farms were thus under-capitalised and burdened with the future obligation to pay the 25 annual instalments for the land they had been granted. Only the newly settled refugees (totally alien to the area they were settled in) profited from a modest, yet real, programme of financial support supervised by the League of Nations. The lowlands ended up with a much higher population and 'agricultural' density than before, while the tendency for cereal yields to fall continued unabated until the early 1930s. Furthermore, the immediate result of the redistribution of the large estates on the number of livestock was negative. With land now fragmented and small farmers adopting cereal rotations chosen according to their perceived needs, the territorial continuity of the winter pasturages was broken. The productive unity of transhumant pastoralism and extensive cereal agriculture in the northern lowlands thus ended, and transhumant shepherds were obliged either to radically reduce the number of their flocks or to change their mode of pastoralism. By and large, the extensive form of animal husbandry was not abandoned and the number of animals did not change (Petmezas, 2009: 364). But it was now small owner peasants or smaller settled pastoralists that held the large majority of the animals.

The land reform of the 1920s had thus swept out all the large landowners in the northern provinces. The stagnation of international demand (and prices) for currants had also encouraged the sale and fragmentation of large and medium-size plantations in the south. Small-owner family farms came therefore to completely dominate the Greek countryside in the late inter-war period. As a rule these smallholders were indebted to the state, to the national or the agricultural banks or to private creditors, and were constantly under-financed (Petmezas, 2009: 365-366). In other words, in the late 1920s, there did not exist in Greece a class of medium or large landowners that was financially and mentally able to adopt (and subsequently diffuse) the necessary but costly and technically demanding innovations and investments.

Thus, the land reform that the Greek Liberals forcibly implemented in the 1920s had initially led to stagnation rather than dynamic productive growth. And yet, in the long run, cereal yields did recover, while agricultural production in the lowlands was transformed due to a large variety of state-supported initiatives undertaken since the 1930s. Effectively, the state was obliged to take the place of the absent capitalist landowners or entrepreneurial farmers in order to provide capital, practical examples and social leadership. All this resulted from piecemeal actions to counter the negative impact of radical reform or market developments, rather than from any overarching strategic policy choices.

V. Crisis and recovery in the inter-war period: autarchy and stagnation

Initially, in the 1920s, Greek governments had tried to break free from the grip of ecological involution by investing in land reclamation and improvement as well as by following a policy of supporting national cereal production. This cereal policy was at once inspired by the Italian (and other Mediterranean) models and imposed by necessity – Greece had experienced the threat of cereal deficit during World War I. Once again the expansion of labour-intensive exports gave a little respite. In the 1920s tobacco, especially the high quality variety produced mainly on the hill slopes of eastern Macedonia, became the main export product. Tobacco accounted for 57 per cent of the total value of all Greek exports in 1929. Perhaps one-seventh of the country's labour force was in one way or another dependent on tobacco production, processing and exporting. Contrary to what happened with currants, the impact of tobacco production and exports spread out beyond the rural areas. In eastern Macedonia and Thrace a large percentage of the local urban workforce was engaged in the processing of tobacco for cigarette production and the export trade (Petmezas, 2009: 365).

An ambitious state project of land reclamation and improvement, drainage, river bank regulation and irrigation in the northern Greek lowlands, which began in 1928 (when Greece once again had gained access to the international financial markets), took more than two decades (separated by the war decade of 1940-1949) to implement. It completely changed the land use and settlement patterns in the basins of the great Macedonian rivers. In the 1930s, a little less than 200,000 hectares were opened to arable culture in Macedonia, thus helping to raise the total Greek agricultural surface to 3.3 million hectares in 1939 (as against 2.4 million in 1929). Even more importantly, the percentage of irrigated fields grew from one per cent in 1929 to 15 per cent in the early 1960s. It is worth noting that the high cost of this land reclamation and improvement was entirely paid from the public purse.

In this same period, thanks to the active support extended by the state-funded Agricultural Bank to rural cooperatives, artificial fertilizers began to be used by farmers. Rural cooperatives became the dominant channel through which state services, and the Agricultural Bank, extended cheap credit, fertilisers etc. in the inter-war period. While less than ten kg of fertilizers per hectare of cultivated land were used in 1922, this index rose to fifty kg per hectare just before the war. New seed varieties were also imported and distributed by state agencies to farmers. Beginning in 1932, cereal yields had finally begun to increase in parallel with the expansion of planted arable land. Between 1930 and 1939 wheat production grew from a record low of 264,000 to a record high of 1,042,000 tons. In the 1920s many autonomous state-sponsored bodies were founded whose task was to control price and production, offer cheap inputs and 'extension' to the small-owner producers – prominent examples here would include the Autonomous Currants Organisation, the Central Office for the Encouragement of Indigenous Wheat, the Cotton Organisation and the Regional Offices of olive oil production. By the end of the 1930s the most important agricultural products were under the guiding influence or control of autonomous state offices.

The result of all this state intervention was a substantial departure from 'traditional' practices and a net growth in output. Wheat tripled its output and doubled its yields (from a low of 450 kg per hectare in 1925 to a high of 1,130 kg per hectare in 1938). Other cereals saw output and yields rise in comparable ways. Cotton, for instance, rose from 11,000 tons in 1929 (the same level as in 1914) to 55,000 tons in 1937. The Greek cotton industry, the chief user of imported cotton, was able to rely on local produce for 75 per cent of its needs in 1937 as against a mere 25 per cent in 1932. In contrast, the production and export of tobacco, the chief labour intensive Greek agricultural product, stagnated in the 1930s. Clearing trade treaties with Germany (along with the USA, the major importer of Greek tobacco) had been agreed upon but they only succeeded in preserving the Greek share in the German market. The volume of exported tobacco did not grow. Stagnation, at relatively high levels of output, characterised the other crucial commodity of Greek labour intensive export agriculture – currants.

State intervention in the 1930s helped producers to adopt new varieties of seed, make better use of fertilizers and even supported rural incomes through guaranteed prices. Yet it proved unable to improve yields, cut costs and standardize product quality. As limited as was the progress made in certain respects, it was nonetheless significant when compared to other countries. Growth continued to be extensive, depending on what land and labour was available. This, of course, meant real output growth and rising incomes for many rural families. But neither productivity nor the quality of the product was in any way affected. The state-sponsored cooperative movement had been able to cut down transaction costs in agricultural credit for banks, but there was very limited improvement in organising cooperative industrial plants. Labour productivity stagnated and Greek agriculture did not improve its position in the international division of labour. It remained a producer of middling quality Mediterranean products (olive oil, wine, fresh fruits and vegetables) clearly distanced from Spain, Italy, and France, while continuing to hold its privileged position vis-à-vis eastern and southern Mediterranean agricultures. Productivity stagnation was partly the result of state intervention since the autarchic agricultural policy discouraged regional and farm specialization and skewed the allocation of productive resources in favour of land-intensive food products. By 1939 Greek agriculture had reached a record high output level but had shown no signs of dynamism (Petmezas, 2009: 366).

VI. Famine, wartime agriculture and the slow post-war recovery

If there were any doubts about the necessary priority of agricultural (*i.e.* food) self-sufficiency in the inter-war period, wartime events would tragically dispel them. The German occupation forces pillaged the available local production sources while the British naval blockade cut off the country from its sources of imported foodstuffs. In the extremely cold war winter of 1941/42, almost a quarter of a million Greeks died indirectly (through diseases, cold, etc.) or even directly from famine. Thus did the extreme dependence of the country on agricultural trade and imported foodstuffs prove to be fatal for the poorer strata of the urban population.

It is clear that the war hit Greece harder than any other Mediterranean country. Apart from the famine it led to mass refugee movements during the occupation of the country and gave rise to the subsequent bloody civil war. Some of these displaced populations never returned to their villages of origin. The year 1940 marks the high point of population density in the countryside. From then on the absolute number of rural inhabitants finally began to fall. The post-war recovery and 'reconstruction' would prove to be a long and difficult process for Greece. In spite of the ambitious industrialization programmes, backed both by right and left-wing parties, many respected civil servants were very pessimistic about the economic prospects for Greece in the post-war world economy. They feared a continuation of the stagnant and autarchic pre-war situation. The loss of export markets for tobacco (Germany) and other labour intensive products was felt as a looming threat to the long-term financial and social stability of the country. The flow of foreign exchange (*i.e.* the balance of current accounts), the income of a substantial part of the agricultural population (especially among smallholders in the hill country of northern Greece) and the employment of those working in the tobacco trade were all perceived to be endangered. Governments in the immediate post-war period followed the inter-war policy of intervention with import quotas, clearing agreements (especially with the new 'socialist' countries), protectionist tariffs, guaranteed prices and income support measures. The Greek state also continued financing the completion of the land improvement and reclamation projects, using the Agricultural Bank and the autonomous agricultural offices to finance and extend services to farmers. The size (small) and composition (fragmented) of farms also began to be addressed as urgent issues.

During the 1950s the country reached the apex of its 'traditional' agricultural structure and development. The cultivated surface had expanded, incorporating the newly drained lands, and all available arable land was being used more intensively. Land in use, which had fallen from 3.4 millions ha in 1939 to 3.1 in 1950, grew again to 3.7 in 1961. Thus in the mid-1950s the pre-war maximum was left behind and land in use reached a maximum of 3.9 million ha in 1965. Ever since it has fluctuated around this level; land in use has never surpassed 4 millions hectares. In 1953, with 1.8 million hectares of land seeded with cereals and rice, wheat cultivation reached its high point. Thereafter this high level fluctuated until 1965 (when it begun to recede), but the yields were rapidly growing and the country became finally self-sufficient in wheat and pulses in 1957. Wheat yields had matched their highest pre-war level (1,139 kg per hectare in 1937) in 1949 (1,109 kg) and again in 1952; and since then they constantly surpassed this level. By the mid 1950s overcrowded Greek agriculture had at last visibly fulfilled the target of inter-war 'productivist' policies: food self-sufficiency through an intensive use of land and labour.

It should be noted that in spite of the negative trends in the global international demand for traditional Greek export products (tobacco, currants, wine and olive oil) a new dynamism is observable. This derived from the growth of the so-called 'new export products' – fresh fruits and vegetables – that were finding an expanding and dynamic market in the northern European countries. The growth of these products depended on investment and the purchase of inputs: *i.e.* the diffusion of irrigation and the use of new equipment, chemical fertilizers, pesticides, etc. The same dynamism was evident in another field of high income elastic products – those based on animal husbandry. Livestock grew in number and greatly improved in quality. Better breeds were adopted and new intensive techniques of animal husbandry were finally diffused in the 1960s. As a result the volume (and value) of livestock products grew rapidly in importance. One can say that this was a clear sign of the true modernization of Greek agriculture.

VII. Agricultural Stagnation in the long inter-war period: a Greek or a Mediterranean phenomenon?

The 'long' inter-war period in Greece (late 1910s-early 1950s), in spite of brief periods of recovery, is one of long-term stagnation, and even regression, on all fronts. Between the years 1911 and 1950 the estimated Total Factor Productivity growth of agriculture in Greece shows a negative rate (-0.57 per cent), which is probably matched by the post-1931 Spanish equivalent. Portugal and Italy had not experienced a similar regression, the former having been spared by staying out of wars (and showing a rather high TFP growth rate in 1927-1962, after a period of slow growth in the first quarter of the century) and the latter having relatively strong inter-war growth, in spite of regression in times of war (see Table 5.1). It is clear that both Greece and Spain paid a higher price during this period. Yet the pattern of causality seems to be different in the two cases. Geopolitical and diplomatic events and the urgent need to engineer social cohesion and ethnic homogeneity in the newly annexed northern provinces forced the Greek state authorities to launch a radical agrarian reform without the financial means to assure highly capitalised viable farms. In the Spanish case diplomatic isolation and the pro-landlord policy of the Franco regime facilitated large landowners in profiting from the particular combination of black market activities and protectionism that allowed higher prices to be charged for a reduced output of low quality products (Barciela, 1986: 393-398). It is also clear that the Italian (Zamagni, 1993: 259-260) and Portuguese (Clar, 2009) owners of large estates profited handsomely from the inter-war policy of protection for cereal cultivators.

All these Mediterranean countries (with the exception of the Po valley in northern Italy) could still be called 'traditional' agricultures. They shared with the cereal producing countries of eastern Europe a very high agricultural population density and were confronted with the same broad problem – low labour productivity in agriculture due to the absence of non-agricultural employment opportunities (in other words latent structural unemployment of their rural population). League of Nations specialists who studied European agricultural economies had concluded that, in view of their agrarian structures and factor endowments, these countries had reached a level of population density that could not be supported and that they were accordingly locked into continuing impoverishment (Moore, 1945). Even earlier Doreen Warriner (1939/1964), in a path-breaking study, had stressed the agricultural stagnation and productivity deadlock of the eastern and southern European periphery. The over-populated agrarian sectors of the periphery were unable to feed their populations and accumulate the necessary surplus to finance the industrial development or/and its technological renovation and structural adjustment that would help the surplus rural population find productive employment in the secondary sector. These countries of the periphery needed either to raise the labour productivity of their agricultural sectors or find alternative non-agricultural employment in the rural or urban areas. Otherwise, these 'involutive' agricultural economies would become fatally impoverished agricultural satellites producing cheap food and raw materials for industrial West Germany, or perhaps become (following the 'Russian path') hotbeds of social revolution.

Warriner's bleak picture proved to be a realistic assessment of southern and eastern Europe in the late inter-war and immediate post-war periods. We may add that rural unrest, due to rural income depression and social strife between landless peasants and big landowners, must be added in the case of those countries that had skipped agrarian reform in the 1920s. Such problems became acute in Spain, Portugal and southern and central Italy. Greece, where small-scale land ownership predominated but where the peasants went through two impoverishing decades of depressed incomes, was not spared either. Social unrest and rural population pressure (with low labour productivity and inadequate family income) continued to be major problems in the agricultural sectors of Mediterranean countries in the 1950s and 1960s. Governments were therefore looking for policies that could reduce population pressure and ease social relations. Irrigation and land improvement and reclamation, as well as the settlement of all available lands by landless peasants, seemed to offer a solution, one that had been tried since the inter-war period, albeit with only relatively minor success (Barciela, 1986: 398-416; Zamagni, 1993: 333-334).

VIII. Agricultural modernization and out-migration in the European Mediterranean countries

Since the mid-1950s the Greek countryside has undergone a series of rapid transformations that completely re-configured relations between the mountain regions and those of the lowlands, between the urban sector and the countryside and between national and international economic structures. These transformations had different effects on the rural population and economy as well as on regional agrarian structures. Small-scale ownership was preserved, but rapid out-migration left more land available for each producer, while those who were able to mechanize their farms could earn considerable supplementary income by renting out their farm equipment. The state-financed policy of land consolidation and land reclamation gave dynamic small farmers the opportunity to rent or purchase more land, especially in the lowlands (Petmezas, 2009: 368). These were the chief beneficiaries of agricultural modernization and growth, and took full advantage of their higher labour productivity. It was these who would also benefit most from the national and later from the European agricultural policy.

What I have called the ecological and institutional constraints on agricultural growth were partly countered in the inter-war and post-war decades (late 1920s-early 1960s) via an active and sustained policy of large-scale public intervention and investment, which begun in the late inter-war era and continued in the first two post-war decades. Public investment in agriculture rose between the late 1950s and the late 1960s, from 22 per cent to 26 per cent of total public spending, while in the same period public spending on agricultural subventions and premiums (*i.e.* mostly agricultural income supports) rose more spectacularly from 2.2 per cent to 8.8 per cent of public spending. Investment and structural policy spending far out-stripped income subvention policy in agriculture in these early years. As time passed, and regional specialization and income divergence came to count for more, the government found itself increasingly engaged in a policy of income support which expanded in parallel with out-migration as successive governments tried to curb the migration of marginalised agricultural populations.

It should be added that public investment had real effects on the modernization of agriculture. Fixed capital in agriculture had finally been formed in substantial amounts, and the available capital per land and labour unit had grown substantially. Both indexes had doubled between 1911 and 1950 and in the next thirty years the available capital per land unit quadrupled while capital per labour unit grew almost sevenfold (Petmezas, 2009: 367). It is interesting to note that while public and private investment in agriculture had been raised, it was especially private investment that had sky-rocketed. A growing part of this private investment was financed with credit supplied by the Agricultural Bank; the credit it advanced (in deflated drachmas per unit of rural labour) grew more than tenfold in 30 years (1950-1981).

Annual fixed capital investment rose from 5 to 16 per cent of the annual aggregate value of agricultural output in the critical years between 1956 and 1978. Irrigation, which enormously stimulated the cultivation of 'new export products' or staple products with growing internal and export demand (such as cotton), rose at an annual rate of 5.8 per cent between 1962 and 1968. In the 1960s an official policy of offering cheap credit through the Agricultural Bank, and the availability of new and innovative equipment (especially the small tractor suitable for working small and fragmented farms on hilly terrain) facilitated smaller farmers to invest in their land. The state also invested heavily (along with private capital and especially with the agricultural cooperatives) in various industrial and commercial consortia, thus offering specialized farmers greater market opportunities. Only after the two oil crises of the 1970s, when Greece was already industrialised, did state spending on income subvention grow more than investment and did private investment became totally dependent on Agricultural Bank loans. At that point out-migration, which had totally emptied a large part of upland Greece, had ceased and successive governments were anxious to retain the remaining rural population in the countryside.

The accelerated agricultural modernization we have described in Greece – which depended very much on a rising international demand for new export products, irrigation, the use of selected seeds, chemical fertilisers and pesticides as well as increased farm mechanization – wasn't very different from that experienced in other Mediterranean countries such as Spain, Portugal and southern and central Italy. Mechanization rose because wages were rapidly increasing as a large segment of the rural population migrated.

Out-migration, income depression and the abandonment of hill and mountain countryside were the other and negative sides of modernization, productivity growth, the capitalisation of farms and social differentiation. In effect, the agro-pastoral populations of the hill and mountain areas sank into a chronic income crisis, while those of the lowlands, delivered from malaria through the abundant use of chemical pesticides, benefited from most of the agrarian policy measures. The majority of the specialized hill farmers who formerly had produced the major Greek export products (oriental tobacco, currants, etc.) had farms which were very small in size; a large part of their family income depended heavily on the prices (and demand) these products could command on international markets. These hill farmers did not have any available alternative to their particular culture mix. Even if the state was willing to assist them financially and technically, they were totally subject to 'uncontrolled market forces'. When international demand for oriental tobacco stagnated and prices fell, a segment of these smallholders gave up (after a decade of misery) and availed of the migration opportunities that offered them a tangible alternative. They were not the only ones. Income crisis was the lot of most hill and mountain populations, especially shepherds. Chronic income crisis, lack of alternative local employment opportunities and the demand for labour in distant markets were responsible for the desertification of the mountain and hill villages. Public policy could only delay this process (through rather timorous income subvention schemes) or orient it towards specific destinations (as, for instance, through bilateral migration agreements with Germany).

In less than twenty years, one fifth of the rural population abandoned the countryside and moved, either to the two largest cities (Athens and Salonica) or emigrated. Sixty per cent of those emigrating from Greece were directly of rural origin. Almost all the natural growth of the Greek population was skimmed off by migration (see Table 5.3).

One should add that those who left were predominantly young males and that those who stayed behind on the mountain and hill villages tended to be the older members of the family, particularly unfit to adopt any innovation.

Table 5.3. Emigration in Greece

p. 1000	*Net emigration rate*	*Out-migration*	*Natural increase*	*In-migration*	*Population change*	*% out-migration to the natural movement*
1955-1961	4.37	4.37	11.75	0.00	7.37	37%
1962-1968	9.46	9.78	9.98	0.31	0.52	98%
1969-1975	3.02	5.81	7.62	2.79	4.60	76%

Source. Petmezas (2005).

In retrospect, one can hardly see how things might conceivably have been different. The Greek urban economy had been structurally unable to absorb sufficient rural labour to reduce the latent unemployment in the countryside and it was also unable to offer substantial markets for agricultural products (or invest in non-agricultural rural activities). Whenever labour productivity rose in the countryside this was related to the evacuation of latent unemployed labour (as in 1900-1920). It is clear that during the long inter-war period (1911-1950) the massive growth of available labour (Δ[L]) depressed productivity in spite of the relatively satisfactory rates of growth for Capital (Δ[K]) and Land (Δ[T]). In the post-war era the high rate of output growth (Δ [O]) depended not only on the growth of capital and (timorously) land but also on the shrinking of available labour (see Table 5.4). European demand for industrial labour was the prime mover for rural depopulation in general and a catalyst for the post-war agricultural modernization of Greece. Most likely this was the case for all the Mediterranean countries.

Table 5.4. Total factor productivity (TFP) growth in Greek agriculture

	Δ (O)	*Δ (K)*	*Δ (L)*	*Δ (T)*	*TFP*	*TFP/Δ (O)*
1860/1911	2.00%	0.07%	0.77%	0.52%	0.64%	0.32
1911/1950	1.60%	0.37%	1.10%	0.70%	– 0.57%	– 0.35
1950/1981	4.31%	0.62%	– 0.36%	0.28%	3.77%	0.87

Source. Petmezas (2005: Table 3).

Italy, in terms of the fundamental trends, preceded Greece by a full decade (see Table 5.5). The high point of Italian out-migration was the 1950s and not the 1960s, though Italy is not a very representative case of rural out-migration since a large part of the rural migrants of underdeveloped southern Italy moved to the heavily industrialized northern part of the country (Volpi, 1989: 63). In consequence, though structurally the phenomenon is the same, the net emigration rate is much lower in Italy than in other Mediterranean countries. Broadly Spain and Portugal followed the Greek pattern, as the greatest number of emigrants left their villages in the 1960s. Of course northern European demand for immigrant labour from the Mediterranean world is a relatively new phenomenon. Before 1960, Latin America constituted the destination of most emigrants (Nicolau, 1989:58; Pais de Morais & de Alarcao, 1975: 25), though this population outflow was insufficient to relieve the Mediterranean countryside of its reserves of latently unemployed labour.

Table 5.5. Estimation of the net emigration rate in various Mediterranean countries

	Population (thousands)	*Net number of migrants (thousands per year)*			*Net emigration rate (p. 1000)*	
Italy						
1942	43,847	1942-1951	63.1	1,44		
1951	47,516	1952-1961	136.5	2,87		
1961	50,624	1962-1971	54.7	1,08		
1971	54,137	1972-1981		0,00		
Spain						
1950	27,977	1947-1962	44	1,57		
1960	30,431	1960-1967	235	7,72		
1970	33,824	1968-1973	202	5,96		
Portugal			*observed*	*estimated*	*observed*	*estimated*
1950	8,441	1941-1950	9,1	12,9	1,18	1,67
1960	8,851	1951-1960	39,7	68,1	4,70	8,07
1970	8,568	1961-1970	38,6	129,9	4,36	14,68

Sources. The Italian data are taken from VOLPI (1989: 63), those for Spain from NICOLAU (1989: 58) and those for Portugal from PAIS DE MORAIS & DE ALARCAO (1975: 25).

The increasing integration of the European economy and the long-term expansion of northern European manufacturing exports provided the necessary impetus that gradually pulled a large number of underemployed peasants from the countryside, without causing any kind of productive regression, with the possible partial exception of Portugal (Lains, 2009: 336-338). One should add that emigrant remittances offered (micro-economically) precious supplementary income to the strained financial situation of the depressed agricultural regions, while they formed (macro-economically) a large part of the invisible resources that closed the large current accounts deficit of the developing Mediterranean economies. It is doubtful whether the modernization of Mediterranean agriculture, and the rise of investment and labour productivity in the countryside in general, would have been possible without the intrinsic relation that developed between the growing urban demand for labour and income elastic agricultural products in northern Europe. The available excess labour of the northern Mediterranean countryside (perhaps five million people) drained away in less than two decades.

As the demand for unskilled labour in the European industrial heartlands relieved population pressure, labour productivity was boosted by public investment in infrastructures (with enhanced transport and communication networks diminishing transaction and transport costs), by land improvement and reclamation as well as by irrigation. These same factors helped reorient agriculture towards high income export products (fresh fruit, vegetables and animal products) with a stable and rising demand in lucrative western markets and, later on, even in local urban markets as incomes grew rapidly here as well. All the while income transfers from migrant remittances were injecting additional fresh capital into the countryside. After out-migration diminished, labour productivity growth and social and regional differentiation in agriculture continued, in spite of the policy of successive governments of supporting incomes among the more vulnerable (but constantly fewer in number and relative size) inefficient family farms. New non-agricultural activities in the countryside (tourism, industry, etc.) began to transform agricultural family labour into multivalent rural labour.

IX. Conclusion

The agricultural sector in Greece has historically been caught in a chronic ecological and institutional deadlock that could only be tackled by the large-scale legal, institutional and financial engagement of the state. The relative political weakness of the large landowners in Greek society, something quite exceptional in the context of other European Mediterranean countries, and the geopolitical and diplomatic entanglements of Greece in 1916-1922, had induced inter-war liberal governments to adopt a radical agrarian reform and to organise a costly policy of land improvement and reclamation. Nonetheless, state intervention that took the form of a turn towards food self-sufficiency (or autarchy) in the late 1930s could do no more than partly challenge the ecological and institutional constraints and timidly induce some modernizing measures. After the war this state policy continued and intensified. Beginning in the 1930s, as a result of policy initiatives, some of the characteristics of the second agricultural revolution found their way to Greece. Of particular significance here were the abundant use of fertilizers, more intensive and more thorough cultivation techniques, improved seeds and animal breeds, better coordination with agronomists and veterinarians and especially the provision of cheaper and more effective credit through the Agricultural Bank (founded in 1928). By virtue of such measures Greece did not miss the second agricultural revolution, as it had missed the first.

And yet it was only the ever increasing inclusion of Greece, along with the other Mediterranean economies, into the re-emerging and ever more integrated European economy that really cleared the way for the modernization of agriculture. These countries achieved integration not only as exporters of labour intensive (traditional and 'new') agricultural products, but also as a major source of unskilled labour for central and northern European industries. Emigrants left the overpopulated Greek countryside, with no negative consequences for agricultural output, and without depressing Greek urban-industrial wages or employment. Thus there is no evidence of immiserating urbanisation in Greece and manufacturing industry did not adopt extreme labour intensive and capital saving techniques. Furthermore, a steady flow of remittances strengthened the incomes of rural households, aggregate demand and the country's balance of current accounts. Thus displacement of population, an extremely painful experience for many emigrants, was to be critical in clearing the way for agricultural modernization and industrialisation in Greece.

Bibliography

BARCIELA, Carlos (1986), 'Introduction', in Ramon GARRABOU, Carlos BARCIELA & José Ignacio JIMÉNEZ (eds), *Historia agraria de la España contemporánea*, v. 3, *El fin de la agricultura tradicional (1890-1960)*, Barcelona, Editorial Crítica, p. 383-453.

CLAR, Ernesto (2009), 'Farm Policy under the Salazar and Franco Dictatorships in Portugal and Spain: Towards an Authoritarian Model of Intervention in Agriculture?', in Nadine VIVIER (ed.), *The State and Rural Societies. Policy and Education in Europe 1750-2000*, Turnhout, Brepols, Rural History in Europe, 4, p. 177-194.

CLAR, Ernesto & PINILLA Vicente (2009), 'The Contribution of Agriculture to Spanish Economic Development, 1870-1973', in Pedro LAINS & Vicente PINILLA (eds), *Agriculture and Economic Development in Europe since 1870*, London, Routledge, p. 311-332.

COLL, Sebastián & BRINGAS, Miguel Angel (2000), 'Estimating Single Input Productivity, TFP and Output from Price Data. An Application to Spanish Agriculture, 1800-1935', *Histoire & Mesure*, 15/3-4, p. 339-353.

FEDERICO, Giovanni (2003), 'L'agricoltura italiana: successo o fallimento?', in Pierluigi CHICCA & Gianni TONIOLO (eds), *Storia Económica d'Italia. 3. Industria Mercati, istituzioni*, Tomo 1: Le strutture dell'economia, Milano, Editori Laterza, p. 99-135.

FEDERICO, Giovanni (2009), 'Agriculture and Economic Development in Italy, 1870-1939', in Pedro LAINS & Vicente PINILLA (eds), *Agriculture and Economic Development in Europe since 1870*, London, Routledge, p. 234-254.

FENOALTEA, Stefano (1993), 'Politica doganale, sviluppo industriale emigrazione: verso una riconsiderazione del dazio sul grano', *Rivista di Storia Economica*, fasc. 1, p. 65-77.

GALASSI, Francesco L. & COHEN, Jon S. (1992), 'La agricultura italiana, 1860-1930: tendencias de la producción y diferencias en la productividad regional', in Leandro PRADOS DE LA ESCOSURA & Vera ZAMAGNI (eds), *El desarrollo en la Europa del sur: España e Italia en perspectiva histórica*, Madrid, Alianza Editorial, p. 139-170.

GEERTZ, Clifford (1963), *Agricultural Involution: The Process of Ecological Change in Indonesia,* Berkeley, University of California Press.

KOSTELENOS, Georgios, PETMEZAS, Socrates *et al.*, (2007), *Ακαθάριστο Εγχώριο Προϊόν (1830-1939) [The Greek Agricultural Economy during the 19th Century]*, Athens.

LAINS, Pedro (2009), 'Agriculture and Economic Development in Portugal, 1870-1973', in Pedro LAINS & Vicente PINILLA (eds), *Agriculture and Economic Development in Europe since 1870*, London, Routledge, p. 333-352.

MOORE, Wilbert E. (1945), *Economic Demography of Eastern and Southern Europe,* Geneva, League of Nations.

NICOLAU, Roser (1989), 'Poblacion', in Albert CARRERAS (ed.), *Estatisticas Historicas de España siglos XIX-XX,* Barcelona, Fundacíon Banco Exterior, p. 53-93.

PAIS DE MORAIS, Joaquim & DE ALARCAO, Alberto (1975), *La Population du Portugal*, Paris, CICRED.

PETMEZAS, Socrates (2000), 'Export-dependent Agriculture, Revenue Crisis and Agrarian Productivity Involution: The Greek Case: 1860s-1930s', *Histoire & Mesure*, XV/3-4, p. 321-337.

PETMEZAS, Socrates (2005), 'The Long Term Development of Greek Agricultural Productivity in a Euro-Mediterranean Perspective (1860-1980)', paper presented in the 6th International Congress of the European Historical Economics Society in Istanbul [http://www.ims.forth.gr/ims/history_studies/agrotiki_oikonomia/PetmezasTPF.pdf]

PETMEZAS, Socrates (2009), 'Agriculture and Economic Development in Greece, 1870-1973', in Pedro LAINS & Vicente PINILLA (eds), *Agriculture and Economic Development in Europe since 1870*, London, Routledge, p. 353-374.

PINILLA, Vicente & AYUDA, María Isabel (2006), *'Horn of Plenty Revisited': The Globalization of Mediterranean Horticulture and the Economic Development of Spain, 1850-1935*, Working Paper, Asociación Española de Historia Económica.

SANZ, Jesús (1987), 'La agricultura española durante el primer tercio del siglo XX: un sector en transformación', in Jordi NADAL, Albert CARRERAS & Carles SUDRIÀ (eds), *La economía española en el siglo XX. Una perspective histórica*, Barcelona, Ariel, p. 237-257.

SIMPSON, James (1995), *Spanish Agriculture: the Long Siesta,* 1765-1965, Cambridge/ New York, Cambridge University Press.

SIMPSON, James & CARMONA, Juan (2003), *El laberinto de la agricultura española. Instituciones, contratos y organización entre 1850 y 1936*, Zaragossa, Prensas Universitarias.

VOLPI, Roberto (1989), *Storia delle populazione italiana dall'unità à oggi,* Firenze, La Nuova Italia.

WARRINER, Doreen (1939/1964), *The Economics of Peasant Farming,* London, F. Cass.

ZAMAGNI, Vera (1993), *The Economic History of Italy 1860-1990,* Oxford, Clarendon, Oxford/New York, Oxford University Press.

6. A 'tendency towards obligation'? The regulation of the international sugar market in the 1930s[1]

Fritz Georg VON GRAEVENITZ

I. Introduction

Writing about the institutional development of agricultural interests on the international level, François Houillier interprets the creation of the International Institute of Agriculture (Rome) in 1905 as a *tendance vers l'obligation*. Before this Institute was brought into being by an inter-governmental agreement, international agricultural meetings were organised around the *Commission Internationale d'Agriculture* as a transnational, non-governmental organisation (Houillier, 1935: 19; Aldenhoff-Hübinger, 2002: 42-70). Hence, Houillier's 'tendency towards obligation' broadly describes a shift from non-governmental to governmental internationalism in agriculture.

The tendency towards obligation in agricultural internationalism appeared in a period characterized by the first wave of globalisation (Torp, 2005: 27). Contextualising the movement of national agricultural interest groups towards the international stage, Madeleine Herren (1994: 1) characterises this nineteenth century internationalism as a 'social analogy' to the globally interlinked economic relations emerging at that time; essentially 'the world is perceived as a sole economic unit and the interdependence of the nations and the national farmers appears more and more clearly' (Houillier, 1935: 20). Having been above all a non-governmental movement establishing a 'dense transnational network' (Herren, 1994: 40), internationalism froze with the outbreak of the First World War. But the mechanisms of internationalism survived the national antagonisms and became the necessary basis for the creation of the economic and social institutions of the League of Nations (Geyer & Paulmann, 2001: 4; Herren, 2001: 144; Clavin & Wessels, 2005).

The initiatives of the League of Nations in international economic cooperation had to cope with an economy very different from the globalized world before World War I. The crisis of international economic interdependence in the interwar years, which peaked in the Great Depression, provoked an 'end of globalisation'

[1] The author wishes to thank Kathrine Lausted Veie, and the two anonymous referees for helpful comments.

according to Harold James (James, 2001). In the 1920s, a massive overproduction of agricultural goods led to the breakdown of worldwide agricultural markets. Unlike the international cross-linking trade relations at the end of the nineteenth century that worked as a growth engine, the growth in global trade in the interwar period lagged behind the growth of production. A further cause of the end of globalisation was an unprecedented increase in national protectionism in the shape of beggar-thy-neighbour tactics in international trade politics (James, 2001: 108).

This paper seeks to transpose Houiller's analysis to the 1930s. Here, I apply his concept of a *tendence vers l'obligation* to a comparable shift in international sugar cooperation from a non-governmental to a governmental agreement between the two World Wars. Taking governmental internationalism as an indicator for a structural change in international relations (Herren, 2001: 135), the interpretation of the term *obligation* in the context of this paper refers to the binding character of sugar agreements beyond the national framework. Accordingly, this case study attempts to contribute to the history of agricultural internationalism in the interwar period.

One key feature of the world sugar market was – and indeed still is – the competition between cane and beet sugar. On the eve of the First World War, both types contributed equally to the world market (Mintz, 1985; Kuster, 1998). Even initially, sugar production from beet was only possible by excluding the lower cost cane sugar from the markets where beet sugar was sold. With the increasing competition, a complex system of export bounties was established in many European countries. The higher prices in the domestic markets encouraged overproduction, and in order to avoid the creation of sugar mountains the excess production had to be exported (Albert & Graves, 1988: 3; Martineau, 1904). This system of export subsidies in the late nineteenth century provoked a dumping price on the world market and could only be countered through an international agreement. The resulting competition in bonuses, with each country attempting to sell its sugar on the world market at the lowest price, meant that national sugar industries derived no advantage from the system (Harney, 1950: 239). On 5 March 1902 the Brussels International Sugar Convention was signed by the principal European countries (Martineau, 1904: 39; Trentmann, 2008: 157). It was a plan conceived against governmental intervention in the free international sugar trade which aimed at ending the system of export bonuses that favoured foreign consumers at the expense of domestic consumers (Wilk, 1939: 862).

This agreement was the first step in the 'International Sugar Régime' (Wilk, 1939). The second step was the so called 'Chadbourne Agreement' of 1931 that was followed by the *International Agreement Regarding the Regulation of Production and Marketing of Sugar* (1937). The Chadbourne Agreement specified export quotas for each of the world's major exporting countries participating in it. This international accord was signed by representatives of the sugar *industries* in the respective countries and amounted to a non-governmental 'gentlemen's agreement'. Even though the Chadbourne Agreement is not understood as being part of the development in international sugar co-operation – Harold James interprets it as a once-off measure which 'expired in 1935 and was not renewed' (James, 2001: 139) – the Chadbourne plan was in fact the core event in the struggle to achieve the regulation of the international sugar market in the years after the First World War. It was replaced by the International Sugar Agreement of 6 May 1937, by which point a 'seemingly hopeless glut in the market has been followed by conditions of shortage' (Bloch, 1939: 141; League of Nations, 1937: 7-19). The sugar agreement of 1937 can be considered to have provided the structures for the international sugar agreements after the Second World War, and therefore to be the direct forerunner of the present-day International Sugar Council in London (Mahler, 1984; Swerling, 1954).

Whether the structural change occurring between the Chadbourne Agreement and its institutional successor in 1937 entailed an obligatory tendency will be considered in two sections. The first of these treats the emerging international discussions in the face of a global crisis that culminated with the signing of the Chadbourne Agreement in 1931. This agreement had to be replaced by the International Sugar Agreement of 1937 as due to its construction, the final results were unsatisfactory. Section two will examine the 1937 international accord. My discussion will conclude with an overall evaluation of the international sugar regime and its tendency towards increasing obligation.

II. The non-governmental agreement of 1931 and the world sugar crisis

II.1. First attempts at establishing a post-war accord: Cuba and the League of Nations

With the outbreak of the First World War, the international sugar trade – and the intra-European trade – was disrupted (Henninger, 1927: 49). Due to the drop in European sugar production between 1914 and 1918, the leading cane sugar producing countries strongly increased their output. In consequence, the share of European beet sugar dropped from roughly a half in 1913-1914 to a meagre fifth of world sugar production in 1919/20 (Mikusch, 1929: 32; International Sugar Council, 1963: 41). The geography of sugar had therefore changed profoundly and to a certain degree had become more international (Chalmin, 1990: 140). Recovering from the Great War by 1925, the increase in European beet sugar production not only inspired intense competition between cane and beet sugar – still evident today – but also caused immense global overproduction that gave rise to sugar mountains. 'Of all the primary commodities, sugar was the most intensely affected by the disturbances of the World War and of the period which followed it' (Guitérrez, 1935: 1).

For the first time after the First World War, the world market saw the appearance of the so-called 'dumped sugar' that caused a free fall in world market prices (Treiber, 1933: 49). Still, sugar production continued to increase because the European sugar exporting industries in particular were well protected by a system of tariff barriers combined with export subventions (Treiber, 1933: 49; Schiller, 1940).

The global crisis of sugar overproduction and consequent falling world market prices could only be countered by concerted international action of the sort familiar to the sugar industry since 1902. But, for a number of reasons, a simple re-establishment of the agreement of 1902 seemed to be out of the question. Firstly, the structure of the world sugar market had completely changed due to the First World War. In view of this an international agreement had to quantitatively go beyond an exclusively European accord. Secondly, the European sugar industries were only surviving under a protectionist umbrella. Removing export subsidies and reducing tariffs on the European continent did not therefore appear to be viable options. Any new international agreement would have to be qualitatively different from the Brussels Convention of 1902.

Since Europe had returned to its former production level by 1925, world market prices began to fall substantially, a fall that seriously threatened the Cuban cane sugar producers (Guitérrez, 1935; Lehmann, 1933: 6). On the initiative of Cuba, delegates of the major sugar-exporting industries in Cuba, Germany, Poland and Czechoslovakia met in Paris in November 1927 to discuss common steps towards an amelioration of the sugar world market. The central idea to emerge in Paris was a limitation of exports so as to release tension on the world market and consequently to support the world market price (Chalmin, 1990: 151). But every attempt to persuade the Dutch owners of the Javanese sugar mills, the second largest global producers of cane sugar, to limit exports ended in failure. Unconvinced of the value of international agreements, the Javanese sugar cartel refused to agree to any accord limiting its sugar production (Rowe, 1930; League of Nations, 1937: 35)[2]. The first attempt to privately regulate the world sugar market thus failed. Without the participation of Java, there

[2] Neither the Cuban nor the Javanese sugar industries were indigenous enterprises. Whereas the Cuban Sugar industry was mainly run by American capital, the Vereenigde Javasuikerproduzenten (VJSP) had its headquarters in Amsterdam. Report of the consul von Bülow in the German consulate general for the Netherlands, Amsterdam, 15.12.1930, Politisches Archiv des Auswärtigen Amtes (PA AA) [Political Archive of the Foreign Ministry, Berlin], R 118211, W 5558, p. 4.

was no effective basis of achieving a significant impact on the world market (League of Nations, 1937: 37).

The resulting standoff on the world sugar market brought the problem to the attention of the League of Nations. The League appointed three internationally recognized sugar experts who submitted a memorandum to the Economic Committee. On the basis of this memorandum, the League of Nations' Economic Committee wrote its final report *La Situation mondiale du Sucre* (League of Nations, 1929). This report promoted co-operation between the sugar industries themselves, only asking the relevant nation-states to provide the necessary policies to stimulate consumption by reducing their excise taxes. Furthermore, the marketing and the consumption of sugar had to be backed up by active propaganda and an international bureau was to be established in order to collect information and to give advice (League of Nations, 1929: 16; Lehmann, 1933: 37).

What is clear, however, is that an inter-governmental co-operation agreement initialised by the League of Nations was not advocated: 'We do not wish to bear any responsibility nor take any initiative in this regard' (League of Nations, 1929: 16). This seemingly cautious statement from Geneva had an impact. On 29 June 1929 the delegates of the sugar industries of Germany, Czechoslovakia, Poland, Belgium, Hungary and Cuba found themselves convened together in Brussels[3]. But again Java boycotted this round of meetings, as its cane sugar production was – in contrast to Cuba's – still profitable at a world market price of 2 cents per English pound (Rowe, 1930: 57). The Javanese competitive advantage was due not alone to the availability of cheap labour, but to the success experienced in breeding a new high-yielding and more disease resistant cane sugar variety (the so-called POJ 2878 launched in 1925). Additionally, due to the sharp competition with Cuba, it seemed to be a better strategy for Java to follow a tough price war strategy and to take advantage of the Cuban trade crisis, than to ameliorate the situation on the world market. The basic aim was to 'kill Cuba as a rival in business'[4].

Cuba accepted this declaration of war. In the summer of 1930 it closed its agency for the centrally organised sale of sugar (*Agencia Cooperativa de Exportacion*). This *Agencia Cooperativa de Exportacion* had been founded on 26 July 1929, when the USA – Cuba's key market for sugar – raised the tariff on Cuban sugar imports and thus limited Cuban export possibilities. The total export of Cuban sugar to the

[3] Letter from the German embassy in Brussels to the German ministry of foreign affairs, 5.7. 1929, PA AA, R 118211, W 2725.

[4] Letter from the German embassy in Lima (Peru) to the German ministry of foreign affairs, 10.3.1930, PA AA, R 118211, W 1527.

USA decreased in the years 1925 to 1930 by 64.9 per cent[5], the decline in 1930 being particularly steep. The Agency controlled and therefore limited the volume of exported Cuban sugar[6]. Following its closure, the world market price fell far below 2 cents per English pound. This shot across the bows of Java, which almost ruined the Cuban sugar industry, had the effect of revitalising the international discussion.

II.2. Chadbourne and the voluntary agreement among sugar-exporting industries

The precarious position of the Cuban sugar industry spurred American and Canadian bankers into action. Having been the controlling shareholders, they feared for their capital invested in the Cuban sugar industry. A consortium for the promotion of Cuban Sugar was established, directed by the New York lawyer Thomas L. Chadbourne. He argued in favour of an international limitation of sugar production, emphasizing at the same time that Cuba must not be the only exporter to make concessions. A possible limitation of Cuban sugar production was related to an international agreement, especially with Java. The Chadbourne plan proposed three measures to prevent the accumulation of surplus stocks and to redress the downward pressure on prices: 'The existing surplus stocks were to be segregated and gradually absorbed by exports; future exports were to be limited to maximum quotas; and production was to be cut to the needs of home consumption and to balance the export contingent not filled by disposing of surplus stocks'[7]. With this plan, Chadbourne hoped to restore the balance between consumption and production in the world sugar industry. Essentially Chadbourne saw his plan initiating a new method for the solution of the problems found on all world markets (Lehmann, 1933: 41).

In order to establish a coherent 'front of cane sugar' (Lehmann, 1933: 40), Chadbourne and the Dutch owners of the Javanese sugar mills commenced pre-negotiations in Amsterdam in early December 1930. After having agreed a bilateral accord and having specified their respective views on how international sugar exports were to be divided[8], the cane and beet sugar interests met in Brussels between 9 and 16 December 1930. Apparently, a common front of cane sugar producers was required to facilitate the general negotiations. In Brussels, Chadbourne was faced with an astonishingly homogenous European front of beet sugar producers, one that had a

[5] Report of the German consulate general of the Netherlands, Amsterdam, to the German ministry of foreign affairs, 2.1.1931, PA AA, R 118212, W 30.

[6] Letter from the German embassy in Havana to the German ministry of foreign affairs, 27.7.1929, PA AA, R 118211, W 3607, p. 1.

[7] Letter from Chadbourne to Rabbethge, 14.12.1930, p. 3, annex to letter from Rabbethge to assistant secretary of state Dr. Ritter, German ministry of foreign affairs, 17.12.1930, PA AA, R 118211, W 5538.

[8] Report of the consul von Bülow, German consulate general for the Netherlands, Amsterdam, 15.12.1930, PA AA, R 118211, W 5558, p. 1.

clear idea about how a possible international agreement was to be achieved. For the European beet sugar producers, Cuba and Java should make most of the concessions in reducing sugar output as they were held to be responsible for the general increase in world sugar production (Die Deutsche Zuckerindustrie, 1930: 1138; Zeitschrift des Vereins der Deutschen Zucker-Industrien, 1930: 266).

While the early cane sugar front talks had taken place in Amsterdam, the representatives of the European sugar industries of Germany, Poland and Czechoslovakia met together on 22 November 1930 in Brussels for preliminary discussions (Die Deutsche Zuckerindustrie, 1930: 1261). The European front was opposed to any international agreement regulating the national tariff and subvention policies and thereby intruding into the space of national sovereignty[9].

In contrast, the Javanese sugar producers demanded a 'genuinely international regulation' *à la* the Brussels Sugar Convention of 1902[10]. Chadbourne tried to include the European producers in the plans to restrict sugar exports agreed between Cuba and Java. The beginning was anything but promising, as conceptions of quotas differed sharply between Chadbourne and the European sugar exporting industries.[11] In particular the German sugar industry, as represented by Erich Rabbethge, proved to be the major obstacle to an international accord by demanding an extraordinarily large sugar export quota compared to the current levels of domestic production and consumption. But an agreement was eventually found when Rabbethge succeeded in imposing the German demands on his future co-signatories. Cuba, along with the other European sugar industries, agreed to a reduction of its quota in line with the German demand that the total export amount fixed in the previous negotiations not be increased.

Having settled one aspect of an international agreement – the export quotas and therefore the foundation of a mechanism for controlling the world's raw sugar supplies – other features remained to be agreed. To make the proposed quota system work, it was necessary to include penalties for non-compliance, to decide a certain price

9 Note for the file of Ritter, German ministry for foreign affairs, 1.12.1930, PA AA, R 118211, W 5281, p. 1f.

10 'Begründung des abweichenden Standpunktes der Ned.Ind.Landbouw-Mij. bezüglich des auf der Amsterdamer Konferenz getroffenen Abkommens zwischen der Visp und der Chadbourne-Kommission', annex 1 to report No. A 1/31, 2.1.31, of the consulate general in Amsterdam, p. 1, annex to the letter from the consulate general of the Netherlands, Amsterdam, to the German ministry of foreign affairs, 2.1.1931, PA AA, R 118212, W 30, p. 1.

11 Letter from Chadbourne to Rabbethge, 14.12.1930, p. 2, annex to letter from Rabbethge of assistant secretary of state Dr. Ritter, German ministry of foreign affairs, 17.12.1930, PA AA, R 118211, W 5538; Letter from the German embassy, Brussels, to the German ministry of foreign affairs, 17.12.1930, PA AA, R 118211, W 5524.

policy for the duration of the agreement, and to institutionalise an administrative body (the International Sugar Council) to oversee the day-to-day implementation of the plan. On 30 March 1930 the parties met again in Paris to come closer to their final goal – the formulation of a contract containing export quotas and related provisions (James, 1931: 490). At a price of 2 cents, a 5 per cent increase in all export quotas would automatically be applied by the International Sugar Council. An additional 5 per cent increase in export quotas was foreseen as soon as the price rose to 2½ cents. Above 2½ cents it was up to the Council to decide further increases in export quotas (Lehmann, 1933: 40)[12].

Table 6.1. Export quotas of the Chadbourne Agreement

Countries of the agreement	*Amount of sugar (in tons) to export without the agreement*	*Export quotas of the Chadbourne Agreement*				
		1931-1932	*1932-1933*	*1933-1934*	*1934-1935*	*1935-1936*
Cuba	2.400.000	655.000	805.000	855.000	855.000	855.000
Java	2.745.000	2.300.000	2.400.000	2.500.000	2.600.000	2.700.000
Germany	887.000	500.000	350.000	300.000	300.000	300.000
Czechoslovakia	756.000	570.817	570.817	570.817	570.817	570.817
Poland	453.000	308.812	308.812	308.812	308.812	308.812
Hungary	118.000	84.100	84.100	84.100	84.100	84.100
Belgium	61.000	30.275	30.275	30.275	30.275	30.275
	7.420.000	4.449.004	4.549.004	4.649.009	4.749.004	4.849.904

Source. TREIBER (1933: 74).

To administer the agreement the permanent International Sugar Council located at The Hague was instituted. Each of the seven countries was allowed three members on the Council. In terms of voting power, Cuba was given 35 votes, Java 30, and Europe 25 (Czechoslovakia 8, Germany 6, Poland 6, Hungary 3 and Belgium 2). A total of 55 votes (60 per cent) was required for making decisions on all important questions concerning price and policy. One of the key policy issues concerned the size of the export quotas; and here an increase rather than decrease in the export quotas was anticipated. Regarding the system of votes, neither of the main blocs – Cuba/Java and Europe – could build a majority on its own. The Council as a purely inter-industrial agency had no powers to adjust governmental policies in the member states. Charged with the simple supervision of the agreement, the Council was

[12] Report from Rabbethge concerning the negotiations in Paris, p. 4, annex to letter from Rabbethge to assistant secretary of state Ritter, German ministry of foreign affairs, 18.4.1931, PA AA, R 118212, W 1843.

granted only auxiliary functions of investigation, advice, and mediations (Guitérrez, 1935: 147; James, 1931: 491; Treiber, 1933: 73).

III. From non-governmental voluntarism to inter-governmental commitment: the International Sugar Agreement of 1937

Figure 6.1. Raw Sugar Production (Chadbourne countries) and Price (New York, cost & freight)

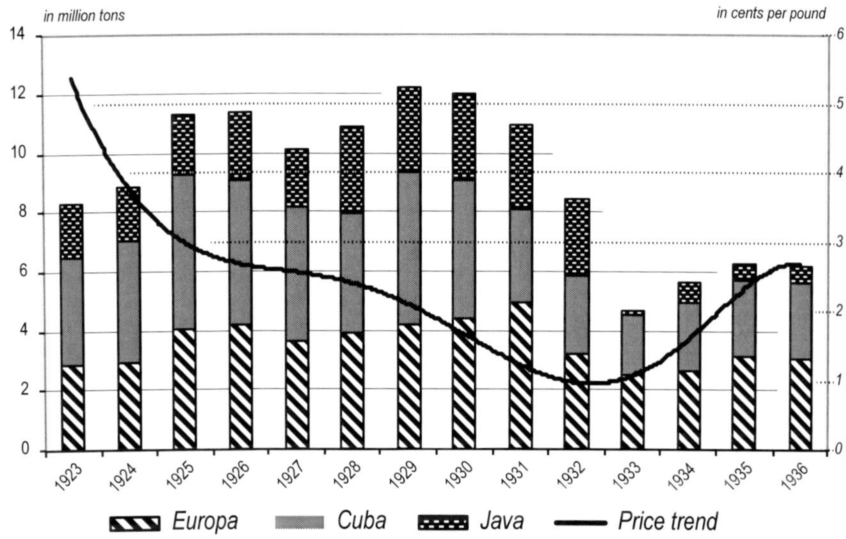

Sources. Statistisches Jahrbuch, 1929: 31* ; 1933: 37*; 1938: 96*; Licht (1937: 168); own calculations.

The project of 1937 was designed to deal with what had proved to be the shortcomings of the Chadbourne Agreement of 1931. Too many outsiders took advantage of Chadbourne's 'gentlemen's agreement' and increased their production[13]. Even so, in terms of its immediate goal of reducing production, the international regime under the Chadbourne Agreement can be considered a 'rousing success' (Mahler, 1984: 714). According to the statistics of the League of Nations, the decrease in worldwide production of 5.2 million tons that had occurred since 1930-1931 was due to the

[13] 'Begründung des abweichenden Standpunktes der Ned.Ind.Landbouw-Mij. bezüglich des auf der Amsterdamer Konferenz getroffenen Abkommens zwischen der Visp und der Chadbourne-Kommission', Anlage 1 zu Bericht Nr. A 1/31 vom 2.1.31 des Generalkonsulats in Amsterdam, p. 1, im Anhang zu Brief des Generalkonsulats der Niederlande, Amsterdam, an das Auswärtige Amt, Berlin, vom 2.1.1931, PA AA, R 118212, W 30, p. 2.

reductions made by the Chadbourne countries (Figure 6.1)[14]. Sugar production in the participating countries was roughly halved in the years after 1931. Therefore, the agreement had a more general impact on the production of its member states, thus showing that international regimes did matter and could make a critical difference.

Of course, these major reductions achieved under the Chadbourne Agreement of 1931 were largely offset by the increased production in the consuming countries not affiliated with the agreement[15]. In consequence, sugar was again a 'commodity in chaos' (Farley, 1935). For this reason the International Sugar Council at The Hague suggested to the committee preparing for the World Economic Conference in London (1933) that the existing agreement of the private sugar interests be complemented by an international sugar convention that would include the sugar-importing countries (League of Nations, 1933). On 3 February 1937 the General Secretariat of the League of Nations convened an international conference for April 1937. Prior to this a meeting had been convened in London in 1934 to explore the possibilities of an extended agreement. This meeting, however, was obliged to suspend its work because at that time there was little chance of inducing the countries concerned to recommend the immediate convocation of a conference (League of Nations, 1937).

The broader and more compulsory character of the agreement that was signed in 1937 became the *sine qua non* for any new international effort at regulating the world sugar market. Opening the conference, the English Prime Minister Ramsay McDonald enunciated the necessary steps in the development of international sugar co-operation:

> It is owing to the decisive influence of State action that this Conference is composed of delegates of Governments not of the producers themselves. Too many factors are involved for the terms of an effective sugar regulation agreement to be settled by producers alone. It is necessary to add to mere regulation of supplies some provision designed to prevent disturbance by further uncoordinated Government intervention (League of Nations, 1937: 26).

The 1937 Conference's beginning was not promising. In the first week, all the invited countries expressed their commitment to reaching a general agreement, but according to the respective delegates it seemed to be difficult if not impossible for each of them to make any concessions that limited exports and production. As in the discussions that preceded the Chadbourne Agreement, Java promoted a *convention*

[14] Letter from the Reichsministerium für Ernährung und Landwirtschaft to the German ministry for foreign affairs, 23.3.1933, PA AA, R 118212, W 2025, p. 2.

[15] Letter from the consulate general of the Netherlands, Amsterdam, to the German ministry for foreign affairs, 2.1.1931, PA AA, R 118212, W 30.

à la Brussels 1902, seeing in the limitation of state intervention the only way to efficiently remedy the problems with the world sugar market. On top of everything, the German delegation again came forward with special demands. With their quotas granted by the Chadbourne Agreement not fully used, the Germans feared that the basis for calculating quotas after 1937 – the export rates from 1935 to 1937 – would not provide a quota allocation adequate to Germany's sugar export plans (League of Nations, 1937: 35).

The experience of the Chadbourne Agreement had shown that the conditions on the world sugar market had first to be stabilized before any export quotas for the sugar exporting countries could be fixed. And to achieve this the non-exporting countries had to be asked not to provoke any contraction of the world market. Autarkic countries should not become exporters and importing countries had to be encouraged not only to augment their non-preferential sugar imports but to refrain from establishing their own sugar industries by resorting to protection via a system of tariffs, export subsidies or import quotas. As the convention enhanced the obligations of non-exporting countries so as to assure the existence of a world sugar market of a certain size, it simultaneously had to prohibit an excessive supply to the world market on the part of the sugar exporters (League of Nations, 1937: 27).

According to the obligatory requirements of the new international accord, the *International Agreement Regarding the Regulation of Production and Marketing of Sugar* was to be joined by almost all sugar-producing countries plus the principal sugar-importing countries. These importers included the United States and Great Britain with their largely preferential markets and that had the ability, in view of their significance as global consumers, to discipline those insubordinate exporting countries that dared to break the rules (El-Gamal, 1941: 365; League of Nations, 1937: 20)[16].

The governmental character of the Convention granted powers of decision and control to a common bureau. The Council was charged with giving effect to the 1937 Agreement and with fulfilling the auxiliary tasks of supplying information and advice as well as undertaking negotiations. It was bestowed with the administrative power to determine the estimates, decisions, and definitions on which the very operation of the flexible plan tied to changing market requirements depended. The Council could even make recommendations for the enforcement of the agreement among members and non-members alike. It is clear therefore that the remit of the Council went beyond

[16] The final agreement of 1937 was signed by Australia, Belgium, Brazil, China, Cuba, Czechoslovakia, the Dominican Republic, France, Germany, Great Britain, Haiti, Hungary, India, The Netherlands, Peru, Poland, Portugal, South Africa, the USSR, the USA, and Yugoslavia. The only non-member states of significance in the world sugar sector were Italy and Japan.

a simple fixing of basic export quotas. The agency was further empowered to regulate the extent of the preferential markets in the United States and the United Kingdom.

Table 6.2. Quotation of the World Sugar Market according to the agreement of 1937

Country	*Basic quota (final rapport)*	*Adjustment for 1937/38 +/-*	*Adjustment 1937/38 (net quota)*
		in metric tons	
Belgium	20.000	- 5.000	15.000
Brazil	60.000		60.000
Cuba	940.000		940.000
Czechoslovakia	250.000	+ 90.000	340.000
Dominican Republic	400.000		400.000
Germany	120.000	- 70.000	50.000
Haiti	32.500		32.500
Hungary	40.000	- 20.000	20.000
Java	1.050.000		1.050.000
Mozambique	30.000		30.000
Peru	330.000		330.000
Poland	120.000	-20.000	100.000
USSR.	230.000	-11.500	218.500
Σ	3.622.500	- 36.500	3.586.000
Reserve for countries which did not export in 1934/36	47.500	- 22.500	25.000
Total	3.670.000	- 59.000	3.611.000

Sources. League of Nations (1937: 62); El-Gamal (1941: 367).

In contrast to the Chadbourne Agreement, the limiting of surplus stock had been made more flexible by fixing both maximum and minimum production levels for cane sugar producers, and by taking a share of these countries' total production as the basis of calculation. This share depended in turn on the flexible amount of exports assigned to the country (League of Nations, 1937: 10).

With a majority of three-fifths in the Council, additional quotas could be proportionally allotted for a maximum period of one year, but only when the market demand exceeded the worldwide sugar supply or vice versa (El-Gamal, 1941: 375). The agreement of 1937 – in contrast to the Chadbourne Agreement – therefore installed a 'governmental machinery [...] [and] the elasticity of basic provisions necessary to adjust the contracting production to still further downward movements of market requirements' (Wilk, 1939: 864). For instance, the International Sugar Council reduced the quotas by 5 per cent during its second reunion in 1938, thereby contributing to a phenomenon not seen on the world sugar market for a long time – a shortage (El-Gamal, 1941: 381 ff.; Bloch, 1939)[17].

IV. Conclusion

The International Sugar Agreement of 1937 concluded an international learning process of ten years duration that had begun with the first Cuban efforts in 1927, progressed through negative price turbulence and difficult negotiations before culminating in 1937 with what was a unique international agreement by the standard of the time. Its uniqueness can be seen in its ambitious aims, the large number of member states that were party to it and in the way it at once added to and complemented the agreements that had preceded it. According to Joseph A. Davis, the director of the Food Research Institute at Stanford, the Sugar Agreement of 1937 was 'the most mature product of thought and efforts [...] that were intensified during the great depression' (Davis, 1946: 203). The universal character of the International Sugar Agreement of 1937 brought the world sugar market into balance by using the power of its International Sugar Council to adjust the export quotas to the needs of the market. The Chadbourne Agreement, in contrast, did not make provision for a possible reduction in export quotas, only an increase[18].

For a number of reasons the progression from the Chadbourne Agreement to the inter-governmental agreement can be characterised as a movement towards obligation. Firstly, an obligatory shift can be seen in terms of its organisational responsibility – the agreement of 1931 was a private accord while that of 1937 was established between states and within a system of international law. Secondly, the 1937 agreement was more comprehensive in its scope and in its memberships: 'essentially different in its form, and both more specific and more elastic in its obligations' (Wilk, 1939: 864). The obligatory tendency of the 1937 agreement manifests itself in the fact that the national concession to the International Sugar Council to reduce the export quotas by 5 per cent was granted for only two years, thereby registering the mistrust of the different countries facing any kind of international obligation (El-Gamal, 1941: 392).

[17] Export Quotas Under the International Sugar Agreement, August 10, 1939. National Archives at College Park/Maryland (NA), RG 59, Department of State, Decimal File 1930-39, 561.35 E I/725, Box 2743.

[18] The resulting rigidity was one of the biggest weaknesses in the Chadbourne Agreement of 1931 (Dye & Sicotte, 2006).

The global cross-linkages of national markets, which became especially obvious during the Great Depression, forced the respective national states into accepting internationally binding obligations. The obligations framed by the sugar agreement of 1937 constituted a prototype of agricultural internationalism that would survive the Second World War without major modifications[19]. In this step towards a more binding agreement, the League of Nations played a central role. By encouraging the Chadbourne Agreement it helped shape and sustain an epistemic community of sugar interests that helped inspire further action in the field of international sugar co-operation. In the context of the Chadbourne agreement, the League of Nations was a facilitator rather than an instigator. By assuming responsibility for the governmental successor of the Chadbourne Agreement, the League of Nations became the Petri dish for future international cooperation.

The extreme difficulties encountered in finding the final agreement within a context of economic crisis are evident on three levels. First and foremost, the general diversity of the positions of the different countries within the world market has to be taken into account. We have seen how the exporting countries differed widely between beet and cane sugar producers. In this regard the Chadbourne plan had decisively paved the way for further measures towards international cooperation, not least by introducing the successful contract-based approach. Having transformed the heterogeneous group of cane and beet sugar-exporting industries into a unified block, the agreement of 1931 became the 'forerunner of a new phase in international economic cooperation' (James, 1931: 481). Another major challenge was to bring into line the different attitudes of the importers intent on maintaining or even augmenting their own indigenous sugar output. The profoundly different preferences of exporters and importers as regards price levels on the world market had also to be balanced. Predictably, exporters tried to promote maximum prices and importers were more interested in securing a moderate world market price for the benefit of their domestic sugar consumers.

On another level difficulties arose from the way compromises in respect to the application of the internationally appointed rules had to be found. Who, for instance, would be obliged to reduce production and who would be obliged to make do with existing levels of domestic production? And how was the reduction to be distributed proportionately among countries?

---

[19] 'Relation of International Sugar Agreement of 6 May 1937 to Any General Principal of Commodity Policy Which May be Agreed by the United Nations'. Enclosure No. 1 to Despatch No. 28462, February 25, 1946, National Archives at College Park/Maryland (NA), RG 59, Department of State, Decimal File 1945-49, 561.35 E I A/2-2546, Box 2492.

What is very clear on another level is that the international negotiations behind the process of restructuring the international sugar market had to take account of domestic economic circumstances across the range of relevant countries. Thus the rate of unemployment, in view of the way the beet sugar sector employed relatively large numbers of workers, gave the sugar-exporting European countries a decisive incentive to stick to the system of tariffs and export subsidies. In addition, sugar exports were commonly seen as an important means of securing foreign exchange (Pennock, 1935, 76). The 1937 agreement therefore had to take as a given national trade policies and to complement these with a set of common international market interventions.

The example of the international sugar regime we have been discussing shows that even during a period of disintegration and excessive beggar-thy-neighbour trade policies, binding international agreements could be established and quantitatively as well as qualitatively upgraded. Consequently, effectual internationalism in the interwar period must be regarded as driven much more by a strategy of complementing national economic policy than by political integrative intentions.

The appearance of this wave of international cooperation during a time of economic de-globalisation is best exemplified by the interventions on the international sugar market in 1931 and 1937. It can also be noted that with the emergence of re-globalisation in the 1980-2000 period, this form of international market intervention came to an end, perhaps to re-emerge at some point in the future – when and if the pendulum swings back again – with an even stronger tendency towards international obligation.

Archival sources

National Archives at College Park/Maryland (NA), RG 59, Department of State, Decimal File 1930-39.

National Archives at College Park/Maryland (NA), RG 59, Department of State, Decimal File 1945-49.

Politisches Archiv des Auswärtigen Amtes, Berlin.

Bibliography

ALBERT, Bill & GRAVES, Adrian (1988), 'Introduction' in Bill ALBERT & Adrian GRAVES (eds), *The World Sugar Economy in War and Depression 1914-40*, London/New York, Routledge, p. 1-25.

ALDENHOFF-HÜBINGER, Rita (2002), *Agrarpolitik und Protektionismus. Deutschland und Frankreich im Vergleich 1879-1914,* Göttingen, Vandenhoek & Ruprecht.

BLOCH, Kurt (1939), 'Impending Shortages Catch Sugar Consumers Napping', *Far Eastern Survey,* 8, 12, p. 140-141.

CHALMIN, Philippe (1990), *The Making of a Sugar Giant, Tate and Lyle 1859-1989*, Chur/ New York, Harwood Academic Publishers.

CLAVIN, Patricia & WESSELS, Jens-Wilhelm (2005), 'Transnationalism and the League of Nations: Understanding the Work of its Economic and Financial Organisation', *Contemporary European History,* 14, 4, p. 465-492.

DAVIS, Joseph S. (1946), 'Experience under Intergovernmental Commodity Agreements, 1902-1945', *Journal of Political Economy*, 54, p. 193-220.

Die Deutsche Zuckerindustrie (*passim*), Berlin, Bartens.

DYE, Alan & SICOTTE, Richard (2006), 'How Brinkmanship saved Chadbourne: Credibility and the International Sugar Agreement of 1931', *Explorations in Economic History,* 43, p. 223-256.

EL-GAMAL, Mohamed Hassan (1941), *Le Problème international du sucre,* Paris, Librairie générale de droit & de jurisprudence.

FARLEY, Miriam S. (1935), 'Sugar – a Commodity in Chaos', *Far Eastern Survey,* 4, 22, p. 172-178.

GEYER, Martin H. & PAULMANN, Johannes (2001), 'Introduction, the Mechanics of Internationalism', in Martin H. GEYER & Johannes PAULMANN (eds), *The Mechanics of Internationalism: Culture, Society, and Politics from the 1840s to the First World War*, London, German Historical Institute; Oxford, New York, Oxford University Press, p. 1-25.

GUTIÉRREZ, Viriato Valladón (1935), *The World Sugar Problem 1926-1935,* London, N. Rodger.

HARNEY, Fritz (1950), *Hundert Jahre Der Deutschen Zuckerindustrie 1850-1950: Verein der Deutschen Zucker-Industrie 1850-1934; Wirtschaftsgruppe Zuckerindustrie (Verein der Deutschen Zucker-Industrie) 1934-1945; Verein der Zuckerindustrie 1945-1950,* Berlin, Verein der Zuckerindustrie.

HART, P. E. (1953), 'Competition and Control in the British Sugar Industry', *Oxford Economic Papers,* New Series 5, 3, p. 317-332.

HENNINGER, Kurt (1927), *Englands Versorgung mit Zucker seit dem Aufkommen der Kontinentalen Europäischen Zuckerindustrie: eine handels- und kolonialpolitische Studie,* Berlin, Winkelmann.

HERREN, Madeleine (1994), 'Modernisierung, Aussenpolitik und Integration im Jahrhundert des Internationalismus', *Historische Mitteilungen,* 7, 1, p. 1-43.

HERREN, Madeleine (2001), 'Governmental Internationalism and the Beginning of a New World Order in the Late Nineteenth Century', in Martin H. GEYER & Johannes PAULMANN (eds), *The Mechanics of Internationalism: Culture, Society, and Politics from the 1840s to the First World War*, London, German Historical Institute; Oxford, New York, Oxford University Press, p. 121-144.

Hobson, Asher (1931), *The International Institute of Agriculture; an Historical and Critical Analysis of its Organisation, Activities and Policies of Administration*, Berkeley, University of California Press.

Houillier, François Jean (1935), *L'Organisation internationale de l'agriculture : les institutions agricoles internationales et l'action internationale en agriculture,* Paris, Librairie technique et économique.

International Sugar Council (1963), *The World Sugar Economy: Structure and Policies,* London, International Sugar Organization.

James, L. Clifford (1931), 'International Control of Raw Sugar Supplies', *The American Economic Review,* 21, 5, p. 481-497.

James, Harold (2001), *The End of Globalization: Lessons from the Great Depression,* Cambridge, Mass., Harvard University Press.

Kuster, Tobias (1998), '500 Jahre Kolonialer Rohrzucker – 250 Jahre Europäischer Rübenzucker', *Vierteljahrschrift für Sozial – und Wirtschaftsgeschichte,* 85, 4, p. 477-512.

League of Nations, Economic and Financial Section (1929), *Memoranda Prepared for the Economic Committee by Dr. H. C. Prinsen Geerligs, Messers F. O. Licht and Dr. Gustav Mikusch*, C.148.M.57.1929.Ii (B), Geneva, League of Nations.

League of Nations (1933), *Report by the International Sugar Council to the Sub-Committee on Sugar of the Economic Sub-Commission II of the Monetary and Economic Conference. London, July 15th*, 1933, Conf.M.E./C.E./90, Geneva, League of Nations.

League of Nations, International Sugar Conference (1937), C.289.M.190.1937.II.B, Geneva, League of Nations, October 27.

Lehmann, Heinrich (1933), *Die Schlechte Lage der Deutschen Zuckerwirtschaft nach dem Kriege und der Versuch ihrer Besserung durch den Chadbourne-Plan,* Bottrop, W. Postberg.

Licht, Franz Otto (1937), *Welt-Zucker-Statistik,* Magdeburg, F. O. Licht.

Mahler, Vincent A. (1984), 'The Political Economy of North-South Commodity Bargaining: The Case of the International Sugar Agreement', *International Organization,* 38, 4, p. 709-731.

Martineau, George (1904), 'The Brussels Sugar Convention', *The Economic Journal,* 14, 53, p. 34-46.

Mikusch, Gustav (1929), 'Memorandum by Dr. Gustav Mikusch, Vienna', *in Sugar. Memoranda Prepared for the Economic Committee by Dr. H. C. Prinsen Geerligs, Messrs F. O. Licht and Dr. Gustav Mikusch*, ed. by the League of Nations, Economic and Financial Section, C.148.M.57.1929.Ii (B), Geneva, League of Nations, p. 29-51.

MINTZ, Sidney Wilfried (1985), *Sweetness and Power: the Place of Sugar in Modern History,* New York, Viking.

PENNOCK, Joop A. (1935), *La Question du sucre en Europe depuis la Guerre Mondiale,* Paris, J.-B. Baillière et fils.

ROWE, John Wilkinson Foster (1930), *Studies in the Artificial Control of Raw Material Supplies. London,* Royal Economic Society.

SCHILLER, Karl (1940), *Marktregulierung und Marktordnung in der Weltagrarwirtschaft,* Jena, G. Fischer.

SOCIÉTÉ des Nations (1929), *La Situation mondiale du sucre,* Rapport du Comité économique de la Société des Nations, C.303.M.104.1929.II, Genève, Société des Nations.

STATISTISCHES Reichsamt (ed.) (1929/1933/1938), *Statistisches Jahrbuch für das Deutsche Reich,* Berlin, Puttkammer & Mühlbrecht.

SWERLING, Boris C. (1954), 'The International Sugar Agreement of 1953', *The American Economic Review,* 44, 5, p. 837-853.

TORP, Cornelius (2005), *Die Herausforderung der Globalisierung. Wirtschaft und Politik in Deutschland 1860-1914,* Göttingen, Vandenhoek & Ruprecht.

TOSI, Luciano (1989), *Alle Origini della Fao. Le Relazioni tra l'Istituto Internazionale di Agricoltura e la Società delle Nazioni,* Milano, Franco Angeli.

TREIBER, Richard (1933), *Die Probleme der Deutschen Zuckerwirtschaft in der Gegenwart unter besonderer Berücksichtigung des Internationalen Chadbourne-Abkommens vom 9. Mai 1931,* Tübingen, A. Becht.

TRENTMANN, Frank (2008), *Free Trade Nation: Commerce, Consumption, and Civil Society in Modern Britain*, Oxford, New York, Oxford University Press.

WILK, Kurt (1939), 'International Affairs: The International Sugar Regime', *The American Political Sciences Review,* 33, 5, p. 860-878.

7. English and Austrian wartime farming, 1939-1945: 'revolution' or something else?

Ernst LANGTHALER

I. Introduction[1]

In the two World Wars, domestic agriculture became highly crucial for feeding the population in Europe, particularly in countries such as Britain that were heavily dependent on food imports in the interwar period (Collingham, 2011). For this reason rural society in general and agricultural factor and product markets in particular experienced enforced state regulation (Tracy, 1989; Trentmann & Just, 2006). While several studies have investigated food production and consumption in the two World Wars in considerable detail, the long-term conditions and consequences of wartime agriculture with regard to rural development in the twentieth century have been rarely discussed (Brassley, Segers & Van Molle, 2012). Recently, studies of rural society in Great Britain before, during and after the Second World War have addressed this topic. These accounts revise the official, somewhat mythical 'story of success' of British wartime agriculture (Murray, 1955; Hammond, 1951, 1954, 1956, 1962; Stamp, 1947) from a more differentiated viewpoint. They also claim that Britain, at least partially and when viewed from a long-term perspective, did experience a 'state-led agricultural revolution'. The suggestion is that agriculture was transformed in the 1939-1945 period from a labour-intensive but otherwise low-input low-output pastoral farming system to an arable 'national farm' with intensive application of land-saving (*e.g.* fertilizers) and labour-saving technology (*e.g.* machinery) acquired from outside the agricultural sector. This 'productivist' food regime that had been established in the Second World War and enshrined in the 1947 Agricultural Act regulated the national agricultural sector at least until the 1970s when Great Britain joined the European Community (Martin, 2000; Short *et al.*, 2007).

This paper aims to contribute to the discussion on the revolutionary character of European wartime farming between 1939 to 1945 by comparing the British case –

1 Research for this paper was funded by the Austrian Programme of Advanced Research and Technology (APART) of the Austrian Academy of Sciences. I would like to thank John Martin for extensively commenting on the first draft of this article. Moreover, Paul Brassley, Peter Moser and Tony Varley gave valuable comments at the COST workshop State Agricultural Policies: Causes, Implementation and Consequences. Finally, the paper profited from the suggestions by two anonymous referees.

with an emphasis on England and Wales – with the case of another territory involved in the Second World War. Austria, as a state annexed by the German Reich in 1938, is an appropriate case for comparison with the UK. Both territories – the former as a part of Nazi Germany, the latter as a member of the Anti-Hitler-Coalition – were subjected to the wartime state regulation of agriculture (as opposed to the pre-1938 German Reich where the expansionist 1936 Four-Years-Plan had already been in force). However, when evaluating the effects of wartime state regulation we have to keep in mind that, contrary to Great Britain, Austria as a province of the German Reich was no nation-state.

Drawing on the concept of 'agrosystem' (II), this paper will focus on upstream and downstream resource flows regulated by individual and collective actors at national, regional and local levels. Despite the author's commitment to actor-centred approaches to the study of agrosystems (Langthaler, 2006), this paper will dwell more on the structural aspects. After a discussion of 'agricultural revolution', agrosystemic regulations at national, regional and local levels in the UK and Austria will be examined (III). This is followed by a comparison, using contemporary statistical data, of agrosystemic resource flows, namely farm inputs, outputs and income (IV). Finally, the notion of 'agricultural revolution' is assessed from a comparative perspective (V).

II. Conceptualising 'agricultural revolution'

Although a case has been made for replacing 'agricultural revolution' with 'more or less rapid change' (Thirsk, 1987: 59), this paper proposes to retain the concept and to use it more properly. My starting point is provided by the agrosystem approach that has yet again attracted the attention of rural historians in recent years (Bieleman, 1999; Thoen, 2004; Langthaler, 2006, 2009b, 2010). An 'agrosystem' can be understood as the 'the theoretical expression of a historically constituted and geographically localized type of agriculture, composed of a characteristic cultivated ecosystem and a specific social production system' (Mazoyer & Roudart, 2006: 51). The concept of 'agricultural revolution' used in this article involves three aspects of agrosystemic transformation: extent, direction and speed. With regard to its *extent*, the transformation must be *fundamental* rather than gradual or, as some would say, 'structural' rather than cyclical (Martin, 2007a: 16). With regard to its direction, the transformation must be *progressive* rather than regressive. And with regard to its speed, the transformation must occur in the *short- or medium-term* rather than in the long-term. Only if all of these requirements are present are we permitted to speak properly of an 'agricultural revolution'; otherwise, we should choose to speak of other kinds of non-revolutionary as against revolutionary agrosystemic transformations.

Undoubtedly, this general definition of agricultural revolution is problematic in itself as a solution to the conceptualisation problem. What, we can ask, do 'fundamental', 'progressive' and 'short- or medium-term' mean exactly? Answers to these questions depend on the relevant historical and geographical contexts. With regard to Europe in the first half of the twentieth century, the three elements of 'agricultural revolution' can be more specifically defined as follows. What 'fFundamental' conjures up here are quantitative and qualitative transformations of, among other things, farm inputs, farm outputs and farm income. 'Progressive' – which is used descriptively rather than normatively here – can be taken to indicate transformations from capital-extensive, diversified and dispersed forms of agriculture to capital-intensive, specialised and concentrated forms, from subsistence to commercial farming and from relative local and regional autonomy to overwhelming state control (Ilbery, Bowler, 1998). 'Short- or medium-term' relates to transformations that occur during a few decades at most.

For comparative purposes two further elaborations of the concept are crucial. Firstly, agricultural development comprises technical as well as *institutional* aspects (Hayami & Ruttan, 1985). Accordingly, we have to differentiate between technical and institutional forms of 'agricultural revolution'. Whereas technical changes are related to quantifiable aspects such as factor substitution, productivity and production growth or rural exodus, institutional changes (which can be captured mostly qualitatively) comprise aspects such as the institutionalisation of property rights, extension services or the formation of farmers' associations. In the event of a considerable time lag between institutional and technical change, two phases of 'agricultural revolution' emerge. Secondly, the scope of an 'agricultural revolution' with regard to a given period and territory can be either total or partial, depending on the prevailing developmental pattern. We can therefore distinguish *general development* (with each farming unit making progress), *unequal development* (some farming units progressing faster than others) and *contradictory development* in the course of which some farming units progress while others are in crisis or even regress. If all farming units regress, there is, by definition, no 'agricultural revolution' only a general crisis (Mazoyer & Roudart, 2006: 50). Taken together, both of these elaborations constitute ideal-types of 'agricultural revolution' that provide a heuristic framework for exploring historically and geographically specific agricultural realities (see Table 7.1).

Table 7.1. Ideal types of 'agricultural revolution'

	General development (1)	*Unequal development (2)*	*Contradictory development (3)*
Technical change only (A)	1A	2A	3A
Institutional change only (B)	1B	2B	3B
Both technical and institutional change (C)	1C	2C	3C

III. Agrosystemic regulation

III.1. State regulation

At the outbreak of war in 1939, the experiences of the First World War food production campaign prompted the British government to extend the state-controlled system of administration that characterised the marketing boards in the 1930s. The Ministry of Agriculture's tried to expand the domestic production of essential and high calorie food, especially carbohydrates such as wheat and potatoes, so as to alleviate Britain's substantial dependence on imported food. Four years later, the 'miraculous result' was already being praised: 'After four years of war the British farmer, who formerly produced less than 40 per cent for home consumption, is now providing 80 per cent of the country's rations [...]' (Stamp, 1943: 523). The newly established Ministry of Food was given responsibility for the distribution of food and feeding stuffs through rationing and price controls. In practice it turned out that the legal separation of powers between the production and distribution sides of the food regime fostered antagonism between the two ministries. The price mechanism, driven partly by state interventions and partly by market forces, was the instrument used to encourage the expansion of food production. Essentially, agricultural commodities were bought at relatively high fixed prices from the producers and sold at relatively low prices to the consumers. The state subsidised the difference between producer and consumer prices. This mechanism of price subsidisation was vulnerable to miscalculations, delayed adaptations and other forms of 'state failure'. The state's main aim was to expand the output of arable farming such as grain, potatoes and sugar beet through the curtailment of livestock farming – except for milk production which was given priority over cattle fattening. To this end, annual ploughing-up campaigns, driven by financial incentives and state coercion, were introduced to convert grassland into arable land. A separate policy for upland farmers was introduced to prevent them abandoning their enterprises (Martin 2000: 36; Pilfold, 2007; Whetham, 1952).

In addition to the co-ordination of production, British wartime controls encompassed the allocation of scarce resources such as labour and capital. Restrictions on the labour market were introduced to limit the recruitment of farm labourers into non-agricultural occupations. Though agricultural wages were raised, they still lagged behind those in the industrial and service sectors. The supply of labour to agriculture was augmented by the contribution of the Women's Land Army, the Voluntary Land Club Movement, school children, prisoners of war (POWs) and other sources of agricultural labour (Clarke, 2007; Moore-Colyer, 2007; Ward, 1988: 34; Grant & Maddren, 1975). Under the Lend-Lease agreement farm machinery was imported from the United States of America (Dewey, 2007). In short, 'in terms of provision of labour and machinery, agriculture received preferential treatment for the allocation of scarce resources at a crucial time' (Martin, 2000: 43).

In comparison with Britain, Germany was more self-sufficient in food and, as an alternative to globalisation, strived for a continental 'greater area economy' (*Großraumwirtschaft*), including the food surplus regions of East and South-East Europe. State protection against adverse market forces in accordance with the National-Socialist 'blood and soil' (*Blut und Boden*) ideology, as codified by the 1933 Reich Hereditary Farm Law (REG), had already been subordinated to the expansionist 1936 Four-Years-Plan (Farquharson, 1976; Grundmann, 1979; Degler & Streb, 2008). The Ministry of Food and Agriculture in close union with the Reich Food Estate (*Reichsnährstand*), the obligatory corporation of producers, manufacturers and distributors of agricultural goods, aimed at raising the degree of self-sufficiency by expanding domestic food production ('battle for production') in order to substitute food imports for imports of raw materials for the booming armaments industry. At the beginning of the war this regulatory framework was complemented at provincial and district levels by food agencies (*Ernährungsämter*) that had a department A for production control (*i.e.* Reich Food Estate) and a department B for food rationing (*i.e.* general administration).

Mindful of the First World War experience of food shortages, the stability of consumer prices was given priority over the rise of producer prices. Thus, the expansion of food production was encouraged by state coercion rather than financial incentives. The conflict between the expansion of food grain production and the closing of the 'gap of fat' (*Fettlücke*) by dairy products and oil seeds could not be solved by domestic agriculture alone. Only through the exploitation of German occupied areas, especially in Eastern Europe, could the provision of German consumers – and, hence, the loyalty of the masses – be ensured until 1944 (Strauss, 1941; Corni & Gies, 1997: 397). Ideological reservations about any massive recruitment of people belonging to 'inferior races' – prisoners of war and civil workers from the German occupied countries – were set aside in the face of the economic need to substitute farm workers conscripted to the army. Millions of foreign soldiers and civilians were therefore directly or indirectly forced to work in German agriculture from 1939 onwards. Domestic labour was extensively recruited before 1942 when Germans, especially youths and women, were obliged to fulfil their official duty (Hornung, Langthaler & Schweitzer, 2004: 107). The emphasis of German agricultural economics shifted from labour to land productivity (Streb & Pyta, 2005). Despite the rhetoric of the 'armament of the village' (*Aufrüstung des Dorfes*), the initial provision of farm technology and credit was reduced from 1942 onwards in favour of military armaments; only a few pioneering projects were implemented in mountainous areas (Langthaler, 2000).

III.2. Regional and local regulation

In order to implement state directives, War Agricultural Executive Committees (WAECs), consisting of eight to ten members personally appointed by the Minister of Agriculture, were established in 1939 throughout each county in the UK. These were complemented by sub-committees whose role was to address specific elements of the agrosystem (machinery, livestock, drainage etc.) and by district committees (Short, 2007a: 159; 2007b; 2012). This decentralised system was partly inspired by the opposition to 'farming from Whitehall' that had characterised the food production campaign of the First World War. The decentralised system of administration, with each county being run by a WAEC composed of a small paid technical staff led by an Executive Officer with administrative experience and a number of unpaid volunteers, was seen as a low cost option. Although the WAEC framework might appear at first glance decentralised and democratic, in practice it was designed to encourage farmers to implement the directives of the central authorities: 'The real power in terms of decision making was held by the nucleus of the main WAECs and their Chief Executive Officers, who exhibited high levels of autonomy and discretion in the way they operated' (Martin, 2000: 45). The functional activities of the WAECs can be divided into general administration, technical advice and sanctions.

Within the first group of activities we find the implementation of wartime regulations, the co-ordination of ploughing-up and drainage works, increasing the production of crops and livestock and the supply of fertilizers, feeding stuffs, machinery, implements and seeds. The second group of functional activities aimed at raising productivity by altering farming methods. And the third group of activities was concerned with imposing sanctions on farmers who failed to comply with the WAECs' directives to plough a proportion of their grassland for arable cropping and also to implement other changes designed to raise output. All this was accompanied by a detailed assessment of the position of individual holdings undertaken by the 1941-1943 National Farm Survey (MAF, 1946; Short *et al.*, 2000). Farmers were graded according to their managerial performance into categories A (achieving more than 80 per cent of potential output), B (achieving 60 to 80 per cent of potential output) and C (achieving less than 60 per cent of potential output). Although this classification produced results that varied considerably between counties, it did reveal a positive correlation between size of holding and managerial performance (Martin, 2000: 43). Sometimes the assessments reflected structural differences affecting productivity rather than managerial differences (Rawding, 2007: 191). What also became apparent was that the evaluation of the farmers' managerial capacities was deeply embedded in the personal relationships to be found in rural communities; typically local committee members had to grade relatives, friends and neighbours (Short, 2007a: 176; 2007b; Waymark, 2007).

Contrary to the official view of the harmonious relationship between the WAECs and the farming communities, tension and conflict between committee members and individual farmers occurred: 'Given the large number of farmers involved, the lack of formal training for WAEC officials and the individualistic spirit of farmers, a degree of friction was inevitable'. In addition to this, 'committee officials saw themselves as functionaries of the mandarins in Whitehall, and often sought to achieve their county targets in a regimented way with little regard for the interests of those farmers who did not comply with their directives' (Martin, 2000: 60). Among the sanctions imposed by the WAECs, the eviction order was the most controversial. With evictions tenancies were terminated and owner-occupiers were compelled to relinquish land and occasionally farmhouses as well. Although only a minority of holdings was directly affected by evictions, the dispossessed area increased year by year and the majority of farmers were affected indirectly by the threat of eviction. In addition to eviction orders, the WAECs possessed an armoury of positive and negative sanctions, ranging from withholding grants and subsidies to controlling the allocation of scarce resources. Among the rare accounts of violent opposition to evictions is the case of George Walden who was shot by the police in the process of being evicted and the cases of a number of other farmers who committed suicide (Martin, 2000: 63, 2007b; Short, 2007a). Neither the courts nor the mass media provided farmers with forums for challenging the WAEC directives being imposed on them. They could only appeal for assistance to the National Farmers Union (NFU) which played a somewhat ambivalent role by virtue of finding itself at once representing the interests of its members and supporting the food-production campaign. What opposition there was tended to be channelled through *ad hoc* organisations such as the Farmers' Rights Association (FRA) and the Farmers and Smallholders Association (FSA) (Martin, 2000: 60).

Immediately after the 1938 *Anschluss*, the regulatory framework of German agriculture was established in Austria. The Reich Food Estate, functioning both as a farmers' organisation and as a state agency, was crucial in linking state directives and farming practice at the provincial, district and local levels. According to the principle of 'leadership', honorary 'peasant leaders' were appointed at each of these three levels. Unlike the local representatives (*Ortsbauernschaften*), the agencies at the district and provincial levels (*Kreis-* and *Landesbauernschaft*) employed three divisions of professional staff. Division I, 'The Man' (*Der Mensch*), was responsible for the mobilisation of the farming community according to the ideology of 'blood and soil' institutionalised by the REG. Division II, 'The Farm' (*Der Hof*), provided material and other assistance for arable and livestock production (machinery, fertilizer, seeds, credit, extension etc.). And division III, 'The Market' (*Der Markt*), organised the obligatory delivery of all products not consumed on the farm at fixed prices and in close cooperation with official marketing boards (*Wirtschaftsverbände*). Through the manipulation of production, manufacturing and distribution as well as the fixing of prices the Reich Food Estate had the means to control the main commodity flows between producer and consumer (Langthaler, 2000; 2008).

In spite of the totalitarian character of the Reich Food Estate, its representatives were in practice embedded in complex relationships, as is shown by the most controversial sanction against farm holders – dispossession. According to the 1933 REG, two categories of farm holders existed: Bauern ('peasants') and Landwirte ('farmers'). Being classified as a *Bauer* depended on the farm's size as well as personal features, namely 'peasant aptitude' (*Bauernfähigkeit*); farm holders who did not conform to these criteria were pejoratively designated *Landwirte*. Although the Reich Food Estate had an organisation for fiduciary farm management (*Landwirtschaftlicher Treuhandverband*), it alone could apply for dispossession. It was the task of the courts – general courts in the case of *Landwirte* and special courts (*Erbhofgerichte*), consisting of professional as well as lay judges nominated by the Reich Food Estate, in the case of *Bauern* – to adjudicate on applications for dispossession. In addition to farm mismanagement and other economic reasons for such applications, moral arguments – such as bad payment behaviour, ideological opposition or illegitimate sexual relations - were also taken into account when making assessments. Because farm holders could appeal against judgements, temporary or permanent dispossessions involved complex and frequently long drawn out judicial proceedings (Münkel, 1996; Hauch, 2006; Langthaler, 2009a).

As regards the number and verdicts of the relevant Austrian legal proceedings, a regional study in the province of Niederdonau reveals that only 5 per cent of all farms under the REG legislation were subjected to evaluations of 'peasant aptitude'. In these cases arguments stressing 'economic ability' were much more important than 'honourable status', therefore indicating the primacy of pragmatism over the 'blood and soil' dogma. The majority (58 per cent) of the proceedings concerning 'economic ability' led to the denial of 'peasant aptitude'; contrarily, only a minority (44 per cent) of the proceedings concerning 'honourable status' resulted in the same judgement. Significantly, evaluations of 'economic ability', which were highly pertinent in the context of the state-led war economy, dominated in absolute terms. Relatively speaking, considerations of 'economic ability' led more often to the dispossession of the farm holder. In contrast, evaluations of 'honourable status' featured in only a minority of all cases and resulted less frequently in dispossessions. In summary, there is evidence that, notwithstanding the REG's ideological intentions, the designation of 'peasant aptitude' was functional in regulating the war food economy. While the direct effect of control by the courts may have been limited due to the small number of proceedings, it is likely that farm owners took careful note of exemplary judgements and were indirectly influenced in their behaviour in consequence (Langthaler, 2010).

IV. Agrosystemic resource flows

IV.1. Farm inputs: Land

In most of the industrialised societies of modern Europe, land becomes an even scarcer resource when agriculture is obliged to compete for it with other sectors of the economy. The competition for land can be expected to become yet more intense in wartime when the state lays claim to land for military and industrial purposes (Foot, 2007). This is certainly true of the UK and Austria between 1939 to 1944 where the total agricultural area declined slightly, by 2.1 and 0.6 per cent respectively. However, this convergence contrasts with the divergent development of the acreage. Whereas the proportion of tilled land increased rapidly from 41 to 62 per cent in the UK, it declined in Austria from 43 to 41 per cent (see Table 7.2). An increasing proportion of the British countryside was ploughed up and, along with the greater use of fertilizer and other inputs, land was farmed more intensively compared to the inter-war period (Stamp, 1947, 1948). Conversely, the relative decline of tilled land led to extensification in Austria. Moreover, intensification and extensification were distributed unevenly across the countryside. In England and Wales there was a strong negative correlation between the proportion of tilled land in 1939 and the changes in the tilled acreage recorded up to 1944. The upland counties in the north, south and west, characterised by small percentages of arable land, showed more significant increases than those of the lowland eastern counties. Austria presents a much weaker but none the less negative correlation. There the alpine districts in the south and west, covered mainly by grassland and forests, were less affected by the relative decline of tilled land than some regional clusters in the mainly flat and hilly land in the north-west, north-east and south-east of the country.

Figure 7.1. Changes in the proportion of arable land in England and Wales, 1939-1944

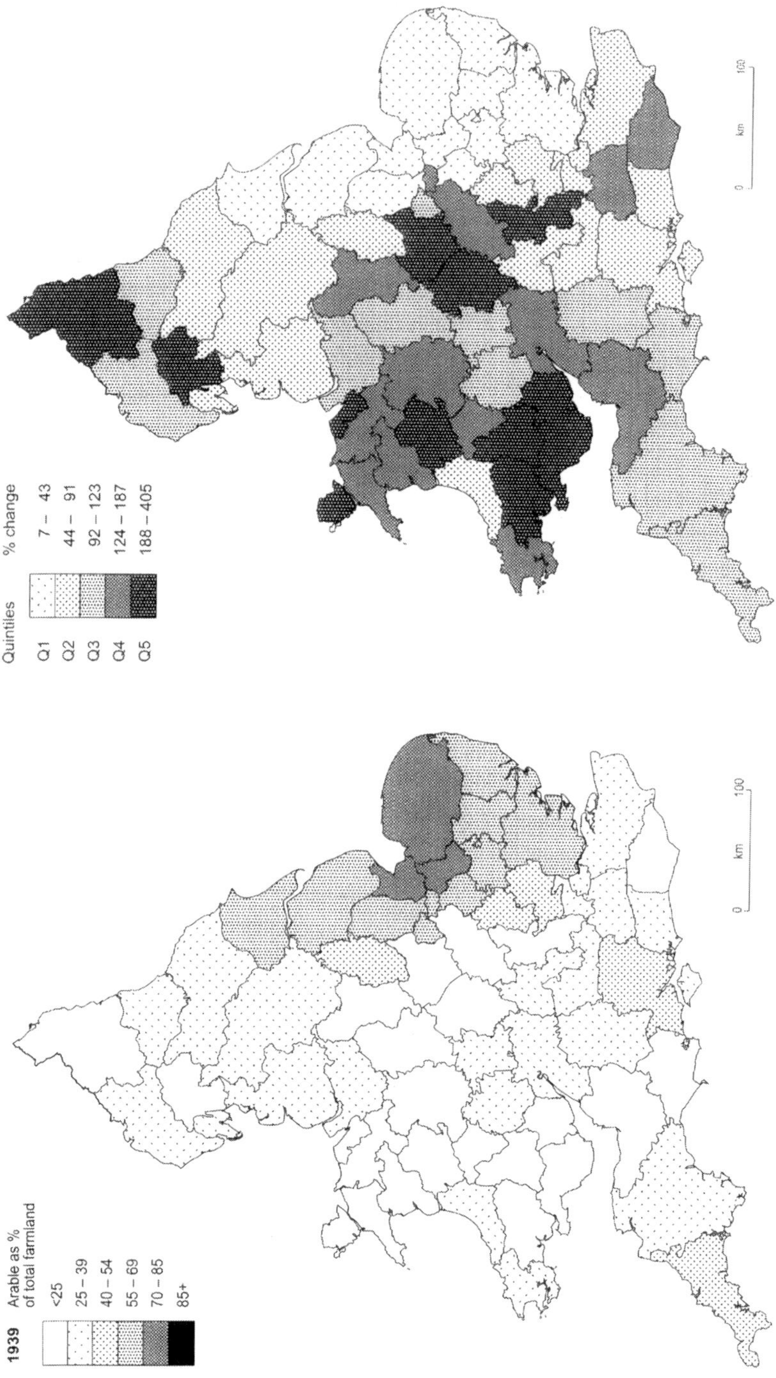

Sources. Short, Watkins and Martin (2007: 7); Martin (2007a: 29).

Figure 7.2. Changes in the proportion of arable land in Austria, 1939-1944

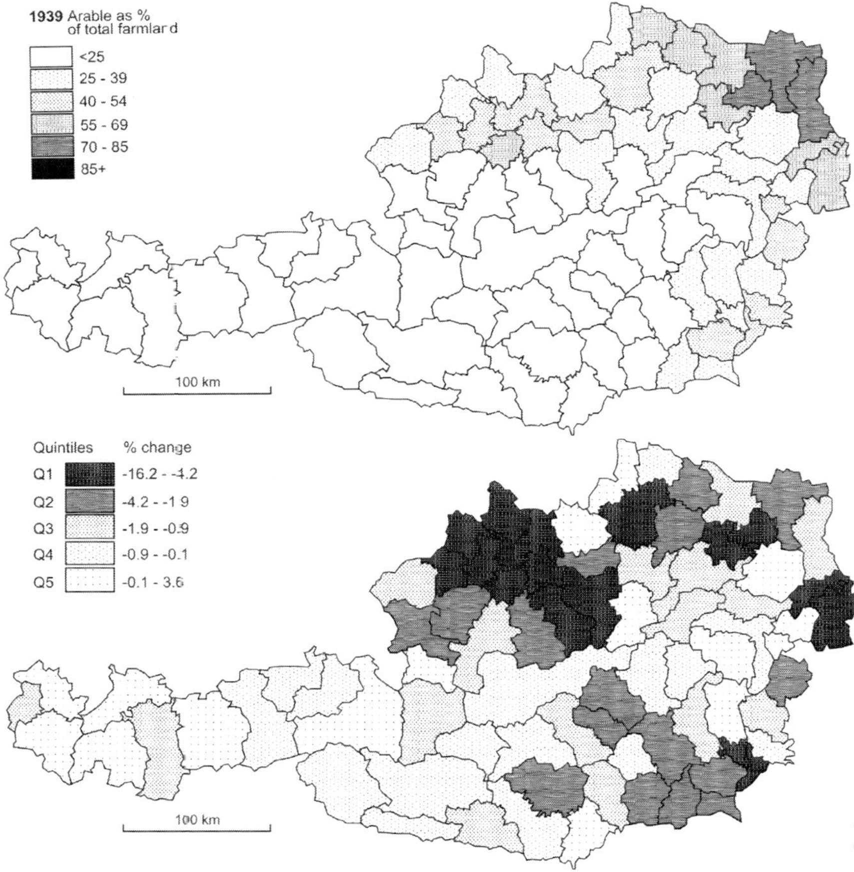

Source. ÖStZA (1948: 2)
GIS design by the author.

Figure 7.2. Changes in the proportion of arable land in Austria, 1939-1944

	United Kingdom			*Austria*		
	Total agricultural area		*Proportion of arable land*	*Total agricultural area*		*Proportion of arable land*
	(1,000 hectares)	*(index)*	*(percent)*	*(1,000 hectares)*	*(index)*	*(percent)*
Pre-war*	12,884	100.5	41.1	4331	104.0	44.6
1939	12,820	100.0	40.7	4163	100.0	43.2
1940	12,719	99.2	45.6	4047	97.2	43.4
1941	12,688	99.0	51.8	4179	100.4	41.9
1942	12,628	98.5	56.1	4136	99.3	41.6
1943	12,569	98.0	60.3	4135	99.3	41.2
1944	12,548	97.9	62.2	4137	99.4	40.8

* UK: 1936-1938, Austria: 1937-1938.

Sources. MURRAY (1955: 373); ÖStZA (1948: X).

The regionally uneven distribution of intensification and extensification might at least partly be explained by the 'law of diminishing returns' (Ellis, 1993: 18). Accordingly, each additional unit of variable input yields less and less additional output (as in the British case) and, *vice versa*, each unit of variable input less causes more and more decline of output (as in the Austrian case). Therefore, assigning additional units of labour and capital to extensively used land in the UK was rational in order to maximise output, since some of the pasture land converted to arable farming was very fertile and capable of producing high yields. In the Austrian agrosystem the law's rationale suggested a reduction of labour and capital in relatively intensively worked agricultural regions. There were however many exceptions to this rule in Austria, indicating that other influences – such as the spatially uneven use of agricultural land for military or industrial purposes – were simultaneously at work. A statistical analysis reveals that the primary – and negative – influence on the change of the proportion of tilled land between 1939 and 1944 in the Austrian countryside was the percentage of the tilled acreage in 1939. In addition, the change in the total farmland area during this period had a secondary – and also negative – influence. What this indicates is that the designation of farmland for non-agricultural uses in Austria 1939 to 1944 affected wasteland, grassland or forests more than arable land[2].

According to the maps above (see Figures 7.1 and 7.2), the proportion of tilled land seems to be a crucial feature of the wartime British and Austrian agricultural development paths. In order to fully assess this feature we must examine not only its

[2] Based on a multiple regression analysis of data of 80 Austrian districts. Source: ÖStZA, 1948: 2.

quantity but also its quality. Whereas the British agrosystem showed a greater relative acreage, the Austrian agrosystem was characterised by higher land use intensity. Although the proportions of grain and potatoes increased in the UK, they never reached Austria's (even if declining) percentages. In spite of increasing rates of fallow and temporary grassland, during the war years arable land was farmed more intensively in Austria than in the UK where these forms of arable land use decreased considerably. However, from 1942 onwards the difference in arable land use intensity – which had equalled 101 in the UK and 118 in Austria in 1939 – became negligible and both agrosystems stagnated around the value of 110 (see Table 7.3)[3]. In short, British and Austrian land use in general and tilled land use in particular, in spite of starting from differing positions, converged to a considerable degree during the wartime period.

Table 7.3. Arable land use intensity in the UK and Austria, 1939-1944

	United Kingdom		*Austria*	
	(intensity value)	*(index)*	*(intensity value)*	*(index)*
pre-war*	100.3	99.6	117.0	99.2
1939	100.8	100.0	117.9	100.0
1940	103.3	102.5	116.4	98.7
1941	108.4	107.6	113.4	96.2
1942	110.0	109.1	111.4	94.5
1943	109.3	108.5	110.5	93.8
1944	108.3	107.4	110.8	94.0

* UK: 1936-1938, Austria: 1937-1938. Intensity coefficients: grain = 1, potatoes = 3, sugar beet = 4, other crops = 1, temporary grass = 0.5, fallow = 0.25

Sources. MURRAY (1955: 373); ÖStZA (1948: X).

IV.2. Farm inputs: Labour

Unlike land, labour is much more difficult to grasp in the available agricultural statistics. In both countries there is a lack of annual data on the agricultural workforce. This contrasts sharply with the importance of farm labour – 'perhaps the input to which most attention has been given' (Brassley, 2007: 42) – during the war. Although the available data are not strictly comparable[4], the main tendencies are

[3] To calculate the arable land use intensity value the area of each crop must be multiplied by the following coefficients: grain = 1, potatoes = 3, sugar beet = 4, other crops = 1, temporary grass = 0.5, fallow = 0.25. The sum of these products is divided by the arable area and then multiplied by 100. See LBG (1948: 88).

[4] The British data come from a calculation on the basis of total numbers of different categories of workers; and the Lower Austrian data are derived from the bookkeeping files of a small sample of middle and large peasant farms.

still evident. Whereas in the UK total workforce units per agricultural area increased by 11 per cent between the pre-war years and 1943-1944, during the same period the Lower Austrian agrosystem faced a considerable decline in the number of rural workers per 100 hectares, up to 47 per cent depending on the type of farming (see Table 7.4). Conscripting rural dwellers to the armed forces as well as maximising farm outputs created a dilemma that could not be solved in principle, but only gradually in practise. In both countries, rural men in uniform were substituted, at least partly, by both male and female workers as well as by labour-saving machinery. The proportion of women from outside the agricultural sector as a percentage of all UK farm workers in 1939 (15.4%) and the proportion of foreign workers in the province of Niederdonau in 1943 (15.9%) were nearly the same. However, the underlying substitution processes differed according to national modes of regulation. In the UK, where the mobilisation of women was pioneered, the main sources of labour were the Women's Land Army, accounting for 10.8 per cent of total workers (Clarke, 2007), and prisoners of war (POWs) accounting for 4.6 per cent (Moore-Colyer, 2007).

These proportions were quite different in Niederdonau where a compromise between pragmatic supporters and dogmatic critics of the use of foreign labour in the Nazi regime saw prisoners of war and civil workers from the German-occupied territories being forced, either indirectly or directly, to work in the territory of the Reich from 1939 onwards. Initially, and prior to 1942, there was no extensive recruitment of domestic labour; from that point on Germans, especially youths and women, became obliged to perform their official duty (Hornung *et al.*, 2004: 107). Yet in 1943 domestic women under official duty accounted for only 1.7 per cent of the agricultural labour force compared to the 14.2 per cent accounted for by the POWs and foreign civil workers that came mainly from German-occupied territories in eastern Europe (see Table 7.5). Our comparison may rest upon questionable data, but the overall difference is quite clear. In the UK, farm workers serving in the army were mainly replaced by the voluntary members of the Women's Land Army; in Austria, it was mainly the POWs and forced civil workers from all over the German-occupied territories who made good the labour shortfall.

Table 7.4. Agricultural labour intensity in the UK and Lower Austria, 1937-1945

	UK	*Lower Austria*							
		Grain-wine farms		*Grain farms*		*Root crop farms*		*Grass-forest farms*	
	*Index of work-force units per agricultural area***	*Workers per 100 ha*	*Index*	*Workers per 100 ha*	*Index*	*Workers per 100 ha*	*Index*	*Workers per 100 ha*	*Index*
pre-war*	100	31	100	21-38	100	19	100	24	100
1939-1940	99	–	–	–	–	–	–	–	–
1940-1941	102	–	–	–	–	–	–	–	–
1940-1942	105	–	–	–	–	–	–	–	–
1942-1943	109	–	–	–	–	–	–	–	–
1943-1944	111	17	55	14-28	53-100	18	95	17	71
1944-1945	112	–	–	–	–	–	–	–	–

* UK: 1937-1939, Lower Austria: 1937. ** Numbers of workers are weighted by coefficients ranging from 1 (regular males, 21 and over) to 0.4 (prisoners of war, not billeted).

Sources. Williams (1954: 334); Murray (1955: 273); LBG (1949: 45).

Table 7.5. Agricultural labour substitution in the UK and Niederdonau, 1943

	Agricultural workers 1939	*Domestic women 1943*		*Foreign workers 1943*	
	Number in 1,000	*Number in 1,000*	*Per cent of 1939*	*Number in 1,000*	*Per cent of 1939*
UK	803.0	87.0*	10.8	37.6***	4.6
Niederdonau	687.9	11.9**	1.7	97.5 (26.5)*** (71.0)****	14.2 (3.9) (10.3)

* Women's Land Army, ** women under obligatory service, *** POWs, **** civil workers

Sources. Murray (1955: 85, 188); Statistisches Amt für die Reichsgaue der Ostmark (1941); Gauarbeitsamt Niederdonau (1943).

IV.3. Farm inputs: Livestock

Compared to labour, livestock and land use are much better documented in the wartime agricultural statistics. Until the first half of the twentieth century land and livestock use were closely interrelated by nutrient transfers *via* fodder and manure. Given that both elements of this integrated agrosystem produced nutrition for human consumption, wartime shortages promoted competition between the cultivation of fodder and food plants. Both countries adopted similar solutions to the resulting problems. While more and more milk cows were fed (UK: +13 per cent, Austria: +4 per cent), the numbers of other livestock declined. In both our cases the strongest decline affected pigs (-58 per cent in the UK), though the Austrian fall-off (-40 per cent) was moderated by the fear of a repetition of the 'murder of pigs' (*Schweinemord*) experienced in the First World War (Corni & Gies, 1997: 402). Milk production was in short given priority over beef and pork production in both countries. Because one unit of animal meat requires five to ten units of fodder crops, this strategy promoted a shift to the production of food crops for human consumption. Total livestock numbers are not automatically accurate indicators of intensity; thus, they have to be related to the agricultural area. Because in both countries the agricultural area declined, increases in livestock numbers resulted in stronger increases in intensity and, vice versa, decreases in numbers resulted in weaker decreases in intensity. What we therefore find is that the boom of dairy cattle breeding amounted to 15 and 5 per cent gains in intensity respectively; and that the decline in pig breeding amounted up to 47 and 40 per cent losses of intensity respectively (see Tables 7.6 and 7.7).

Table 7.6. Livestock intensity in the UK, 1939-1944

	Dairy cattle		*Other cattle*		*Sheep*		*Pigs*		*Poultry*	
	number per 100 ha	*index*	*number per 100 ha*	*index*	*number per 100 ha*	*index*	*number per 100 ha*	*index*	*number per 100 ha*	*index*
pre-war (1936-1938)	31	101	37	94	200	95	35	101	592	102
1939	30	100	39	100	210	100	34	100	580	100
1940	31	103	40	104	207	99	32	94	560	97
1941	31	104	39	100	175	84	20	59	489	84
1942	33	110	39	99	170	81	17	50	458	79
1943	34	113	39	101	162	77	15	42	404	70
1944	35	115	41	105	160	76	15	43	439	76

Source. Murray (1955: 373).

Table 7.7. Livestock intensity in Austria, 1939-1944

	Dairy cattle		*Other cattle*		*Sheep*		*Pigs*		*Poultry*	
	number per 100 ha	*index*	*number per 100 ha*	*index*	*number per 100 ha*	*index*	*number per 100 ha*	*index*	*number per 100 ha*	*index*
pre-war (1938)	28	95	31	94	7	95	66	97	209	101
1939	30	100	33	100	8	100	68	100	206	100
1940	31	104	33	99	8	111	54	80	181	88
1941	30	102	30	89	8	111	49	72	158	77
1942	31	104	30	90	10	127	43	63	136	66
1943	31	106	30	90	11	141	45	67	139	67
1944	31	105	30	91	11	146	41	60	127	62

Source. ÖStZA (1948: XIV).

IV.4. Farm inputs: Technology

The data on technical inputs are as scarce as the labour data, especially for Austria. Despite the lack of data comparability, massive gains in technical farm inputs are indicated both for the UK (Short *et al.*, 2007; Martin, 2007a; Dewey, 2007) and for Austria (Langthaler, 2000). Regarding mechanical technology, one of the few comparable indicators available for both countries is the change in farm power supplies. Between 1939 and 1946 total horsepower increased by 154 per cent in the UK and by 80 per cent in Austria. Strikingly, total power per agricultural area in 1939 was similar in both countries, before the UK began to extend its lead up to 1946. In comparing agrosystems we should keep in mind that in Austria a considerable (albeit unquantifiable) amount of machinery was lost due to damages and requisitions at the end of the war (Sandner, 1948: 71). Even if horsepower more than tripled in both countries during the war years, tractors were the main source of power in the UK (1939: 56 per cent, 1946: 81 per cent) as opposed to Austria (1939: 9 per cent, 1946: 16 per cent) (see Table 7.8). With regard to biological-chemical technology, data on the use of mineral fertilizer provide comparative insights. In both countries the wartime application of nitrogen, phosphate and potash increased, even though these materials became ever scarcer due to the needs of the armaments industry. Compared to Austria the application of mineral fertilizer in Britain was more effective given that much of the grassland ploughed up had been laid down in grass for a couple of generations. Whereas the application of mineral fertilizer continued in post-war Britain, it fell off in Austria. The comparison of the amount of fertilizer use per agricultural area between 1938 and 1945 leads therefore to rather ambivalent results – the application of nitrogen and phosphate was higher in the UK; more potash was applied in Austria (see Table 7.9). All in all, this rough comparison indicates an accelerated use of labour-saving and land-saving technologies in both countries, thereby following a secular trend towards the substitution of labour and land with capital (Hayami & Ruttan, 1985). While the rates of technical change in Austria were considerably lower, the UK experienced a 'resurgence of progressive, high-input arable farming' (Martin, 2007a: 16) after the decline of 'high farming' in the late nineteenth century.

Table 7.8. Supply of mechanical power in agriculture in the UK and Austria, 1939-1944

	United Kingdom		*Austria*	
	1939	*1946*	*1939*	*1946*
Stationary power (horsepower)	854	911	525	866
Tractor power (horsepower)	1075	3995	49	165
Total power (horsepower)	1929	4906	574	1031
Proportion of stationary power (per cent)	44.3	18.6	91.5	84.0
Proportion of tractor power (per cent)	55.7	81.4	8.5	16.0
Index of stationary power	100	107	100	165
Index of tractor power	100	372	100	337
Index of total power	100	254	100	180
Total power per agricultural area (horsepower per 100 hectares)*	15.0	39.1	13.8	24.9

* The value for 1946 is related to the agricultural area of 1944.

Sources. Murray (1955: 274); Sandner (1947: 72).

Table 7.9. Application of mineral fertilizer in the UK and Austria, 1939-1944

	United Kingdom			*Austria*		
	Nitrogen	*Phosphate*	*Potash*	*Nitrogen*	*Phosphate*	*Potash*
pre-war* (1000 tons)	–	–	–	5.9	12.8	7.5
1938-1945 (1000 tons)	136.8	268.3	87.1	19.5	18.9	42.1
post-war** (1000 tons)	164.6	358.7	107.2	13.8	16.9	12.8
1938-1945 (tons per hectare)	10.8	21.2	6.9	4.7	4.5	10.1

* Austria: 1930-1938, ** UK: 1945-1946, Austria: 1945-1950.

Sources. Murray (1955: 259); Meihsl (1961: 744).

IV.4. Farm outputs: Arable production

In view of the crucially important challenge of maintaining food supply in wartime, rich data relating to farm outputs exist for both countries. However, they only indicate the official commodity flows between producers and consumers; what was sold and bought privately on the 'black market' can hardly be measured. According to the official figures, neither the UK nor Austria achieved any significant increase in yields per hectare. In the UK grain yields oscillated by up to 10 per cent in some years, but there was no marked trend towards higher yields. Potatoes and sugar beet yields even fell from year to year from 1940 onwards. In Austria all crop yields declined, except for 1943 which featured an above average grain harvest (see Tables 7.10 and 7.11). It is worth noting that the wartime changes were not indicative of the very rapid and unprecedented increases in crop yields that took place in the post-war period. The critical difference between both agrosystems prior to 1945 was not yields per hectare but total *yields*. Ploughing up grassland and dedicating more acreage to grain and potatoes in the UK allowed total production to increase by 47 to 96 per cent between 1939 and 1944. Only sugar beet harvests stagnated. In contrast, in Austria the decline of arable land in general, and of the planted area of most food crops, caused considerable losses (30 to 53 per cent) in total output (see Tables 7.12 and 7.13).

Table 7.10. Yield per hectare of selected crops in the UK, 1939-1944

	Wheat		*Barley*		*Oats*		*Potatoes*		*Sugar beet*	
	Tons per ha	*Index*	*Tons per ha*	*Index*	*Tons per ha*	*Index*	*Tons per ha*	*Index*	*Tons per ha*	*Index*
pre-war (1929-38)	2.26	97	2.06	93	2.03	98	16.56	91	21.25	83
1939	2.33	100	2.21	100	2.07	100	18.29	100	25.45	100
1940	2.27	97	2.07	94	2.13	103	19.03	104	23.97	94
1941	2.23	96	1.95	88	2.06	99	17.54	96	22.98	90
1942	2.56	110	2.37	107	2.16	104	17.79	97	22.98	90
1943	2.50	107	2.31	105	2.10	101	17.54	96	22.49	88
1944	2.45	105	2.23	101	2.03	98	15.81	86	19.03	75

Source. MURRAY (1955: 374).

Table 7.11. Yield per hectare of selected crops in Austria, 1939-1944

	Wheat		*Barley*		*Oats*		*Potatoes*		*Sugar beet*	
	Tons per ha	*Index*	*Tons per ha*	*Index*	*Tons per ha*	*Index*	*Tons per ha*	*Index*	*Tons per ha*	*Index*
pre-war (1937-38)	1.76	101	1.82	103	1.64	113	16.00	112	25.85	89
1939	1.75	100	1.78	100	1.45	100	14.28	100	28.97	100
1940	1.34	77	1.54	86	1.39	96	13.68	96	21.29	73
1941	1.51	87	1.57	88	1.29	89	13.91	97	24.81	86
1942	1.32	76	1.46	82	1.26	87	12.87	90	23.46	81
1943	1.63	94	1.65	93	1.43	99	10.49	73	20.67	71
1944	1.41	81	1.41	80	1.23	85	10.06	70	19.33	67

Source. ÖStZA (1948: 86).

Table 7.12. Total output of selected crops in the UK, 1939-1944

	Wheat		*Barley*		*Oats*		*Potatoes*		*Sugar beet*	
	1,000 tons	*Index*	*1,000 tons*	*Index*	*1,000 tons*	*Index*	*1,000 tons*	*Index*	*1,000 tons*	*Index*
pre-war (1929-38)	1651	100	765	86	1,940	97	4,873	93	2,741	78
1939	1,645	100	892	100	2,003	100	5,218	100	3,529	100
1940	1,641	100	1,104	124	2,892	144	6,405	123	3,176	90
1941	2,018	123	1,144	128	3,247	162	8,004	153	3,226	91
1942	2,567	156	1,446	162	3,553	177	9,393	180	3,923	111
1943	3,447	210	1,645	184	3,064	153	9,822	188	3,760	107
1944	3,138	191	1,752	196	2,953	147	9,096	174	3,267	93

Source. Murray (1955: 375).

Table 7.13. Total output of selected crops in Austria, 1939-1944

	Wheat		*Barley*		*Oats*		*Potatoes*		*Sugar beet*	
	1,000 tons	*Index*	*1,000 tons*	*Index*	*1,000 tons*	*Index*	*1,000 tons*	*Index*	*1,000 tons*	*Index*
pre-war (1937-38)	460	103	295	103	454	126	3356	121	1089	88
1939	447	100	286	100	360	100	2765	100	1235	100
1940	286	64	280	98	343	95	2605	94	875	71
1941	342	76	233	81	285	79	2602	94	832	67
1942	275	62	222	78	272	76	2252	81	744	60
1943	344	77	215	75	302	84	1772	64	666	54
1944	293	66	181	63	251	70	1751	63	576	47

Source. ÖStZA (1948: 52).

IV.5. Farm outputs: Livestock production

What has been said about the segmentation of arable product markets between legal and illegal domains applies even more to livestock product markets. The official figures for milk production in both agrosystems differed only slightly. It is striking that in most years the Austrian performance, where total output as well as output per milk cow are concerned, was even better than the British which was adversely affected by the decline in imported feeding stuffs. Only the proportion of milk sold to manufacturers or consumers was much higher in the UK than in Austria (see Table 7.14). The officially recorded numbers of slaughtered animals, which are comparable only with reservations[5], show variation between our two cases. The indices of slaughtered cattle and calves in the UK exceeded those of Austria; in the case of pigs, however, the relationship was reversed. In both agrosystems the numbers of slaughtered calves – increasing in the UK and stagnating in Austria – performed much better than those of pigs which fell most of all (see Tables 7.15 and 7.16).

Table 7.14. Milk production and marketing in the UK and Austria, 1939-1944

	United Kingdom					*Austria*				
	Total production	*Production perdairy cattle*		*Proportion of milk sold*		*Total production*		*Production per-dairy cattle*		*Proportion of milk sold*
	1,000 tons	*Index*	*kg*	*index*	*per cent*	*1,000 tons*	*Index*	*kg*	*index*	*per cent*
pre-war*	8,323	101	2,111	99	–	–	–	–	–	–
1939	8,277	100	2,130	100	72.8	1,983	100	1,604	100	49.9
1940	7,515	91	1,899	89	77.7	1,925	97	1,542	96	56.7
1941	7,309	88	1,833	86	81.7	1,945	98	1,551	97	59.6
1942	7,744	94	1,844	87	81.4	1,947	98	1,537	96	59.5
1943	8,001	97	1,851	87	81.3	2,020	102	1,570	98	61.5
1944	8,071	98	1,846	87	82.6	1,888	95	1,452	91	61.0

* UK: 1936-39, Austria: -

Sources. MURRAY (1955: 375, 380); HAMMOND (1956: 798); ÖStZA (1948: 188).

[5] While the British data cover only slaughtered animals at the collecting centres, the Austrian data encompass animals slaughtered on and off the farm.

Table 7.15. Registered slaughtered farm animals* in the UK, 1940-1944

	Cattle		*Calves*		*Sheep and lambs*		*Pigs*	
	1,000 heads	*Index*	*1,000 heads*	*Index*	*1,000 heads*	*Index*	*1,000 heads*	*Index*
1940	1,889	100	861	100	10,961	100	5,331	100
1941	1,863	99	1,062	123	8,452	77	3,428	64
1942	1,672	89	1,126	131	8,158	74	1,752	33
1943	1,744	92	1,363	158	7,665	70	1,588	30
1944	1,866	99	1,357	158	6,830	62	1316	25

* Purchases for slaughter at collecting centres
Source. HAMMOND (1962: 794).

Table 7.16. Registered slaughtered farm animals* in Austria, 1940-1944

	Cattle		*Calves*		*Sheep and lambs*		*Pigs*	
	1,000 heads	*Index*	*1,000 heads*	*Index*	*1,000 heads*	*Index*	*1,000 heads*	*Index*
1940	420	100	668	100	199	100	1,361	100
1941	436	104	705	106	79	40	1,497	110
1942	372	88	738	111	65	32	1,311	96
1943	341	81	699	105	70	35	1,146	84
1944	246	59	660	99	38	19	1,042	77

* Slaughters on and off farm.
Source. ÖStZA (1948: 174).

IV.6. Farm income

How did arable and livestock production affect the income of farmers? The answer to this question must, of course, take account of agricultural product prices. Prices rose considerably higher – up to 124 per cent from 1939 to 1944 – in the UK compared to Austria, where only sugar beet, cattle and pig prices increased by up to 31 per cent during the same period. Thus, the incentive to produce for the official market was probably stronger in the UK than in Austria, where threatening sanctions did not prevent the emergence of a 'black market' (see Table 7.17). The extent to which these different price levels affected farm incomes can be measured by the 'cash net income' for the UK and by the 'net yield' (*Reinertrag*) for Austria. What the British 'cash net income' indicates is the difference on a cash basis between gross income and gross expenditure, excluding adjustments for valuation differences between the beginning and end of the year. In the UK cash net income was three to five times higher in 1943-1944 compared to 1937-1938, though there were individual differences according to the type of farming and farm size. This figure is remarkable: 'Farmers collectively saw significantly greater increases in their disposable income than entrepreneurs or managers outside the agricultural sector' (Martin, 2000: 58). The Austrian 'net yield' measures difference on a cash basis between gross yield (*Rohertrag*) and expenditure (*Aufwand*), including assumed family labour wages. In the province of Lower Austria the indices, on the basis of 1937, were scattered over an extremely wide range in 1943/44. Although the available data are not strictly comparable, they do point to the conclusion that the vast majority of UK farmers benefited in economic terms while Austrian farmers were more split between winners and losers during the war period (see Tables 7.18 and 7.19).

Table 7.17. Agricultural price index in the UK and Austria, 1939-1944

	United Kingdom							*Austria*						
	wheat	*oats*	*potatoes*	*sugar beet*	*cattle*	*pigs*	*milk*	*wheat*	*oats*	*potatoes*	*sugar beet*	*cattle*	*pigs*	*milk*
pre-war*	100	100	100	100	100	100	100	100	100	100	100	100	100	100
1939	111	159	93	118	125	130	120	95	98	95	94	98	106	100
1940	148	195	131	154	138	139	154	92	101	88	91	98	103	100
1941	154	195	149	161	145	150	173	92	101	100	100	98	111	100
1942	171	208	148	204	154	154	180	92	105	103	105	98	123	100
1943	189	211	148	201	154	154	188	93	105	107	114	109	130	100
1944	198	224	152	192	159	155	193	92	100	107	114	128	131	100

* UK: 1936-1938, Austria: 1938.

Sources. Murray (1955: 381); LBG (1949: 16).

Table 7.18. Cash net income* per farm by farming type in England and Wales, 1937/38-1943/44

	1937/38	*1943/44*	
	Pounds	*Pounds*	*Index 1937-1938 = 100*
Arable farming types	285	1,545	542
Intermediate farming types	252	1,086	431
Grassland farming types	196	661	337

* Cash net income: difference on a cash basis between gross income and gross expenditure, excluding adjustments for valuation differences between the beginning and end of the year.

Source. Murray (1955: 382).

Table 7.19 Net yield (*Reinertrag*)* per hectare by farming type in Lower Austria, 1937-1943/44

	1937	*1943-1944*	
	Schilling	*Schilling*	*Index 1937-1938 = 100*
Wine farms	196	5,077	2,586
Grain-wine farms	109	135	123
Grain farms	95	87	91
Root crop farms	158	60	38
Grass-forest farms	67	4	6

* Net yield (*Reinertrag*): difference on a cash basis between gross yield (*Rohertrag*) and expenditure (*Aufwand*), including assumed family labour wages.

Source. LBG (1949: 140).

V. Conclusion

Do the wartime agrosystemic developmental paths taken by the UK and Austria between 1939 and 1945 deserve the attribute 'revolutionary' in the sense of a fundamental, progressive and short to medium-term transformation? For the British case, an answer to this question has recently been given as follows:

> In terms of both internal and external changes, the rapidity of land use change, the degree and lasting duration of state support and control, the adoption of mechanization, and the resultant impact of farming communities, it was undoubtedly revolutionary. And since the outcome was to establish agriculture thereafter as a key element in receipt of preferential treatment within national strategic planning, it could also be seen as the most important turning-point of the twentieth century. In these terms, it was an agricultural revolution. (Short *et al.*, 2007: 15).

For the Austrian case, much more guarded verdicts that speak of 'regressive modernisation', 'forced modernisation' or 'proto-modernisation' have been offered. My own claim of a 'selective modernisation' points to the ambivalent transformation of Austrian rural society during the Nazi era (Langthaler, 2000: 372). Can these various evaluations be validated from a comparative perspective?

Due to different criteria of data generation or lack of relevant data, some of our statistical comparisons pose unsolvable problems. Nevertheless, they can be shown to make sense in most cases in the context of additional evidence[6]. To begin with the indicators of agrosystemic resource flows, there is evidence of revolutionary changes occurring in British agriculture between 1939 and 1945. A large proportion of grassland was converted into tilled land, thereby increasing land use intensity. The intensity of dairy cattle farming was raised considerably. Land-saving and labour-saving technologies were widely applied. Total arable output was raised, particularly in monetary terms and agricultural prices and incomes were strongly boosted. These changes were fundamental insofar as they continued into the post-war era, progressive insofar as they raised the intensity of farming, and short-term insofar as they occurred during the space of half a decade (Brassley, 2000; Short *et al.*, 2007; Martin, 2000). The 'revolutionary' effect of wartime farming on subsequent agricultural development was already being discussed as soon as one year after the end of the war. 'I venture to suggest that the farming community has emerged from the war a far more efficient force in a competitive world than it was in 1939' (Stamp, 1947: 53). However, not all indicators point to an 'agricultural revolution' in the UK, thus suggesting that the official history of wartime heroism on the land needs to be revised to some degree (Murray, 1955). The most obvious exception is crop yields per hectare that stagnated during the wartime period and began to rise only in the post-war era. A non-revolutionary, even regressive tendency, is indicated by the measure of milk production per cow as well.

Concerning Austrian agriculture, there are only a few indicators that might be labelled 'revolutionary' or at least progressive – in particular the increasing use of mechanical and biological-chemical technology that continued into the 1950s and beyond (Sandgruber, 2002: 204, 323). The contradictory changes in farm income presaged socio-economic differentiation in post-war rural society (Krammer & Scheer, 1978). In view of the mainly regressive change in farm inputs and outputs, there was definitely no overall 'agricultural revolution' in Austria between 1939 and 1945.

Alongside these quantifiable indicators some more qualitative aspects of agrosystemic regulation must also be considered. Farming communities and the state were more closely interlinked in both cases via dense relations of support and control, therefore institutionalising planning procedures at national, regional and local levels. In the UK, the WAECs served as the crucial link between state agricultural planning and the farming communities. In Austria, the Reich Food Estate as a hybrid of state agency and farmers' organisation fulfilled a similar intermediate function. The differences between the two political-economic systems – liberal democracy in the UK and fascist dictatorship in German-annexed Austria – became blurred due

[6] For an overview of British and Austrian agricultural statistics, see Britton & Hunt (1951); ÖStZA (1979).

to similar measures of agricultural regulation. The example of dispossession in the event of farm mismanagement also reveals system-specific differences. Whereas in Britain farmers' movements such as FRA and FSA provided formal and collective expressions of rural unrest, under the Nazi regime this occurred rather informally and individually by everyday forms of peasant resistance that lay beyond the domain of the Reich Food Estate (Hanisch, 1990). This is not to deny that in both countries mechanisms of state regulation promoting more productive farming styles – a 'new morality' oriented towards 'short term economic advantages and unquestioning compliance' (Martin, 2000: 61) – were institutionalised in rural society. To what extent the wartime institutions set the scene for the post-war formation of the 'productivist' food regime in the UK and Austria has been discussed elsewhere (Martin & Langthaler 2012).

In terms of the typology outlined above (see Table 7.1), the British agrosystem experienced in part a 'revolutionary' change of institutions and technology, mainly characterised by general and unequal modes of development, in the course of the Second World War. In contrast, 'revolutionary' changes of the Austrian agrosystem were confined to the institutional framework. With regard to technology, only mechanical and chemical inputs reveal unequal and contradictory progress; most other indicators point to a general crisis. To cut a long story short, the so-called 'agricultural revolution' in wartime Britain was both institutional and technical to a considerable – but by no means total – degree (types 1C and 2C). In German-annexed Austria, several institutional but hardly any technical changes occurred that deserve to be labelled 'revolutionary' (types 2B and 3B). If we choose to retain the general concept of 'agricultural revolution', we should therefore properly differentiate between its context-specific varieties.

Bibliography

Bieleman, Jan (1999), 'Farming System Research as a Guideline in Agricultural History', in Bas J. P. van Bavel & Erik Thoen (eds), *Land Productivity and Agrosystems in the North Sea Region (Middle Ages-twenntieth Century). Elements for Comparison*, Turnhout, Brepols, Corn Series, 2, p. 235-250.

Brassley, Paul (2000), 'Output and Technical Change in Twentieth-century British Agriculture', *Agricultural History Review,* 48/1, p. 60-84.

Brassley, Paul (2007), 'Wartime Productivity and Innovation, 1939-45', in Brian Short, Charles Watkins & John Martin (eds), *The Front Line of Freedom. British Farming in the Second World War*, Exeter, British Agricultural History Society, p. 36-54.

Brassley, Paul, Segers, Yves & Van Molle, Leen (eds) (2012), *War, Agriculture, and Food: rural Europe from the 1930s to the 1950s*, London/New York, Routledge.

BRITTON, D. K. and HUNT, K. E. (1951), 'The Sources and Nature of Statistical Information in Special Fields of Statistics: The Agricultural Statistics of the United Kingdom', *Journal of the Royal Statistical Society,* Series A (General), 114/1, p. 59-98.

CLARKE, Gill (2007), 'The Women's Land Army and its Recruits, 1938-50', in Brian SHORT, Charles WATKINS & John MARTIN (eds), *The Front Line of Freedom. British Farming in the Second World War*, Exeter, British Agricultural History Society, p. 101-116.

COLLINGHAM, Lizzie (2011), *World War Two and the Battle for Food*, London, Penguin.

CORNI, Gustavo & GIES, Horst (1997), *Brot, Butter, Kanonen. Die Ernährungswirtschaft in Deutschland unter der Diktatur Hitlers*, Berlin, Akademie Verlag.

DEGLER, Stephanie & STREB, Jochen (2008), 'Die verlorene Erzeugungsschlacht: Die nationalsozialistische Landwirtschaft im Systemvergleich', *Jahrbuch für Wirtschaftsgeschichte*, 2, p. 161-181.

DEWEY, Peter (2007), 'The Supply of Tractors in Wartime', in Brian SHORT, Charles WATKINS & John MARTIN (eds), *The Front Line of Freedom. British Farming in the Second World War*, Exeter, British Agricultural History Society, p. 89-100.

ELLIS, Frank (1993), *Peasant Economics. Farm Households and Agrarian Development,* second edition, Cambridge, Cambridge University Press.

FARQUHARSON, John E. (1976), *The Plough and the Swastika. The NSDAP and Agriculture in Germany 1928-1945,* London/Beverly Hills, Sage.

FOOT, William (2007), 'The Impact of the Military on the Agricultural Landscape of England and Wales in the Second World War', *in* Brian SHORT, Charles WATKINS & John MARTIN (eds), *The Front Line of Freedom. British Farming in the Second World War*, Exeter, British Agricultural History Society, p. 132-142.

GAUARBEITSAMT Niederdonau (1943), *Der Arbeitseinsatz im Gau Niederdonau,* Neue Folge, vol. 1 ff., Vienna.

GRANT, Ian & MADDREN, Nicholas (1975), *The Countryside at War*, London, Jupiter.

GRUNDMANN, Friedrich (1979), *Agrarpolitik im Dritten Reich: Anspruch und Wirklichkeit des Reichserbhofgesetzes,* Hamburg, Hoffmann und Campe.

HAMMOND, Richard James (1951), *Food,* vol. 1: *The Growth of Policy* (History of the Second World War), London, H. M. Stationery Off.

HAMMOND, Richard James (1954), *Food and Agriculture in Britain 1939-45. Aspects of Wartime Control,* Stanford, Stanford University Press.

HAMMOND, Richard James (1956), *Food,* vol. 2: *Studies in Administration and Control* (History of the Second World War), London, H. M. Stationery Off.

HAMMOND, Richard James (1962), *Food,* vol. 3: *Studies in Administration and Control* (History of the Second World War), London, H. M. Stationery Off.

HANISCH, Ernst (1990), 'Bäuerliches Milieu und Arbeitermilieu in den Alpengauen: Ein historischer Vergleich', in Rudolf G. ARDELT & Hans HAUTMANN (eds), *Arbeiterschaft und Nationalsozialismus in Österreich*, Wien, Europaverlag, p. 583-598.

HAUCH, Gabriella (2006), '"Deutsche Landfrauen" – zwischen Angeboten und Zumutungen: Reichsnährstand – Tätigkeitsprofile – Landwirtschaftsschulen – Reichserbhofgesetz', in Gabriella HAUCH (ed.), *Frauen in Oberdonau. Geschlechtsspezifische Bruchlinien im Nationalsozialismus*, Linz, OÖLA, p. 147-190.

HAYAMI, Yujiro & RUTTAN, Vernon W. (1985), *Agricultural Development. An International Perspective*, second edition, Baltimore/London, John Hopkins University Press.

HORNUNG, Ela, LANGTHALER, Ernst & SCHWEITZER, Sabine (2004), *Zwangsarbeit in der Landwirtschaft in Niederösterreich und dem nördlichen Burgenland*, Vienna/Munich, Oldenbourg.

ILBERY, Brian W. & BOWLER, Ian (1998), 'From Agricultural Productivism to Post-productivism', in Brian W. ILBERY (ed.), *The Geography of Rural Change*, London, Longman, p. 57-84.

KRAMMER, Josef & SCHEER, Günter (1978), *Das österreichische Agrarsystem, 2. Teilbericht: Probleme, Perspektiven und Alternativen*, 2 vols., Vienna, Fachverlag für Wirtschaft und Technik.

LANGTHALER, Ernst (2000), 'Eigensinnige Kolonien. NS-Agrarsystem und bäuerliche Lebenswelten 1938-1945', in Emmerich TÁLOS *et al.* (ed.), *NS-Herrschaft in Österreich. Ein Handbuch*, Vienna, öbv & hpt, p. 348-375.

LANGTHALER, Ernst (2006), 'Agrarsysteme ohne Akteure? Sozialökonomische und sozialökologische Modelle in der Agrargeschichte' *in* Andreas DIX and Ernst LANGTHALER (eds), *Grüne Revolutionen: Agrarsysteme und Umwelt im 19. und 20. Jahrhundert* (Jahrbuch für Geschichte des ländlichen Raumes 3), Innsbruck, Studien-Verlag, p. 216-238.

LANGTHALER, Ernst (2008), 'Nahe und entfernte Verwandtschaft: Agrar-korporativismus in Niederösterreich', in Stefan EMINGER & Ernst LANGTHALER (eds), *Niederösterreich im 20. Jahrhundert, vol. 1: Politik*, Vienna/Cologne/Weimar, Böhlau, p. 687-719.

LANGTHALER, Ernst (2009a), 'From Capitalism to 'Neo-Feudalism'? Property Relations, Land Markets and the Nazi State in Austrian Rural Society, 1938-1945', in Rosa CONGOST & Rui SANTOS (eds), *The Social Embeddedness of Property Rights to Land in Europe*, Turnhout, Brepols, Rural History in Europe, 5, p. 165-186.

LANGTHALER, Ernst (2009b), *'Ökotypen'* in Friedrich JAEGER (ed.), *Enzyklopädie der Neuzeit*, vol. 9, ed. Stuttgart, Metzler, p. 419-423.

LANGTHALER, Ernst (2010), 'Regional Agrosystems, Labour Markets and the Nazi State: Farm-Work Relations in Niederdonau, 1938-1945', in Erich LANDSTEINER & Ernst LANGTHALER (eds), *Agrosystems and Labour Relations in European Rural Societies*, Turnhout, Brepols, Rural History in Europe, 3, p. 13-29.

LBG Land- und Forstwirtschaftliche Landes-Buchführungsgesellschaft (1948), *Die Lage der Landwirtschaft des Bundesjahres Niederösterreich im Vergleichsjahr 1937*, Vienna.

LBG Land- und Forstwirtschaftliche Landes-Buchführungsgesellschaft (1949), *Die Lage der Landwirtschaft der Bundesländer Niederösterreich,* Oberösterreich, Steiermark und Kärnten im Berichtsjahr 1946/47, Vienna.

MAF = Ministry of Agriculture and Fisheries (1946), *National Farm Survey of England and Wales. A Summary Report,* London, H.M. Stationery Off.

Martin, John (2000), *The Development of Modern Agriculture. British Farming since 1931,* Houndmills/Basingstoke/New York, St. Martin's Press.

Martin, John (2007a), 'The Structural Transformation of British Agriculture: the Resurgence of Progressive, High-input Arable Farming', in Brian Short, Charles Watkins & John Martin (eds), *The Front Line of Freedom. British Farming in the Second World War*, Exeter, British Agricultural History Society, p. 16-35.

Martin, John (2007b), 'George Odlum, The Ministry of Agriculture and 'Farmer Hudson'', *Agricultural History Review*, 55/2, p. 229-250.

Martin, John & Langthaler, Ernst (2012), 'Paths to Productivism: Agricultural Regulation in the Second World War and its Aftermath in Great Britain and German-Annexed Austria', in Paul Brassley, Yves Segers & Leen Van Molle (eds), *War, Agriculture, and Food: Rural Europe from the 1930s to the 1950s*, London/New York, Routledge, p. 55-74.

Mazoyer, Marcel & Roudart, Laurence (2006), *A History of World Agriculture from the Neolithic Age to the Current Crisis*, New York, Monthly Review Press.

Meihsl, Peter (1961), 'Die Landwirtschaft im Wandel der politischen und ökonomischen Faktoren', in Wilhelm Weber (ed.), *Österreichs Wirtschaftsstruktur. Gestern – heute – morgen,* vol. 2, Berlin, Duncker und Humblot, p. 551-837.

Moore-Colyer, Richard J. (2007), 'Prisoners of War and the Struggle for Food Production, 1939-49', *in* Brian Short, Charles Watkins & John Martin (eds), *The Front Line of Freedom. British Farming in the Second World War*, Exeter, British Agricultural History Society, p. 117-131.

Münkel, Daniela (1996), *Nationalsozialistische Agrarpolitik und Bauernalltag,* Frankfurt am Main/New York, Campus.

Murray, Keith Anderson Hope (1955), *Agriculture (History of the Second World War),* London, H.M. Stationery Off.

ÖStZA = Österreichisches Statistisches Zentralamt (1948), *Ergebnisse der Landwirtschaftlichen Statistik in den Jahren 1937-1944 (Beiträge zur österreichischen Statistik,* 3), Vienna, Carl Ueberreuter.

ÖStZA = Österreichisches Statistisches Zentralamt (1979), *Geschichte und Ergebnisse der zentralen amtlichen Statistik in Österreich 1829-1979 (Beiträge zur österreichischen Statistik,* 550), Vienna, Österreichisches Statistisches Zentralamt.

Pilfold, Will (2007), 'Defending Farmland: Dudley Stamp, the Ministry of Agriculture and Fisheries, and Rural Land Utilization Planning, 1939-47', in Brian Short, Charles

WATKINS & John MARTIN (eds), *The Front Line of Freedom. British Farming in the Second World War*, Exeter, British Agricultural History Society, p. 194-203.

RAWDING, Charles (2007), 'The Treatment of 'Failing farmers' in South-west Lancashire during the Second World War', in Brian SHORT, Charles WATKINS & John MARTIN (eds), *The Front Line of Freedom. British Farming in the Second World War*, Exeter, British Agricultural History Society, p. 179-193.

SANDGRUBER, Roman (2002), 'Die Landwirtschaft in der Wirtschaft – Menschen, Maschinen, Märkte', in Ernst BRUCKMÜLLER *et al.* (eds), *Geschichte der österreichischen Land- und Forstwirtschaft im 20. Jahrhundert, vol. 1: Politik*, Gesellschaft, Wirtschaft, Vienna, Carl Ueberreuter, p. 191-591.

SANDNER, R. (1948), 'Die Landmaschine in der österreichischen Landwirtschaft', *Die Bodenkultur. Österreichs Zentralorgan der Landwirtschaftswissenschaften*, 1/1, p. 69-84.

SHORT, Brian (2007a), 'The Dispossession of Farmers in England and Wales during and after the Second World War', in Brian SHORT, Charles WATKINS & John MARTIN (eds), *The Front Line of Freedom. British Farming in the Second World War*, Exeter, British Agricultural History Society, p. 158-178.

SHORT, Brian (2007b), 'War in the Fields and Villages: The County War Agricultural Committees in England, 1938-45', *Rural History*, 18, p. 217-244.

SHORT, Brian *et al.* (2000), *The National Farm Survey 1941-1943. State Surveillance and the Countryside in England and Wales in the Second World War*, Wallingford/New York, CABI.

SHORT, Brian, WATKINS, Charles & MARTIN, John (2007), "The Front Line of Freedom': State-led Agricultural Revolution in Britain, 1939-1945', in Brian SHORT, Charles WATKINS & John MARTIN (eds), *The Front Line of Freedom. British Farming in the Second World War*, Exeter, British Agricultural History Society, p. 1-15.

SHORT, Brian (2012), 'The Social Impact of State Control of Agriculture in Britain, 1939-1945', in Paul BRASSLEY, Yves SEGERS & Leen VAN MOLLE (eds), *War, Agriculture, and Food: Rural Europe from the 1930s to the 1950s*, London/New York, Routledge, p. 172-192.

STAMP, Laurence Dudley (1943), 'Land Utilization in Britain 1937-1943', *Geographical Review*, 33/4, p. 523-544.

STAMP, Laurence Dudley (1947), 'Wartime Changes in British Agriculture', *The Geographical Journal*, 109/1-3, p. 39-54.

STAMP, Laurence Dudley (1948), *The Land of Britain. Its Use and Misuse*, London, Longmans-Green.

STATISTISCHES AMT FÜR DIE REICHSGAUE DER OSTMARK (1941), *Die land- und forstwirtschaftlichen Betriebe im Reichsgau Niederdonau nach den Ergebnissen der im Deutschen Reich am 17. Mai 1939 durchgeführten landwirtschaftlichen Betriebszählung*, Vienna, Carl Ueberreuter.

STRAUSS, Frederick (1941), 'The Food Problem in the German War Economy', *The Quarterly Journal of Economics*, 55/3, p. 364-412.

STREB, Jochen & PYTA, Wolfram (2005), 'Von der Bodenproduktivität zur Arbeitsproduktivität. Der agrarökonomische Paradigmenwechsel im Dritten Reich', *Zeitschrift für Agrargeschichte und Agrarsoziologie*, 53, p. 56-78.

THIRSK, Joan (1987), *England's Agricultural Regions and Agrarian History*, 1500-1750, London, Macmillan.

THOEN, Erik (2004), "Social Agrosystems' as an Economic Concept to Explain Regional Differences', in Bas J.P. VAN BAVEL & Peter HOPPENBROUWERS (eds), *Landholding and Land Transfer in the North Sea Area (late Middle Ages-twentieth Century)*, Turnhout, Brepols, Corn Series, 5, p. 47-66.

TRACY, Michael (1989), *Government and Agriculture in Western Europe 1880-1988*, New York, Harvester Wheatsheaf.

TRENTMANN, Frank & JUST, Flemming (eds) (2006), *Food and Conflict in Europe in the Age of the Two World Wars*, Houndmills/Basingstoke/New York, Palgrave Macmillan.

WARD, Sadie (1988), *War in the Countryside*, 1939-45, London, Cameron.

WAYMARK, Janet (2007), 'The War Agricultural Executive Committee in Dorset: State-directed Farming or the Preservation of the Landed Estate?', in Brian SHORT, Charles WATKINS & John MARTIN (eds), *The Front Line of Freedom. British Farming in the Second World War*, Exeter, British Agricultural History Society, p. 143-157.

WHETHAM, Edith H. (1952), *British Farming*, 1939-49, London, T. Nelson.

WILLIAMS, H. T. (1954), 'Changes in the Productivity of Labour in British Agriculture', *Journal of the Agricultural Economic Society*, 10/4, p. 55-71.

8. The Iberian dictatorships and agricultural modernisation after the Second World War

Dulce FREIRE & Daniel LANERO TÁBOAS

I. Introduction

Where the relationship between the state and agriculture is concerned, the three decades that followed World War II in Europe proved to be especially significant. Two dominant historical processes can be identified: the consolidation of the democratic model of the welfare state and the rapid modernisation of the agrarian sector. As Portugal and Spain were ruled between the 1930s and the mid-1970s by fascist dictatorships – the New State in Portugal (1933-1974) and Francoism in Spain (1939-1975) – they clearly did not conform to the constitutional arrangements of liberal democracy. But to what extent did Portugal and Spain under fascism embrace the rapid modernisation of the agrarian sector? This is the issue addressed in this chapter.

While there was no question of regime change in 1945, the defeat of the fascist regimes of Italy and Germany forced the governments of Portugal and Spain to adapt to the new international political and economic realities. To this end they undertook internal reshuffles, changing the governing elites and the direction of certain policies. New strategies were devised towards the external world that sought to keep up with trends in diplomatic relations and the move towards greater economic integration. Recognition by other countries and international organisations was a key factor in ensuring regime survival in Iberia.

Despite state repression and the limitations imposed by peripheral spatial location, closer external relationships promoted receptiveness (at least among the Iberian technical and economic elites) to the sorts of innovations that were driving European agricultural modernisation. In line with increased state intervention in the economy and society, which had been obvious since World War I, technical and economic specialists supported modernising measures to assist the Iberian countries overcome their economic and social backwardness.

How well were the policy measures drawn up for agriculture able to deliver on the sort of modernisation being championed by the specialists? The specialist view of agricultural modernisation, in line with the wider consensus of the 1950s, favoured increasing the productivity of land and labour. Such intensification would depend on the use of fossil fuels, a greater application of external inputs (capital /machinery, chemical fertilizers, pesticides, etc.), the reduction in rural manual labour and the enlargement of urban consumer markets.

If the post-war model of agricultural modernisation demanded strong state intervention and was associated with profound social, economic and institutional change, how did the Iberian regimes embrace and implement it? What adjustments were made in public bodies to provide a new administrative framework for stimulating agriculture? How were the political measures designed and applied?

These questions will be discussed in a number of sections, the first of which will consider the general economic trends relating to state intervention in agriculture. We will then review the organisation and functioning of public and corporate services relevant to agriculture. From here we will proceed to analyse the measures taken to promote land restructuring, irrigation and technical assistance, measures viewed by external and internal agronomists and economists as indispensable to agricultural modernisation. Although the productive success of this model of agricultural modernisation is widely acknowledged (Federico, 2005), an adequate explanation of the factors underlying its mixed fortunes in Iberia has yet to be provided.

II. Agriculture, economy and policy: a general portrait

A current debate in Iberian historiography – to date more advanced in Spain (Baptista, 2004) – concerns the backwardness of agriculture and the evaluation of agriculture's contribution to economic convergence with the Iberian countries' European neighbours. As shown in Table 8.1, general trends in rural population size and the contribution of Gross Agricultural Product (GAP) to GDP were broadly similar in both countries. During the twentieth century, the high percentages of the economically active population engaged in Iberian agriculture and the GAP's significance were constantly invoked as evidence of vast economic and social problems. Of course, delayed change did not mean that no progress was being made. Recently a consensus has developed that Portuguese and Spanish agriculture, despite their competitive disadvantages (both ecological and institutional), narrowed the gap between them and central Europe (Clar & Pinilla, 2009; Lains, 2009). Yet general patterns cannot disguise important chronological and regional variations in processes of agricultural change and economic growth.

Table 8.1. Active agricultural population as a proportion of the total active population and Gross Agricultural Product (GAP) as a proportion of Gross Domestic Product (GDP)

Year	*Portugal*		*Spain*	
	Agricultural population (%)	*GAP (%)*	*Agricultural population (%)*	*GAP (%)*
1910	57(a)(c)	32(b)	55.7	35
1920		29	57.2	39
1930	49(c)	30	45.5	29
1940	53(c)	29	50.5	
1950	49(c)	31	47.6	31
1960	42	20	36.6	21
1970	26	12	22.8	11
1980	17	9	14.4	7
1990	12	8	10.0(d)	

Notes. (a) 1911; (b) 1915; (c) includes fisheries; (d) 1991.

Sources. Tortella (1994: 227, 242); Soares (2005: 162)

During the first half of the twentieth century, with the exception of the two world wars, the performance of Portuguese agriculture was comparable with other sectors of the national economy (Soares, 2005: 159). The average annual rate of growth of agricultural output, which Lains has estimated at 0.35 per cent between 1902-1927 and 2.36 per cent between 1937-1951 (Lains, 2003), exceeded the equivalent French rates in the same periods. Similarly agricultural output grew in Spain in the decades leading up to the civil war. The 55 per cent growth recorded between 1900 and 1931, with a cumulative annual increase of 1.37 per cent, was greater than that of France (0.58 per cent per annum over the same period). Land productivity also increased (by 53 per cent between 1900-1931), as did labour productivity (by 76 per cent) (Jiménez Blanco, 1986; GEHR, 1983). There were of course short periods, as during the civil war and the 1940s, when productivity either stagnated or regressed, but the overall tendency since the end of the nineteenth century was clearly one of growth (Simpson, 1995; 1997).

For the four agricultural regions of Spain (North, Mediterranean, Andalusia and Interior) that Simpson (1995) identifies it was the first two, and particularly the Mediterranean region, that saw the most vigorous economic growth. In response to the late nineteenth century crisis, the Mediterranean region created an irrigation infrastructure and intensified production of goods with a high commercial value (such as vegetables and dried and fresh fruits). In the Northern region there was increased specialisation on cattle farming (both meat and dairy), associated with the intensification of crop rotation. However, in the two other Spanish regions, which accounted for 72 per cent of the agricultural land area, land and labour productivity remained stubbornly below the national average. Although the regional growth patterns are less well documented for Portugal, there are indications that similar trends towards specialisation occurred in some regions (Serrão, 2009). As in Spain, however, much of the territory showed low levels of labour and land productivity.

Ever since the nineteenth century, technical specialists had argued that natural and institutional characteristics were among the greatest obstacles to the modernisation of Iberian agriculture. Guided by the assumptions of agronomy, and inspired by the example of the industrially advanced countries (especially the USA), agronomists desired that both private initiatives and public policies create the conditions for the maximum exploitation of available resources. To that end they argued that it was necessary to change the agrarian structure, extend the area devoted to irrigated crops, introduce laboratory-improved plant and animal species, intensify the use of fertilizers and pesticides and replace labour with technology. Some of these proposals were implemented in both countries before the 1950s, but what greatly fostered rural change from then on was the increasing domestic and international demand for labour, raw materials and food.

The data in Table 8.1 are suggestive of the rapid and profound changes that affected the Iberian countries in the last decades of dictatorship. But did agriculture modernise in the same way in the two countries? An analysis of agricultural output can provide one answer to this question. Between 1954 and 1966 Portuguese agricultural production increased at an average rate of 1.2 per cent per annum compared to an average 3.8 per cent increase in Spain (OECD, 1969: 13). And despite changes in agricultural practices, differences in overall agricultural growth persisted in the two countries (Clar & Pinilla, 2009; Lains, 2009).

The composition of agricultural production reveals further differences. The livestock sub-sector was the most successful, growing in response to the changing dietary habits that accompanied rapid urbanisation. Both countries registered a significant relative decline in cereals used in the making of bread. While Spain experienced a growth in the production of animal fodder (Clar, 2005), Portugal's livestock industry was based on the importation of large quantities of animal feed (Soares, 2005). Overall, traditional crops (like cereals for bread and vines) shrank proportionately more in Spain than in Portugal. In both countries we find a diversification of production and the adoption of crops dependent on irrigation (for example, citrus fruits, horticulture, sugar beet and maize), though the trend was more pronounced in Spain[1]. Contemporary experts subjected the different indicators published nationally and internationally to close analysis[2]. For contemporary Portuguese agronomists and economists the emerging trends were further proof that the New State, unlike the European democracies and even the Franco regime, was not adopting the most appropriate policies to modernise agriculture.

The Spanish data indicate that pre-civil war levels of production and productivity were reached again during the 1950s in response to changing government policies. The changes began when the agronomist Rafael Cavestany held the agriculture portfolio between 1951 and 1957. The modernisation model Cavestany favoured is well summarised in his declaration: 'Fewer farmers, more agriculture'. Once it came to be implemented, Cavestany's programme brought about a rapid increase in both agricultural output and exports. In the decades that followed, the model adapted itself without major difficulties to the many challenges of modernisation. The economic and social development plans (*planes de desarrollo económico y social*) of the 1960s — the first running from 1964-1967 and the second from 1967-1971 — reinforced the basic patterns established since the 1950s, thus consolidating the agricultural sector's capacity to respond to the expanding demands of both domestic and external markets.

Many Portuguese researchers believe that the New State made no real effort to help agriculture benefit from post-war modernisation possibilities, despite Portuguese technicians going abroad to be educated. Early in the 1950s development plans (*planos de fomento*) were outlined that contemplated agricultural modernisation (settlement, irrigation, product specialisation, mechanisation, etc.), but these plans, even when enacted in legislation, were rarely implemented (Baptista, 1993; Rosas, 1994). Rather than extending intensive production, extensive agriculture was allowed to continue. There was a change in the system of guaranteed prices for certain products in the early 1960s, though these reform measures too had little impact. It was therefore not surprising that the dictatorship's final development plan in 1974 insisted on the 'urgent need to provide agriculture with conditions of recovery that can no longer be delayed' (Presidência do Conselho, 1974: 333).

1 The quantification of global agricultural output (percentages, calculated at current prices) indicates that Spanish vegetable production amounted to 80.5% of global output between 1950 and 1955, 73.0% between 1961 and 1965 and 65.6% between 1971 and 1975. The equivalent livestock percentages for the same dates were 19.4%, 26.9% and 34.4% (Clar & Pinilla, 2009: 317). Turning to Portugal, vegetable output as a proportion of global agricultural output amounted to 52.5% between 1950 and 1954, 50.7% between 1960 and 1964 and 51.8% between 1970 and 1973; for the same dates livestock production was recorded at 31.5%, 32.9% and 31.3% of global agricultural output (Lains, 2009: 343).

2 Besides the annual Economic Surveys, the OECD commissioned more specific studies, such as the Mediterranean Regional Project on education in the 1960s (OECD, 1966) and a slightly later study that focused on agriculture (OECD, 1969).

Clar (2009) suggests that compared to Portugal the process of agricultural modernisation in Spain began ten years earlier (in the 1950s) due to the strong pressure exerted on the Franco regime by international economic bodies. The Spanish Government, politically sensitive to international opinion and free from domestic agrarian opposition, committed itself to promoting policies that accelerated the modernisation of agriculture. In contrast, the New State was anxious to maintain the support of a conservative agrarian lobby. Yet Clar's interpretation is not entirely satisfactory in explaining the different agricultural paths taken by the two countries. For one thing Portugal was also subject to international pressure and[3], while rural support was important to the regime, this did not mean that the State did not take unpopular decisions (as in its support for industrialisation). What still remains to be explained as well is how the Franco regime reconciled the new agrarian policy with economic interests, and with the power of the large landowners to exert political pressure.

Given that the modernisation of agriculture aims to meet the demands of consumers, we must also assess the power of market forces. Here the differences are striking. Not only did Portugal have a weak domestic market historically (Lains, 2003), but also it retained its colonial empire until 1974. While Spanish agricultural exporters had to respond to free market conditions, the Portuguese were tied to the protected markets of the colonies. Economic relations with the colonies made for the production of surplus goods and for a poor ability to compete on the open market (as with wine, for example). They also made expansion difficult in some of the modern products becoming increasingly popular in Spain. Certain crops (sugar beet, maize, cotton, etc.) competed with colonial agricultural output in Portugal, while others (barley, hops and oil seed plants) competed with traditional wine and olive oil production. In both countries, protectionism concentrated on certain products (wheat, wine, olive oil). The factors of production may have increased, but were they used efficiently? Agronomists would say 'no' due to the poor implementation of reforms. As industrialisation and urbanisation intensified, Portuguese agriculture failed to respond adequately to rising market demand, thus forcing the increased imports of basic foodstuff (Soares, 2005).

While attempts at agricultural modernisation may have been influenced by the international context, domestic political and economic considerations were also of great importance. An analysis of how agricultural policies were devised and

[3] Portugal was a founder member of NATO in 1949 and EFTA in 1959; it also joined the FAO in 1946, the OEEC in 1948, the OECD in 1960 and the UN in 1955. For its part Spain joined the FAO in 1951, the UN in 1955 and the OECD in 1960.

implemented in our two cases promises therefore to clarify the key factors accounting for the different prospects modernisation proposals faced in Spain and Portugal.

III. Agriculture's institutional framework

How did the existing institutional arrangements impact on agricultural modernisation? Independent ministries for agriculture pre-dated the Iberian dictatorships. Once formed, the autocratic regimes inspired by the fascist *Corporazioni* model of the 1930s began to create new organisations along both production and spatial lines, a process that continued into the post-war period.

The Portuguese Ministry of Agriculture established in 1918 survived, with some minor changes, until the 1940s. In 1940, however, agriculture was added to commerce and industry within the Ministry of the Economy. This change, which the government claimed was necessary in view of war economy demands, remained in force until the end of the dictatorship. Those who supported a speedy modernisation of agriculture believed that the persistence of this arrangement became an obstacle to change (Caldas, 1978). A basic difficulty was that industrial and commercial interests seemed more capable of influencing the state's decisions within the Economy 'super-ministry'.

Yet despite these criticisms and the presentation of alternatives, frequently based on international examples, the New State refrained from introducing significant changes. Up to 1958 agriculture was represented by an under-secretary of state and thereafter by a secretary of state. There were, reflecting the changes of the 1930s, three administrative departments (directorates general): Agriculture, Livestock and Forestry. There was also a Committee for Internal Colonisation (*Junta de Colonização Interna*) created in 1936 with its own directorate general and which acquired financial and juridical independence in 1946. Each of these central departments had its own separate regional and sub-regional offices.

Spain was an example that the Portuguese agronomists could easily evoke. Francoism would retain its inherited Ministry of Agriculture (created in 1931) with only some minor denominational alterations. The ministry's internal structure, with its *direcciones generales de agricultura, ganadería y montes* (general directorates of agriculture, livestock and forestry), and provincial directorates (*direcciones provinciales*), did not differ greatly from Portugal's Secretariat of State for Agriculture. During the lifetime of the regime some important organisational changes were made, as for exemple when Rafael Cavestany established the General Directorate for Coordination, Credit and Training (*Dirección General de Coordinación, Crédito y*

Capacitación), a body that in certain ways was the predecessor of the Agricultural Extension Service (*Servício de Extensión Agraria*) of 1955. Prior to this technical assistance had been provided by the *Servicio Nacional de Concentración Parcelaria* (1952). Cavestany's innovations were aimed at facilitating the implementation of the new policy guidelines for agriculture in the 1950s.

Alongside state administrative structures, the creation of hierarchical corporatist systems during the 1930s occurred concomitantly in Iberia with the repression and destruction of agricultural trade unions and other associations. The Spanish corporatist organisation, which appeared in 1942, implemented a plan conceived in the civil war that called for the complete destruction of existing unions and non-religious agricultural and co-operative societies. *Falange* pressure even prompted the partial destruction of bodies operating within the orbit of the Catholic Church[4]. Francoism's brutal repression had one clear goal: the political demobilisation of the rural population. The approach in Portugal, where political mobilisation appears to have been significantly weaker, was not as extreme, though there too the idea was to weaken local initiatives, destroy social networks and prevent the persistence of solidarities not favoured by the state.

In theory, the New State's view was that corporatist organisations should emerge from civil society (Lucena, 1976). In reality, however, they emerged only when and where the state determined. The new corporatist organisations, as was equally true of cultural, sporting, educational, welfare, technical and economic associations, remained subject to strict political and administrative control. In Spain, on the other hand, where syndicalist structures were imposed from above, there was never any question of corporatist organisations emerging from civil society. This would have conflicted with the principles of totality, unity and hierarchy that were intrinsic to the *Falange's* national-syndicalist organisational policy.

Both the Francoist Agrarian Trade Union Organisation (OSA – *Organización Sindical Agraria*) and the Portuguese New State's Agriculture Corporation (OCL – *Organização Corporativa da Lavoura*) were ideologically conceived as vertical syndicalist structures that sought cross-class integration (with strongly paternalist connotations in Portugal) within the rural communities. Viewed as a re-creation of the (supposed) 'natural order' of rural society, the new corporatist order amounted to

[4] *Falange*, the only political party recognised by the Francoist regime, succeeded in converting itself into a powerful bureaucratic machine that mobilised support for the regime and controlled important institutions. Relatively speaking, the *Falange's* political and administrative presence was greater than that of the National Union (UN – *União Nacional*), the single party recognised by the Portuguese political authorities. The National Union exercised no control over corporatist organisations and had but weak and inconsistent political functions.

the doctrinaire defence of elite interests in either case. The actual creation of the new corporatist structures generated multiple tensions between private agents and several state bodies and other entities. Nor did they enjoy the state's full support, something that gave rise to delays and contradictions.

The activities of the New State's OCL were closely linked to those of the economic co-ordination bodies (OCE: *Organismos de Coordenação Económica*), organisations that had no equivalents in Spain. In fact, the Portuguese corporatist system was controlled *via* these OCEs, membership of which was obligatory. As public organisations responsible to the secretary of state for commerce (within the economics ministry), the OCEs were appointed to oversee the main productive sub-sectors. Guided by common basic principles, their main duties were to intervene in domestic markets where they exercised a powerful influence over pricing policies. Their activities in the external market were designed to protect the privileged access to colonial markets. From the 1950s on, OCE activity concentrated heavily on encouraging the creation of new cooperatives that remained under state supervision.

In Portugal, as in Spain, the agricultural corporatist structures functioned on three territorial administrative levels: the local, the provincial-regional and the national. In Spain the local level revolved around the Brotherhood Union of Farmers and Ranchers (*Hermandades Sindicales de Labradores y Ganaderos*). As with the Portuguese Agricultural Societies (*Grémios da Lavoura*), the Spanish Brotherhoods were responsible for the distribution of factors of production and collaborated with other organisations in the fixing of prices in local markets. Mostly the *grémios* were nothing more than extensions of the heavily bureaucratised Portuguese state. The Spanish Brotherhoods also assumed responsibility for the management of social security and benefits that formed part of the system of rural social welfare. In Portugal the limited functions of social welfare assumed by the state did not become a responsibility of the *grémios*, but of the *Casas do Povo* (People's Halls), corporatist bodies responsible for providing medical and social assistance at the local level.

The regional level of agriculture's syndicalist organisation was more diffuse in nature. We find in Spain the Agrarian Trade Union Official Chambers (*Cámaras Oficiales Sindicales Agrarias*) functioning as intermediaries between the state and the farmers in implementing agricultural policies (for example, in relation to agricultural credit or new crops). The official desire to convert these bodies into real commercial enterprises, specialising in the distribution and promotion of inputs, generated strong opposition from agricultural specialists as well as from Catholic co-operativism. Something similar took place in the New State with the few (but highly dynamic) Federations of Agricultural Societies (*Federações de Grémios da Lavoura*), which from the 1950s were led by agrarian technicians, agricultural businessmen and some local authorities. Despite the obstacles that the state erected, the federations brought about a swift integration of economic activities and demonstrated entrepreneurial flair in several sectors (Lucena, 1980; Larcher Graça, 1999).

The organisations that came to occupy the peak of the corporatist structure made a late appearance in our two cases. The Portuguese Agriculture Corporation (*Corporação da Lavoura*) was created only in 1956; the Spanish National Brotherhood (*Hermandad Nacional*) followed in 1962. In the event, the syndicalist structures were left incomplete. Portugal lacked both the first (local) and the intermediate (district/regional) levels, whereas in Spain, even though all municipalities possessed corporatist organisations, in some towns and in every province these were frequently inoperative, inefficient and excessively bureaucratic (Lanero, 2010).

If in the 1940s there was a predominance of *Falangists* and Catholics in the institutional framework of Spanish agriculture, from the early 1950s the conception and application of modernising agricultural policies were almost exclusively controlled by agricultural and forestry engineers and by some large agricultural companies (Gómez Benito, 1996). While Portuguese specialists also sought to influence the direction of government policy, they were less successful than their Spanish counterparts. In part this was due to the political weight of the Corporate Chamber (*Câmara Corporativa*), the dictatorship's consultative body, in which interests linked to traditional metropolitan agriculture, commerce and the colonies dominated (Rosas, 1994). This body ratified the most important decisions and often opposed the most innovative measures. And even when technocratic proposals were enacted, the public and corporate bodies frequently lacked the capacity to implement the legislation (Baptista, 1993).

IV. Agrarian restructuring and the expansion of irrigation

For decades Portuguese and Spanish experts argued that inequality in Iberian land ownership between the north and the south was a fundamental cause of social and economic problems. This problem persisted, despite ameliorative attempts since the end of the nineteenth century by the state and individuals. An OECD report of 1969, in arguing for land reform, stated that 'farm structures are far from adequate in southern Europe' (OECD 1969: 18)[5].

---

5 In Portugal the average size of arable farms was 5.1 hectares; in Spain it was 7.1 hectares. Another problem resulted from the number of fields, with the average farm having six fields in Portugal and fourteen in Spain (OECD, 1969: 18).

Resistance to agrarian reform had been important to the rise of the authoritarian regimes during the 1930s. After World War II agronomists highlighted the prospective technical and economic benefits of land reform that would result in economically viable farms. Solutions based on the creation of economically efficient family farms were promoted, as well as the provision of education in the various types of 'group agriculture'. Proposals for agrarian reform that called for the large southern estates to be divided and the northern smallholdings to be restructured into economically viable peasant farms were seen as inseparable from the need for increased irrigation. How effective in our two cases did these proposals prove to be in practice?

The legislative framework of Spanish settlement policy during the 1950s, based on a model in use in the western United States, was set out in the law on the settlement and distribution of property in irrigated areas (*ley sobre colonización y distribución de la propiedad de las zonas regables*) of 1949. Quite distinct from the rather limited Italian Fascist-inspired settlement *bonifica integrale* projects of the 1940s that had relied on private ventures, state action was central to the Spanish initiative. It fell to the National Settlement Institute (INC – *Instituto Nacional de Colonización*) to encourage technical transformation in agricultural practice in places with potential and where the law had decreed irrigation. Typically the state expropriated for itself the 'excess' marginal land that landowners chose not to include within their irrigation zones (Barciela & López Ortiz, 2000).

The Franco regime did not seek to undermine the large estate owners with these initiatives; indeed, the landowners benefited in two significant ways. Firstly, the INC subsidised the cost of irrigating arid land and paid the landowners for the expropriation of their poorest land. And secondly, the regime's settlement legislation scrupulously respected private property rights; expropriations were subject to numerous limitations and provided 'exemplary' compensation. From 1950 onwards the policy strove to make the interests of the large estate owners compatible with the more efficient cultivation of the best land. Between 1951 and 1960 c. 200,000 hectares were thus settled and irrigated. Western Andalusia, where the greatest concentration of large estates was found, became the region with the largest irrigated land surface in Spain.

Simultaneously, the Franco regime invested in other elements of settlement policy. Legislation passed in 1946 allowed for additional funded activities and increased the resources placed at the farmers' disposal. State assistance thus became available for numerous improvements to the rural infrastructure, such as the collection and distribution of the water required to turn small barren fields into well-watered land (Barciela, 1990). More than one million hectares of previously marginal land were improved between 1941 and 1971.

If Spanish settlement policies were closely bound up with irrigation, the same was not true of Portugal. While Rafael Duque was minister of agriculture in the New State he created the Committee for Internal Colonisation in 1936[6], which was authorised to sub-divide and parcel land; it was further empowered to provide finance for capital and labour while implementing local improvements and offering technical assistance to farmers. Law No. 1949 of 1937 sought to develop irrigation projects and committed the state to expropriating poorly utilised land that would benefit from irrigation. A total of 50,000 hectares of land, including alluvial plains and drained marshlands, was irrigated between the 1940s and 1960s. During this same period, only about ten tenant farms were established over an area of 22,500 hectares (Baptista, 1993). These farms were located in barren areas, as in the largely infertile mountain areas where small estates predominated. The state tended to expropriate communal and public land, taking great care not to upset the owners of the large estates who employed every means to stop the agricultural reform initiative from impinging on their property. In fact, most of the expropriations envisaged in the legislation never occurred.

Faced with the inadequacy of earlier measures, the experts proposed a new set of solutions in the 1950s. The ambitious Alentejo Irrigation Plan (*Plano de Rega do Alentejo*), which covered an area of 170,000 hectares, was published in 1957. The plan's implementation was to be slow (indeed, it is still being implemented today); it was not integrated with the settlement project and its impact was very localised. The Portuguese experts, faced with the crisis of traditional agriculture in the early 1960s, prepared new legislation that called for changes to the structure of land ownership and for ending water wastage in the south. The project was submitted to the Corporate Chamber but, when published in 1962, it had been expunged of the agrarian reform and irrigation proposals central to the agricultural modernisation model. In fact, the irrigation public works were little more than a way of occupying the excess agricultural labour that the clandestine opposition (linked to the Communist Party) was successfully mobilising against the dictatorship (Freire, 2007).

Just as the division of large estates was intended to create economically efficient farms, so too was the consolidation of the smaller farms. The chief goal was to correct what was regarded to be the distortion caused by the multitude of smallholdings. The plans sought to ensure that each landowner had enough relatively similar land to permit intensive cultivation of the fewest possible fields.

[6] Duque, as the New State's last minister of agriculture (1936-1940) and its first minister of the economy (1940-1944), supported modernising policies and (particularly during the 1930s) made some attempts to implement them.

The measures to consolidate smallholdings into larger units had no more success in Portugal that those aimed at the division of large estates. Only two consolidated areas were created in the northwest of the country during the 1960s, affecting less than 400 hectares (Baptista, 1993: 47). The situation was very different in Spain where the first experimental law for consolidating smallholdings was introduced in 1952, a year that also saw the creation of the National Service for Farm Consolidation (SNCP – *Servicio Nacional de Concentración Parcelaria*). More proposals were published during the years that followed, particularly the definitive farm property consolidation law of 1955 (revised in 1962). However, in Spain too the results were very modest, prompting the government to reinvigorate consolidation programmes during the 1960s through the introduction of 'rural organisation' plans. As well as proposing farm consolidation and land redistribution, these plans offered guidance as to which crops were suitable for each zone, the planning and exploitation of natural resources, the development of collective agriculture and the promotion of agri-industries, etc. While in many cases the results were not as positive as was hoped, a total of four million hectares were consolidated between 1953 and 1975.

The objectives set by agronomists were often left unrealised (Scott, 1998; Federico, 2005). In both Spain and Portugal the greatest tensions were experienced with interventions in the South, though plans for irrigation and land restructuring were more successful in Andalusia than in the Alentejo where efforts to improve specialised production in irrigated areas failed. Here the different political and financial capacities of the two states proved telling. The large Portuguese estates reacted unfavourably to the new opportunities and deliberately under-utilised the new infrastructures, despite the state granting them several enticements, including an exemption from the irrigation tax introduced as a means of amortising the government's expenditure. All the Portuguese agronomists could do was to praise the Spanish irrigation projects and lament the fact that the Portuguese experience had not followed suit.

V. Technical assistance

In the years under review the Iberian countries were not at the vanguard of applied agricultural research. Yet despite living under repressive regimes and experiencing many material limitations, the agronomists were able to follow the scientific and technological developments emanating from foreign research centres in the more 'advanced countries', particularly the United States, the Netherlands and France. Diffusion of imported innovations still required prior research to determine how suited they were to local conditions; and this required the creation of scientific research units.

The separation of education and research was among the changes that both the New State and the Franco regime made to the organisation of the research centres. Contemporary witnesses noted nonetheless that the research units were subject to severe budgetary limitations and that there was a lack of co-ordination between them (Fernández Prieto, 2007; Bernárdez Sobreira, 1999). Apparently, the way in which scientific research was organised did little to encourage the dissemination of results or the provision of technical assistance to the farmers. The view of the research units as little more than ivory towers is especially true of Spain, where the institutions charged with promoting research and applying the innovations to regional agriculture had been both diverse and dynamic in the pre-Franco period (Fernández Prieto, 1992; 2007; Bernárdez & Cabo, 1996; Lanero, 2010).

From early in the 1950s American specialists arriving in Portugal and Spain under the auspices of the Marshall Plan[7], began to influence research and technical assistance. The specialists best positioned to design agricultural extension programmes were trained in the United States and, in some cases, at the International Agricultural Centre in Wageningen, which from early in the 1950s was collaborating actively with North American agronomists in Western Europe. The sort of agricultural modernisation associated with such cooperation fared better in Spain than in Portugal.

The establishment of the Agrarian Extension Service (SEA – *Servicio de Extensión Agraria*) in 1955 brought several North American agricultural extension specialists to Spain[8]. The SEA's main goal was to promote the modernisation of agriculture through the development of cooperatives, animal nutrition and the use of fertilisers, for instance. Its work centred on areas dominated by the large estates with settlement projects already underway and on the smallholding areas in which consolidation projects were anticipated. Predictably the creation of an autonomous SEA, with its own budget and goals, led to a dispute over its competence with the regime organisations, particularly those controlled by the *Falange*. Countering the *Falange*'s influence was no easy task for the SEA. Nonetheless the SEA had 44 agencies by 1957 and 135 by 1961. Throughout the 1960s it continued to expand and by 1971 had 755 county agencies, reporting to 50 provincial agencies under the control of 11 regional centres.

---

7 Although Spain was outside the Marshall Plan, it did sign co-operation agreements with the USA (September 1953). Under the terms of the 'Food for Peace' programme (Agricultural Trade Development and Assistance Act, 1954. Public Law 480), the United States' economic mission provided technical assistance.

8 The Minister for Agriculture, Rafael Cavestany, favoured the US model of agricultural extension, having observed it at first hand prior to the SEA's creation.

Another phase began in 1962 when the SEA, at a time when farm mechanisation was proceeding rapidly, began to take on a rural development remit that was much broader than technology transfer. By 1971 the SEA, employing many different types of specialists (including sociologists skilled in group dynamics), had therefore assumed a more local orientation closer to the farmers (Sánchez de la Puerta; 1997).

The stress laid on development via stimulating collective activity did not appeal to the dictatorship. It was therefore not an accident that, during the regime's final years, agricultural extension agents and other specialists were important to the emergence of clandestine political and union organisations intent on overcoming social inequality and other rural problems. In this way SEA activity had the effect of contributing to the outbreak of social, environmental and economic conflicts between 1960 and 1975 (Lanero, 2008; Cabana, Lanero, 2009).

Under the auspices of the United States Technical Assistance and Productivity Programme (promoted by the Economic Cooperation Administration of the European Recovery Plan), American specialists arrived in Portugal in the late 1940s and early 1950s (Rollo, 2007). With the support of Vitória Pires, an agronomist and under-secretary of state for agriculture (1950-1958), they encouraged the education of Portuguese experts and the establishment of several experimental and extension research centres[9]. However, some projects lacked continuity, and their impact on agricultural modernisation has yet to be sufficiently assessed.

Even though the training of agronomists was relatively up to date, it turned out that the technical assistance to farmers provided by public services was ineffective. Such assistance required co-ordination between regional public services and local corporate bodies that were unevenly spread across the country and often overwhelmed by the daily bureaucratic requirements of distributing state aid. Another problem was that candidates for direct assistance or training were selected more for political reasons, a practice that favoured the largest landowners or their clienteles. It appears as well that the normative advice of agronomists did not meet the immediate needs of the farmers (Freire, 2007).

To remedy some of the many evident failures, a group of renowned agronomists (some with close ties to the regime and occupying public positions of responsibility) proposed in 1957 that the Calouste Gulbenkian Foundation create a centre for the study of agrarian economics (CEEA – *Centro de Estudos de Economia Agrária*). For almost 30 years, this privately funded centre promoted research and debates that –

[9] Only in 1977, after the 1974 revolution had overthrown the New State, was a law introduced to expressly create a public service for agricultural extension.

among other things – sought to provide official services and economic agents with the knowledge needed to make agricultural modernisation projects viable (Freire 2007). During its early years, often in collaboration with corporatist and official bodies, researchers and technicians also strove to develop extension programmes in specific areas, even if the implementation of these almost always proved to be problematical. In fact, the nature of the CEEA kept it even more isolated from institutional and political power structures than was the case with the SEA in Spain.

VI. Conclusions

The post-war decades were crucial to the advance of an agricultural modernisation model based on maximising income from land and labour. Our study has shown how it became possible to promote a measure of modernisation in Iberian agriculture – more successful in Spain than in Portugal – without profoundly altering the state's institutional structure. The fate of Iberian modernisation policies depended on two essential factors: the flexibility and independence of the institutional framework and the willingness and capacity to implement modernising measures.

Very conspicuously the Portuguese central state deprived agriculture of the necessary institutional independence during what proved to be the crucial years for modernisation. The large-scale agricultural interests, linked to traditional cultivation, had sufficient power to control state agricultural policy. The dictatorship could never dispense with the political support of the traditional elites and so found itself unable to foster the structural changes that agricultural modernisation required.

In Franco's Spain, the ministry of agriculture resisted the *Falange's* attempt to contest its control of agricultural policy in the 1940s so effectively that even the *Falangist* syndicates came under ministerial control. For their part, the large-scale agricultural interests remained linked to the political power centres through the channels that corporatism provided.

Franco's government considered it necessary to create a body – the SEA, closely linked with North American experts – that promoted new agricultural methods at local level and that was independent of the ministry of agriculture. While the SEA made some notable progress, the same cannot be said of Portugal's CEEA. This body, despite being a prestigious research organisation, never managed to influence either state policy or the production practices of farmers.

These institutional realities had a big bearing on the ability of specialists to design and apply modernisation methods and on the opportunities available to agronomists in our two cases. In Spain, in line with the general technocratic redirection of economic policy during the 1950s, the regime trusted the experts to conceive and pursue programmes of agricultural modernisation. Far from opposing them the large estate owners viewed many of the measures – particularly land settlement – as a state-subsidised opportunity to capitalise and develop their estates. Fortuitously, the specialists saw the medium and large estates as ideally suited to the development of an economically efficient agricultural sector. And the rural exodus beginning in the 1950s helped defuse the potential for social conflict on the land, making the industrialisation of agriculture even easier.

In contrast, the Portuguese agricultural specialists did not have the political and institutional independence that their Spanish peers enjoyed. While there were plans (sometimes enacted in legislation) for modernisation, the more ambitious proposals were opposed by the influential traditional agrarian elites. This was so despite their knowledge that modernisation would deliver more economic efficiency in a context of increasing mechanisation and intensified rural exodus in the 1960s. Changes in Portuguese agriculture were therefore stimulated more by the growth of other sectors and by international developments than by specific state measures.

Bibliography

Baptista, F. Oliveira (1993), *A política agrária do Estado Novo,* Porto, Edições Afrontamento.

Baptista, F. Oliveira (2004), 'Espanha e Portugal, um século de questão agrária', in Dulce Freire, Ines Fonseca & Paula Godinho (eds), *Mundo rural: transformação e resistência na Península Ibérica (século xx)*, Lisboa, Edições Colibri/CEEP, FCSH-UNL, p. 15-51.

Barciela López, Carlos (1990), 'La colonización agraria en España, 1939-1951', in Teresa Pérez Picazo & Guy Lemeunier (eds), *Agua y modo de producción*, Barcelona, Crítica, p. 98-120.

Barciela López, Carlos & López Ortiz, María Inmaculada (2000), 'La política de colonización del franquismo: un complemento de la política de riegos', in Carlos Barciela López and Joaquín Melgarejo Moreno (eds), *El agua en la Historia de España*, Alicante, Publicaciones de la Universidad de Alicante, p. 323-363.

Bernárdez Sobreira, Antonio (1999), *Planificación agraria na Galicia da autarquía (1939-1955),* Santiago de Compostela, Xunta de Galicia.

Bernárdez Sobreira, Antonio & Cabo Villaverde, Miguel (1996), 'Ciencia y dictadura: la investigación agronómica en Galicia durante el primer franquismo', *Noticiario de Historia Agraria*, 12, p. 119-139.

Cabana, Ana & Lanero, Daniel (2009), 'Movilización social en la Galicia rural del Tardofranquismo (1960-1977)', *Historia Agraria*, 48, p. 111-132

Caldas, Eugénio de Castro (1978), *A agricultura portuguesa no limiar da reforma agrária,* Lisboa, Centro de Estudos de Economia Agrária/Fundação Calouste Gulbenkian.

Clar, Ernesto (2005), 'Del cereal alimento al cereal pienso. Historia y Balance de un intento de autosuficiencia ganadera: 1967-1972', *Historia Agraria,* 37, p. 513-544.

Clar, Ernesto (2008), 'Farm Policy under the Salazar and Franco Dictatorships in Portugal and Spain: Towards an Authoritarian Model of Intervention in Agriculture?', in Nadine Vivier (ed.), *The State and Rural Societies. Policy and Education in Europe 1750-2000*, Turnhout, Brepols, p. 177-194.

Clar, Ernesto & Pinilla, Vicente (2009), 'The Contribution of Agriculture to Spanish Economic Development, 1870-1973', in Pedro Lains & Vicente Pinilla (eds), *Agriculture and Economic Development in Europe since 1870*, London, Routledge, p. 311-332.

Federico, Giovanni (2005), *Feeding the World: an Economic History of Agriculture. 1800-2000,* Princeton and Oxford, Princeton University Press.

Fernández Prieto, Lourenzo (1992), *Labregos con ciencia: Estado, sociedade e innovación tecnolóxica na agricultura galega (1850-1939)*, Vigo, Xerais.

Fernández Prieto, Lourenzo (2007), *El apagón tecnológico del franquismo: Estado e innovación en la agricultura española del siglo xx,* Valencia, Tirant lo Blanch.

Freire, Dulce (2007), *Portugal e a terra. Itinerários de modernização da agricultura na segunda metade do século xx,* Dactilografado/Dissertação de Doutoramento, Lisboa, Faculdade de Ciências Sociais e Humanas-Universidade Nova de Lisboa.

Gómez Benito, Cristóbal (1996), *Políticos, burócratas y expertos: un estudio de la política agraria y la sociología rural en España (1936-1959)*, Madrid, Siglo xxi.

Grupo de estudios de historia rural (GEHR) (1983), 'Notas sobre la producción agraria española, 1891-1931', *Revista de Historia Económica*, año I/2, p. 185-252.

Jiménez Blanco, José Ignacio (1986), 'Introducción', in Ramón Garrabou, Carlos Barciela & José Ignacio Jiménez Blanco (eds), *Historia agraria de la España contemporánea. 3. El fin de la agricultura tradicional (1900-1965)*, Barcelona, Crítica, p. 9-24.

Lains, Pedro (2003), 'New Wine in Old Bottles: Output and Productivity Trends in Portuguese Agriculture', *European Review of Economic History,* vol. 7/1, p. 43-72.

Lains, Pedro (2009), 'Agriculture and Economic Development in Portugal', in Pedro Lains & Vicente Pinilla (eds), *Agriculture and Economic Development in Europe since 1870*, London, Routledge, p. 333-352.

Lanero Táboas, Daniel (2008), 'Una relación compleja: comunidad rural, conflicto socio ambiental y organizaciones políticas en la Galicia de la transición. El caso de As Encrobas', en *XII Congreso de la Sociedad Española de Historia Agraria (SEHA)*, Córdoba, 2008.

LANERO TÁBOAS, Daniel (2010), *Historia dun ermo asociativo. Labregos, sindicatos verticais e políticas agrarias en Galicia baixo o franquismo*, Santiago de Compostela, 3c3 Editores y Fundación 10 de Marzo de CCOO.

LARCHER GRAÇA, Laura (1999), *Propiedade e agricultura: evolução do modelo dominante de sindicalismo agrário en Portugal,* Lisboa, Conselho Económico e Social.

LUCENA, Manuel de (1976), *A evolução do sistema corporativo português,* vol. 1, O Salazarismo, Lisboa, Perspectivas & Realidades.

LUCENA, Manuel de (1980), 'Sobre as federações de grémios da lavoura: breve resumo sobre o que fizeram e deixaram de fazer', *Análise Social,* vol. 16/64, p. 713-744.

OECD (1966), *The Mediterranean Regional Project. Portugal,* Paris.

OECD (1969), *Agricultural Development in Southern Europe: Report Adopted by the Committee for Agriculture and by the OECD council,* Paris.

PRESIDÊNCIA DO CONSELHO (1974), *IV Plano de Fomento: 1974-1979,* Tomo I, Lisboa, Imprensa Nacional-Casa da Moeda.

ROLLO, Fernanda (2007), *Portugal e a reconstrução económica do pós-guerra: o Plano Marshall e a economia portuguesa dos anos 50,* Lisboa, Ministério dos Negócios Estrangeiros.

ROSAS, Fernando (1994), *O Estado Novo,* in José MATTOSO (dir.), *História de Portugal,* vol. 7, Lisboa, Editorial Estampa.

SÁNCHEZ DE LA PUERTA, Fernando (1997), 'Estado y agricultura: La extensión agraria', in Cristóbal GÓMEZ BENITO & Juan Jésus GONZÁLEZ RODRÍGUEZ (eds), *Agricultura y sociedad en la España contemporánea*, Madrid, CIS – Ministerio de Agricultura, Pesca y Alimentación, p. 913-947.

SCOTT, James (1998), *Seeing like a State: How Certain Schemes to Improve the Human Conditions have Failed,* Yale University Press, New Haven.

SERRÃO, José Vicente (2009), 'Land Management Responses to Market Changes. Portugal, Seventeenth-eighteenth Centuries', in Vicente PINILLA (ed.), *Markets and Agricultural Change in Europe, from the 13th to the 20th century*, Turnhout, Brepols, Rural History in Europe, 2, p. 47-73.

SIMPSON, James (1995), 'Spanish Agricultural Production and Productivity, 1890-1936', in Pablo Martín ACEÑA & James SIMPSON (eds), *The Economic Development of Spain since 1870*, Aldershot (U.K.), Edward Elgar, p. 182-220.

SIMPSON, James (1997), *La agricultura española (1765-1965): la larga siesta,* Madrid, Alianza Editorial.

SOARES, F. Brito (2005), 'A agricultura', *in* Pedro LAINS and Álvaro Ferreira DA SILVA (eds), *História Económica de Portugal*, vol. 3, Lisboa, Imprensa de Ciências Sociais, p. 157-183.

TORTELLA, Gabriel (1994), *El desarrollo de la España contemporánea: historia económica de los siglos XIX y XX,* Madrid, Alianza.

9. Conflicts and compromises between the Hungarian socialist state and the peasantry: contextualising the 'Hungarian agricultural miracle'

Zsuzsanna VARGA

I. Introduction

The character of the 'Hungarian Agricultural Miracle' of the 1970-1980s is well illustrated by the fact that, while in other socialist countries there was still a food shortage at the end of the 1980s, agriculture in Hungary was able to satisfy consumer demand at home as well as to export food abroad[1]. By the 1980s Hungary's production of meat and cereals (as measured in per capita terms) ranked among the top five countries in the world[2].

Since agricultural policy played a crucial role in attaining this outcome, it is necessary to understand the character of this facet of state policy. The bulk of recent research on the social history of post-war Hungary has argued that the dictatorship was effectively able to strip society of its autonomy. By implication social groups have tended to be seen as passive victims of socialist despotism. This picture, however, is far too one-sided. To understand the socialist agricultural policies in Hungary a new approach is required. My own research, which takes as its point of departure the notion of 'the limits of dictatorship', follows an approach that allows us to investigate the degree to which the communist party and the socialist state influenced social and economic processes and how these in turn were influenced by peasant resistance[3]. As crucial as it is to consider the issues of how the power of the socialist state affected peasants, it is just as important to investigate how peasants responded to state power.

1 This paper was supported by the János Bolyai Research Scholarship of the Hungarian Academy of Sciences.

2 Measured in terms of per capita cereals, Hungary ranked fifth in the world in 1985. In terms of wheat production, Hungary came second after Canada, and in terms of meat production it took fourth place after Denmark, Holland and Australia. These data, published by the Central Statistics Office of Hungary (KSH), are based on calculations made by the Food and Agriculture Organisation of the United Nations (FAO). See: A magyar mezőgazdaság nemzetközi összehasonlításban. [Hungarian agriculture in international comparison] (Budapest: KSH, 1987: 25-28).

3 The phrase 'the limits of dictatorship' is borrowed from the title of an edited volume dealing with the DDR. See: BESSEL & JESSEN (1996).

Only by considering both sides of the equation can we hope to better understand the nature of the 'the limits of dictatorship' and thus of socialist society in Hungary[4].

Based on newly opened archival sources at the national, county and local levels, this study presents new empirical evidence relating to the conflicts and compromises that emerged between the party-state and rural society in socialist Hungary. My discussion begins by concentrating on the early conflicts that emerged between the state socialist regime, intent on achieving the collectivization of agriculture, and farmers committed to preserving their private property rights in land. Special attention is paid to the forms assumed by peasant protest and to family survival strategies as these manifested themselves during the Stalinist collectivization campaigns of the first half of the 1950s. The second section of the chapter asks how the relationship between rural society and the political leadership changed as a result of the revolution in 1956. The nature of the 'dialogue' between the state and the agricultural cooperatives during the final phase of collectivization in 1959-1961 is explored. The divergence of the Hungarian collective farms of the 1960s from the Soviet model provides the subject of the third section. Here the goal is to investigate the interplay between the initiatives coming from 'below' and 'above' in the shaping of Hungarian agricultural policy; and to point to the differences with other socialist countries.

II. The socialist state and the peasantry during the first half of the 1950s

Shortage of food, rationing and standing in long queues had not been unusual in Hungary in the 1950s. These phenomena were primarily the product of the Soviet Union preparing for a Third World War and therefore seeking to exert absolute control over the Eastern European countries in 1948-1949 (Berend, 1996; Crampton, 1997). After assuming political power, the local communist leaders were then instructed to launch the Sovietization of the economy. This entailed eliminating private enterprise in industry and trade, bringing companies under state control and fitting them into an economy planned according to socialist principles.

The situation was somewhat different in agriculture, where private property continued to play a decisive role. The official intention here was to achieve the 'socialist reorganisation of agriculture' by means of collectivisation. On 3 March 1949, the Political Committee (PC) of the Hungarian Workers' Party (HWP) passed a resolution declaring that the transition from small peasant farming to large-scale

[4] For a sample of this literature on socialist agricultural transformation in the Soviet Union and East-Central Europe, see: Bauerkämper (2002); Bokovoy (1998); Fitzpatrick (1994); Iordachi & Dobrincu (2009); Lampland (1995); Schöne (2005); Swain (1985); and Viola (1996).

collectivised farming was to be accomplished in the course of the first five-year plan (1950-1954)[5].

In reality, the role the party and the state assigned to the new cooperatives (collective farms) was that of making it possible for the profit generated by agriculture to be siphoned off into other spheres of the national economy, predominantly heavy industry. It seemed far easier to appropriate the profit of a few thousand collective farms than hundreds of thousands of peasant farms.

As things turned out, a major conflict emerged between the socialist state, trying to push through collectivization, and private farmers committed to preserving their property rights. The reason why this conflict proved to be particularly sharp in Hungary, a country of large estates in the interwar period, was that the land reform of 1945 had brought a radical change to the structure of landed property. As large and middle-sized estates were redistributed, the number of small landowners grew significantly. Prior to the land reform almost 46 per cent of the agricultural population were landless proletarians and 47 per cent were peasants working small-scale farms in 1941. After the land reform this proportion changed to 17 and 80 per cent respectively. Land distribution also had a significant effect on the entire social structure. The proportion of peasants working small-scale farms within society reached 43 per cent in 1945. (Szakács, 1998: 257-267). It is therefore not surprising that only a small group of landless peasants and dwarf holders were attracted to collective farming in the early 1950s. The great mass of the newly created class of land-owning peasants had no intention whatsoever of giving up their land.

By way of response to this situation the representatives of the governing socialist leadership sought to set limits to commodity production as well as to market relations; and they launched an attack on private property, especially landed property (Szakács, 1998: 257-298; Varga, 2004: 221-252). The first victim of this policy was the wealthy *kulak* stratum of the peasantry[6]. Partly due to their financial independence, emotional attachment to the land and influence over other sections of the village population, it was also these wealthy peasants that put up the greatest resistance to collectivisation. The measures used by the state in the campaign against the *kulaks* included economic and administrative pressure as well as the use

[5] Magyar Országos Levéltár [Hungarian National Archive, hereafter MOL] M-KS 276. f. 53. cs. 22. ő.e. Jegyzőkönyv a Politikai Bizottság (PB) 1949. március 3-i üléséről. [Minutes of the PC of HWP] 3 March 1949. Agenda 2. Javaslat a mezőgazdaság ötéves tervére. [Proposition on the agricultural part of the first five-year plan].

[6] MOL M-KS 276. f. 54. cs. 14. ő. e. A Titkárság előterjesztése a 'kulák' meghatározására [Proposition of the Secretariat on the definition of the [*kulak*] 14 October 1948.

of physical force, involving arrests, imprisonment and deportation to labour camps (Nagy, 1999: 41-97).

Although the strength of the wealthy peasantry was shattered as a result of the state's agrarian campaign, the broad mass of the peasantry could neither be forced nor persuaded to abandon private farming[7]. Only the landless and some of the poorest peasants showed any interest in collective farming. The vast majority was reluctant to join farming cooperatives voluntarily; what they had seen and heard of Soviet-style cooperatives had aroused a profound distrust[8].

It was in the Great Plain, predominantly in Békés, Csongrád, Hajdú and Szolnok counties, that the first collective farms were formed. Those who joined the Soviet-type cooperatives were obliged to surrender their tools of production and animals to the collective farm and to carry out all work tasks (apart from work on household plots) as members of brigades and work teams. The constitutions of the cooperatives strictly defined the operational scope of both the organisation and its membership (Szakács, 1979: 95-141; Wädekin, 1977: 95-116).

By giving up individual farming the peasants knew that they would lose their personal independence. A significant number of them therefore opted to accept the role of overburdened individual farmer rather than join a cooperative. The authorities increasingly became convinced that peasants would only join collective farms if they became entirely impoverished. While it is true that it was predominantly the burdens on *kulak* farms that increased enormously from 1949 on, the burdens imposed on all peasant farms rose dramatically from 1952 on[9].

---

7 MOL 276. f. 53. cs. 81. ő. e. Jegyzőkönyv a PB 1951. augusztus 23-i üléséről. [Minutes of the PC of HWP] 23 August 1951. Agenda 1. Gerő Ernő tájékoztatója a mezőgazdaság helyzetéről [Ernő Gerő's Memorandum on the Situation of Agriculture].

8 In the first two attempts at collectivisation, three types of producer cooperative group were distinguished (types I, II and III). The difference between these three types of cooperative groups was based on the amount of communal production and the manner in which the harvest was shared. On type I farms the members sowed jointly, but harvested individually, and kept their harvested produce for themselves. On type II farms, the members cut the harvest on an individual basis but then pooled it. Type III groups operated on an entirely cooperative basis, distributing the surplus in relation to the amount of work performed. There was also an independent producer cooperative which was organised along the same lines as the type III cooperative groups described above. During the third collectivisation campaign (1959-1961) this latter type became the dominant one because of its very close resemblance to the Soviet 'artel'-type cooperative.

9 MOL M-KS 276. f. 53. cs. 83. ő. e. Jegyzőkönyv a PB 1951. szeptember 20-i üléséről. [Minutes of the PC of HWP] 20 September 1951. Agenda 1. A Mezőgazdasági és Szövetkezeti Osztály javaslata a termelőszövetkezeti gazdaságok megerősítésére [Proposition of the Agricultural and Cooperative Department on the strengthening of collective farms].

Up to December 1950, a total of 2,185 cooperatives were formed and this number rose to 4,625 by the close of 1951 and to 5,110 by December 1952 (Fazekas, 1976: 76-77). Usually only one person per family joined the cooperative; other family members left the land and went to work in industry. The main industrial investments of the first five-year plan gave rise to ample employment opportunities outside agriculture. A common occurrence was even for members of the cooperatives themselves to work illegally in industry, construction or on state farms in order to provide a living for their families (Lampland, 1995: 144-231; Nagy, 1995: 647-661). Between 1951 and 1952 this practice became quite common as the compulsory delivery quotas of the cooperatives increased significantly, thus reducing the already meagre members' share even further.

Farmers forced to join the cooperatives seldom surrendered their work ethic and diligence to the collective farm. The official figures[10] suggest that at least 40 per cent of all members did not perform any collective work; and in 1951, and especially 1952, tens of thousands of members actually left the cooperatives. Around 10 to 12 per cent of the type I and II cooperative groups existed only formally and were not involved in actual production (Fazekas, 1976: 71-74). It is no wonder therefore that labour discipline on collective farms was soon viewed officially as a major problem.

Not only was the appropriate working morale missing on these farms, but they lacked almost every other element of efficient production (such as machines, professional skills, organisation, etc.) as well. This absence indicates that only a tiny portion of the investment provided for under the first five-year plan actually ended up in agriculture.

The subjugation of village residents to arbitrary and coercive measures did not result in open resistance, but its negative consequences grew by leaps and bounds. Hundreds of thousands abandoned agricultural work, more and more land went uncultivated, agricultural output declined – as did the productivity of the land. Thus while there was a catastrophic decline in private peasant farming, the newly established and continuously expanding cooperative farm sector was incapable of compensating for the shortfall in production (Pető & Szakács, 1985: 204-217). By the turn of 1952-1953 the situation in the agrarian sector had become threatening in its consequences for the whole of Hungarian society.

---

10 A common problem in state socialist countries stems from the questionable reliability of the surviving statistical sources. The adulteration and distortion of statistical data was most common in the 1950s. In addition, there were collections of data classified for internal use only that better reflected the real situation. This existence of a dual set of statistical sources persisted even after 1956 in the Kádár regime. Faced with such a situation, all the historian can do is to try to compare the published statistics with those found in the classified sources that were meant for official use only.

The changes that followed Stalin's death in March 1953 at once aroused expectations and lessened the danger of a violent explosion of resentment. In Hungary these changes were associated with the person of Imre Nagy. The 'New Course' he announced in July 1953 involved a re-assessment of Stalinist agrarian policies and their partial correction. The subsequent directives of Nagy's government significantly reduced the peasantry's tax burdens and compulsory deliveries. They also decreased the uncertainty involved in agricultural production. And, what was even more important, they allowed peasants to leave the collectives legally[11].

These changes were welcomed, especially in the countryside where their impact was most directly felt. But soon it became obvious that the rejoicing was premature. Hardly two years had passed when, in the spring of 1955, Nagy was forced from office and Mátyás Rákosi regained power. Rákosi and his supporters quickly reverted to the pre-1953 policies. For the agrarian population this meant an increase in taxes and deliveries as well as the resumption of the campaign to drive people into the collectives. Not surprisingly this new turn of events caused huge disappointment in the countryside.

The individual and collective grievances of the farming population, as mentioned already, had not led to open rebellion before 1953. Active resistance did not manifest itself after the re-establishment of the old Stalinist order in 1955 either. In light of this we can ask what kind of survival strategies were the farmers able to develop inside and outside the cooperatives[12]? It is clear that they availed themselves of various means of protesting at an individual level (Varga, 2004: 251-258). For example, they made their deliveries late or failed to fulfil them completely. With renewed determination they resisted the demands to enter the collectives. Even if they signed the declaration of intent to join, they often kept postponing the act of joining for months and when eventually they became members did everything possible to avoid collective work. It appears that in the summer of 1956 in the western counties – especially in Zala, Vas, Baranya, Somogy and Győr-Sopron where agriculture was traditionally more productive – attempts to leave the collectives became frequent.

The revolution and fight for independence started on 23 October 1956 with student demonstrations and subsequent armed clashes in Budapest and other big cities. A few

[11] Az MDP Központi Vezetőség 1953. június 27-28-i határozata. [Resolution of the Central Leadership of HWP, June 27-28, 1953] in A Magyar Dolgozók Pártja határozatai, 1948-1956. [Resolutions of HWP, 1948-1956] (Budapest: Napvilág, 1998:. 188-206).

[12] The survival strategies resorted to within and outside the cooperatives are well documented at local level. See: Zala Megyei Levéltár [Zala County Archive, hereafter: ZML] MDP Zala M. Biz. ir. 2.fcs. 50.ő.e, 63. ő.e., 64 ő.e; 1.f.cs. 26. ő.e.; Hajdú-Bihar Megyei Levéltár [Hajdú-Bihar County Archive, hereafter: HBML] 24. f. 1. fcs. 9. ő.e. 2. fcs. 13. ő.e. 18. ő.e.

days later the revolution arrived in Hungary's villages and the village communities succeeded in organizing themselves within a very short time (Szakolczai & Varga, 2003; Szakolczai, 2006). The government quickly assessed the peasants' demands. There may have been differences of emphasis in the phrasing of these demands, but in essence they were the same: an end to forceful collectivization, the restoration of the expropriated lands, the abolition of compulsory deliveries and the scaling down of taxes.

III. 'Dialogue' and Informal Bargaining after 1956

The fate of the 1956 revolution was to teach definitive lessons both to civil actors and to the socialist regime in Hungary (Romsics, 1999; Tőkés, 1996). The Hungarian people had to realize that they could not count on the help of the western countries and that the country was destined to remain as part of the Soviet block for the foreseeable future. At the same time Hungary's new leadership realized that they had to avoid the sort of abuses that had led to the revolution of 1956. The most important lesson the leaders of the Hungarian Socialist Workers' Party (HSWP) drew from the revolution was that the building of socialism could not come at the expense of the living standards of the rural and urban masses. As János Kádár, the chief secretary of the Central Committee of the HSWP, pointed out at the 2 December 1956 meeting of the Party's Provisional Central Committee: 'Our policy on economic questions, and especially on questions of redistribution, is that the primary aspect should be the gradual rise in living standards'[13].

This was in contrast to the situation prevailing before 1956 when the forced development of the heavy and military industries was predicated on depressing the production of consumer goods. The new policy of raising the standard of living in the early years of the new regime could only succeed if the supply of foodstuffs was increased since a large proportion of income was still being spent on food (Valuch, 2001: 321-330).

The aim to supply cheap food to the masses made the increase of agricultural output a critical issue for the regime. This involved, among other measures, offering incentives to private producers who had the greatest potential to increase overall agricultural production. To settle things between the state and the agrarian population, the severe tensions generated by the former agrarian policy had first to be eased.

[13] A Magyar Szocialista Munkáspárt ideiglenes vezető testületeinek jegyzőkönyvei I. 1956. november 11. 1957. január 14. [Minutes of the Provisional Leadership Bodies of the HSWP, vol. I. November 11, 1956. January 14, 1957.] (Budapest: InterArt, 1993: 151).

Of the new measures the abolition of compulsory deliveries was the most significant, as it occasioned a sharp break with state socialist strategy[14]. Compulsory deliveries had been considered an integral part of the socialist system and Hungary was the first socialist country to take the step of abandoning them (Wädekin, 1982: 65). Instead of using economic compulsion, the state was now establishing commercial relations with the agricultural producers, peasants and cooperatives, and trying to give them market incentives to produce more food. The significance of all this was that market forces were being applied, albeit to a limited extent, in one of the main sectors of the post-1956 Hungarian economy[15].

As a result of these measures, two-thirds of the existing collectives dissolved themselves and several hundred thousand private agricultural units started functioning in their place. The general trend was for formerly landed peasants to leave the collectives and for those who originally had little or no land to remain on. The dissolution rate of cooperatives varied greatly by county and region. The area that saw the most change was the south-western region (the counties of Zala, Somogy, Baranya and Vas) and the counties least affected were those of Csongrád, Bács and Heves in eastern Hungary[16].

At the end of 1956 it was not only those people who said 'no' to the regime of *kolkhozes* who left the collectives but also those who, lacking other economic opportunities, had felt compelled to stay on. It soon became apparent that the collectives that had survived found that their members wanted them to function differently from the way they had functioned up to then. This amounted to a wholesale abandonment of the Soviet-style model of collectivised agriculture (Varga, 2001: 17-57). A report from the county of Hajdú-Bihar, written at the end of 1956, announced that: '... agricultural cooperatives wish to follow the route of free farming and intend to introduce a new order that grants them total autonomy from machine stations and the bank. It is their most sincere wish that there should be no fixed patterns or a unified model charter at all, but merely a loose framework'[17].

---

14 Törvények és rendeletek hivatalos gyűjteménye, 1956 [Official Collection of Laws and Ordinances, 1956], (Budapest: KJK, 1957: 62, 68-69, 263-265).

15 MOL M-KS 288. f. 28/1957/13. ő. e. Az Élelmezésügyi Minisztérium előterjesztése az MSZMP Politikai Bizottságához az új felvásárlási rendszer eredményeiről és tapasztalatairól [Food Ministry submission to the PC of HSWP on the achievements and experiences with the new central purchasing system], 19 November 1957.

16 MOL 288.f. 28/1957/11.ő.e. A Földművelésügyi Minisztérium jelentése a termelőszövetkezeti mozgalom helyzetéről [Report of the Ministry of Agriculture on the situation of collective farms], 4 February 1957.

17 Hajdú-Bihar Megyei Levéltár [Hajdú-Bihar County Archive] XXIII. 2. 4.k. Jegyzőkönyv a Hajdú-Bihar megyei Tanács Végrehajtó Bizottságának üléséről [Minutes of the County Council in Hajdu-Bihar], 4 December 1956.

In the first half of 1957 it seemed that the Hungarian Socialist Workers' Party (HSWP) expected Hungarian agriculture to become diverse in its organisation in the long run. By the end of 1958, however, the agrarian policy of the HSWP took a sudden turn in response to a change affecting the 'fraternal countries'. Acting on a motion coming from the Soviet Union, the socialist bloc countries now resolved upon completing a process that would transform the entirety of small-scale farms into large-scale production units (Pittaway, 2004: 98-103).

The winter of 1958-1959 proved to be a difficult time for the HSWP. First of all, the new acceleration of collectivization meant that the earlier promises made to the peasantry had to be broken. Secondly, the Party had to cope with the fact that the memory of the forced collectivization campaigns of the past was still vivid among both the peasants and party functionaries. The latter had not forgotten the failures of these campaigns. But the new collectivisation campaign was important as the Kádár regime felt obliged to demonstrate its ability to rule Hungary according to the wishes of the Kremlin.

The new wave of collectivization was preceded by a great debate within the HSWP in which two positions soon emerged. The leaders of the Ministry of Agriculture held highly dogmatic views, and on every question insisted on a strict implementation of the pre-1956 policy. In contrast, the Party's agricultural department now vehemently criticised the pre-1956 agrarian policy. Working alongside the Central Committee, Lajos Fehér, who headed the Department of Agriculture when setting out the HSWP's agrarian political line, professed a desire to avoid past errors and to respect peasant interests (Sipos, 1991: 188-197). Amidst the preparations for the re-launching of collectivisation, representatives of dogmatic left-wing views came to the fore in agrarian policy.

In the winter of 1958-1959, at the commencement of the new collectivization campaign, a mere 13 per cent of the total tillage area had been under cooperative ownership. By March 1961 this proportion had grown to nearly 70 per cent. And the number of cooperative members had increased from 200,000 to 500,000 in the first three months of 1959. It rose further to 900,000 in the winter of 1959-1960 before reaching an historic high point of 1,200,000 in the winter of 1960-1961. This meant that while about 80 per cent of farmers were managing individual farms at the beginning of the reorganization, about 75 per cent of them had become cooperative members by the end of the campaign[18].

---

18 Mezőgazdaságunk a szocialista átszervezés idején, 1958-1962 [Our Agriculture in the Time of Socialist Reorganization, 1958-1962] (Budapest: KSH, 1963: 141-153).

It came as another bitter blow to the Hungarian peasantry that the authorities tolerated private farming only provisionally. However, the mechanisms of defence and the strategies of survival developed in the early 1950s underwent a quick revival in the course of the third wave of collectivisation that occurred in the winter of 1958-1959[19]. Farming families rediscovered their interest in satisfying subsistence needs outside the cooperatives and, if the industrial sector offered any possibility of employment, the best worker in the family was quick to avail of it while his wife or an elder member of the family joined the cooperative in his place (Kovács, 2013). It was typical for members to work in the cooperative only on the understanding that this granted them the right to have household plots. Even if the head of a family had not been able to work in the industrial sector, the family was apt to seek out other employment opportunities outside the cooperative. Day-labour opportunities were thus exploited, as were those that involved working illegally in construction or on nearby state farms where labour could expect to be paid at once (Valuch, 2001: 189-200).

All these tendencies find expression in a number of telling statistics. In the course of the third period of collectivisation, the number of agricultural workers decreased by 350,000. Their proportion within the total occupied population went from 42.5 per cent in 1959 to 35 per cent in 1962. During the same period the age and gender profiles of the collective farmers went from bad to worse. Whereas half of the members had been younger than 40 years in 1958, by the end of 1961 only one quarter was in this age-group while 36 per cent were over 60 years. It is not surprising therefore that mean age within the cooperatives should go from 41 years in 1958 to 52 years in 1961. Considering the fact that most of the workers who quit the agrarian sector were men, the proportion of women grew significantly during the period of reorganization – from 26 per cent in 1958 to 38 per cent at the close of 1961[20].

Another feature of the third wave of collectivisation was that discipline quickly became a problem for the authorities. It is evident that a significant segment of registered farmers (25-40 per cent), due to a lack of personal interest, did little or no work at all for the cooperative they formally belonged to. The average amount of work done by cooperative members decreased each year. Altogether, the number of work units per farming family dropped from 390 in 1958 to 301 in 1959 and

[19] MOL M-KS 288. f. 28/1961/2. ő. e. Jelentés a Magyar Rádió és Televízió levelezési rovatához érkező levelekről, elsősorban a mezőgazdaság szocialista átszervezésével kapcsolatban [Report on the Letters Sent to the Correspondence Department of the Hungarian Television and Broadcast, Mainly on the Socialist Reorganization of Agriculture], June 1961.

[20] MOL M-KS-288. f. 17/5. ő. e. A Központi Statisztikai Hivatal feljegyzése a paraszti családok és a paraszti népesség számának alakulásáról [Record of the Central Statistical Office on the Change of the Number of Farmer Families and Agricultural Population in General], November 1960.

then to 169 in 1960[21]. In addition to this, family members had not been drawn into cooperative work.

The adverse effect of the problems associated with the supply and motivation of labour was not offset by any swift mechanization of the cooperatives (Pető, Szakács, 1985: 156-164). In consequence, food imports were required to maintain the welfare commitments made by the Kádár regime. Even though Hungary had been a substantial grain exporter in the interwar period (Szuhay, 1998: 217-218), the country's needs could not now be supplied from domestically grown grain. Following collectivisation, imports of bread grains had therefore to be increased, even though consumption was falling. Between 1959 and 1962, an annual average of 227 thousand tons of bread grains had to be imported[22].

Early in 1960, on his return from a trip to Moscow, János Kádár had the following to say about the problems being created by the grain output deficit:

> The situation in connection with the production of grain is that in our [socialist] camp [...] not only can we make no impact on world markets [...] but we can supply our own internal needs only with difficulty.[...] In this connection the opinion was expressed [in Moscow] [...] that the first duty of every member of the socialist camp is to assure its own grain supplies. In the name of the Soviet delegation, comrade Khrushchev made it absolutely clear to us that they were not willing to take over all the growing themselves for an indefinite period of time and offer their grain to every other country[23].

The economic and social problems arising in the agrarian sector became so severe that the leadership was compelled to initiate corrective measures. In a symbolic gesture, the dogmatic minister of agriculture, Imre Dögei, was dismissed from office. His successor was Pál Losonczi, the head of a cooperative that had displayed a spirit of enterprise. Furthermore, at the session of the Political Committee held on 16 February 1960, the party conveyed the essence of what was envisaged by way of correction in its declaration that:

21 Work units were both units of measurement of various kinds of work and a measure of evaluating the work accomplished that served as a basis for remuneration. The income of each cooperative member was thus determined by the number of work units performed or completed. Mezőgazdaságunk a szocialista átszervezés idején, 1958-1962 [Our Agriculture in the Time of Socialist Reorganization, 1958-1962] (Budapest: KSH, 1963: 147).

22 MOL M-KS 288. f. 5/354. ő. e. Jelentés az MSZMP Politikai Bizottsága részére a magyar-szovjet gazdasági kapcsolatokról [Report to the Political Committee of the HSWP on Hungarian-Soviet Economic Relations], 15 December 1964.

23 A Magyar Szocialista Munkáspárt Központi Bizottságának 1959-1960. évi jegyzőkönyvei [Minutes of the Central Committee of the HSWP, 1959-1960] (Budapest: MOL, 1999: 420).

> The Political Bureau considers it necessary that the remuneration of the workers according to the locally proven methods receive the widest possible dissemination [...] What constitutes the best way of doing this has to be decided by the general meeting of the cooperatives in every case[24].

With this decree the regime accepted the validity of locally produced solutions that were designed to take advantage of the cooperative membership's interest in material profit and that clearly departed from the Soviet *kolkhoz* model. Once the party leadership accepted the idea that the agricultural cooperatives should apply methods adapted to their own conditions, the cooperative membership was given a 'bargaining' position from which it could seek leave to keep more stock on the family farm, practise a form of share-cropping on collective farms, receive their premium in kind and so on.

That this 'correction' was controversial within ruling circles is clear from a far-reaching debate during a session of the Politburo. What really concerned the majority of the party leadership was the fact that the cooperatives, in being permitted to assign more importance to individual incentives, were departing from the regulations of the Soviet-style statutes, thus weakening the influence of the central administration over them. Despite the split in the party's upper leadership, the political change of direction in 1960 was significant, as it introduced a new structure of incentives, at least provisionally and until such time as the cooperatives gained more strength.

IV. Departing from the Soviet Model

On 17 February 1961, the Hungarian Socialist Workers' Party's (HSWP) Central Committee issued a communiqué to inform the public that 'a turning point has been reached in the transformation of Hungarian agriculture: following industry, Socialist relations in agriculture have also become dominant'[25]. The statistical data reflect the turnaround as follows: in 1961 there were 271 state farms, approximately 4,200 cooperatives, and almost 165,000 registered individual farms. The cooperatives owned almost 70 per cent of the country's arable land, and employed three-quarters of the total number of agricultural workers[26].

---

24 MOL M-KS-288. f. 5/170. ő. e. Jegyzőkönyv a Politikai Bizottság 1960. február 16-ai üléséről [Minutes of the PC of the HSWP, February 16, 1960].

25 A Magyar Szocialista Munkáspárt határozatai és dokumentumai 1956-1962 [Resolutions and Documents of the Hungarian Socialist Workers' Party, 1956-1962] (Budapest: Kossuth, 1973: 497).

26 Mezőgazdaságunk a szocialista átszervezés idején, 1958-1962 [Our Agriculture in the Time of Socialist Reorganization, 1958-1962] (Budapest: KSH, 1963: 25-27).

Formally, therefore, the collectivization appeared to be a success, but in practice serious problems still persisted. For the next six years (1960-1965), agricultural production levels hardly reached the average for 1958-1959. There were supply difficulties for several years because the collectivization resulted in a sharp fall in national livestock herds – sheep are the sole exception here – and the condition of livestock deteriorated[27]. For several years, burdened as they were by initial difficulties and shortages of equipment and labour, most cooperatives failed to demonstrate the advantages of large-scale production.

During the different collectivization campaigns the political decision-makers had all found themselves in a tight spot before long. The experience of 1956, and their living-standard commitments, forced them into their first compromises with the peasantry. A certain kind of 'dialogue' had started between the representatives of the state and the cooperatives. Most of the party leadership, however, considered this to be no more than a transitional situation. Yet the historical sources reveal that this 'dialogue' had not been terminated after the completion of collectivization; if anything, it had become stronger and more widespread.

Of importance to this development was the emergence of a new mediator between the state and the farmers: the new presidents of the cooperatives. The conception of the cooperative president as head of farm management affairs would undergo significant change. Whereas in the 1950s the party had practically appointed the 'elected' presidents of the cooperatives by sponsoring loyal party members, in the course of the final phase of collectivization the party organizations effectively withdrew and allowed the local communities elect their own cooperative heads. This development resulted in well-experienced and widely respected local farmers taking over the leadership of many cooperatives. This 'natural selection', which the party leadership had acceded to, produced a distinctive group within the socialist economic elite (Hann, 1980; Kunszabó, 1974; Swain, 1985). In the 1960s the presidents of the cooperatives had to deal with two sources of pressure: the expectations of organizations of the party and state on the one hand and those of the cooperatives that had elected them on the other. Under such pressures, it became almost inevitable that the cooperative presidents should come to mediate relationships between the state and the peasant population.

When it came to the implementation of the emerging pragmatic agrarian policy, the greatest problem was caused by the fact that the local measures regarding remuneration and work organization were largely inconsistent with the Stalinist idea

27 Mezőgazdaságunk a szocialista átszervezés idején, 1958-1962 [Our Agriculture in the Time of Socialist Reorganization, 1958-1962] (Budapest: KSH, 1963: 68-79).

of *kolkhoz*. Since the Hungarian party leadership did not wish to confront the basic doctrines of the Soviet model, the local measures had been authorized only in practice without being recognised officially or given legal standing.

Each year, after the Minister of Agriculture had analyzed the experiences of the year that had just passed, he went on to make suggestions for possible changes in the remuneration system and in the distribution of income within the cooperatives for the coming year. In these suggestions he was able to offer concrete solutions for remuneration and work organization that had not been authorized officially but were nonetheless permitted in the everyday practice of the cooperatives. Early in 1961 the Minister of Agriculture announced his first proposals regarding the remuneration system[28]. His suggestion was that premiums be paid to members only if and when production plans had been fulfilled[29]. It turned out, however, that this suggestion did not contain sufficient incentives for the membership and as a result a different policy for the paying of premiums was suggested for 1962.

Parallel to these developments the Party's opinion of sharecropping also changed. By 1961 this practice had gone from being illegal to being tolerated, because the Party acknowledged that it was well suited to serving both the interests of the cooperative members and the state[30]. Thus an old tradition, one that had its origins in feudal times, was able to gain formal acceptance under socialist rule. Why this was so can again be attributed to the triumph of pragmatism over dogma.

From 1962 on, as the practice of making cash payments to members became universal, a means of remuneration became accepted that had little to do with the original system of allocating work units to individuals. The Stalinist dogma that cooperatives should distribute their profits solely by reference to work units was thus abandoned.

The plans for members' remuneration in 1962 were further influenced by the fact that the three-year ban on members withdrawing from the collectives was coming to an end. It could reasonably be expected that many of the people who had entered since 1959 – as they had done so under duress – would use this opportunity to leave.

---

28 Javaslatok a jövedelemelosztás és munkadíjazás formáira a termelőszövetkezetekben [Suggestions for Forms of the Distribution of Income and the System of Remuneration in Agricultural Cooperatives], in: Mezőgazdasági Értesítő [Agricultural Gazette], 2 (1961: 9-14).

29 MOL M-KS 288.f. 28/1962/3. ő.e. Jelentés az FM Kollégiuma részére a termelőszövetkezetek 1961 évi zárszámadásáról [Report to the Ministry of Agriculture on the annual accounts of cooperatives for the year 1961], 28 March 1962.

30 MOL 288.f. 28/1962/4. ő.e. Tájékoztató jelentés a PB számára a termelőszövetkezetek 1961 évi zárszámadásáról [Informative report to the Political Committee on the annual accounts of cooperatives for the year 1961], February 1962.

As well as accepting the practice of sharecropping on a more widespread basis, the new system of remuneration was meant to decrease the number of departures – and that is exactly what happened[31]. What the local archival material also reveals is that more people left the cooperatives in those counties where sharecropping was not tolerated.

The emerging tendency was for a growing number of local initiatives and practices to move from the category of being 'tolerated' to being 'supported', thus extending the scope and local autonomy of the agricultural cooperatives year by year[32]. Even if this special 'dialogue' was significantly extending the scope of the agricultural cooperatives, legal recognition and regulation of the new forms of remuneration and work organization did not inevitably follow. On the contrary, a large gap between the *de jure* and *de facto* situation of the cooperatives had opened up in the first half of the 1960s.

In this situation the standing of the local cooperatives, and their ability to innovate within the structures that impinged on them, was a decisive factor. The majority of the Hungarian Socialist Workers' Party (HSWP) administration consisted of former members of the Hungarian Workers' Party who strictly adhered to the original principles laid down in the 1950s. One of these principles stated that the predominant form of work organization was the brigade and that the main form of socialist remuneration had to be tied to the system of work units. A dogmatic acceptance of such principles led to an insistence that 'everything that is not expressively authorized is prohibited'. In contrast, as we have seen, party and government leaders came to accept and let it be publicly known that 'everything that is not expressively prohibited is authorized'. Given the mixed messages emanating from the state, it is not surprising that the pragmatic agrarian policy propagated by the government should result in a great deal of local variation in the first half of the 1960s. Vast differences in the application of the agrarian policy, not only between counties but also between districts, are observable.

31 MOL 288.f. 28/1962/9. ő.e. Megyénkénti jelentések a termelőszövetkezetek anyagi ösztönzése területén történt változások hatásairól [Reports from counties on the impact of the changes in remuneration of cooperatives].

32 MOL 288.f. 28/1963/37., 38., 39. ő.e. A megyei párt és tanácsi vezetőkkel folytatott megbeszélések a mezőgazdasági termelés és irányítás fontosabb kérdéseiről [Discussions with county party and local county leaders on the essential issues of agricultural production and management]; 288.f. 28/1963/13. ő.e. Jelentés a Földművelésügyi Minisztérium Kollégiumához a termelőszövetkezetek jövedelemrészesedési és munkadíjazási tapasztalatairól [Report to the Ministry of Agriculture on the experience of the distribution of income and remuneration in cooperatives].

In the dissemination of the informal profit-sharing and remuneration practices the press played an important role[33]. What also was of considerable significance was the way the members of the Party's Agitation and Propaganda Department organized local seminars for party and government functionaries to familiarize them with new and changing remuneration and management practices. The official desire to embrace and accommodate change is well captured in the lecture delivered by Lajos Fehér[34] at the Political Academy of the HSWP on 20 December 1961. In this talk he made it clear that:

> The initiatives aiming at the expansion of the distribution of income in cooperatives deserve the attention and the close study of our party organizations. They should help these initiatives, which comply with the principle of socialist distribution and better represent the interests of cooperative members, to spread rapidly in all cooperatives. [...] We shall take firm steps against inflexibility, incomprehension, conservatism; against efforts that are concerned about the 'prestige' of work units and which actually stunt the proper expansion of the system of work units. (Fehér, 1961: 21).

If the gap between the *de jure* and the *de facto* position of the cooperatives was to be bridged, it had to be accepted that the new informally permitted practices were good for both the cooperatives (in terms of membership and output) and for society (in terms of the availability of greater quantities of better quality food). Eventually, in the course of the economic reform process of 1966-1967, the *de facto* realities in the cooperative sector were given legal recognition. At one level this recognition endorsed the local initiatives that took the form of a range of remuneration and management practices. They also, at another level, cleared away several restrictions still remaining from the former *kolkhoz* model and that constituted obstacles on the path to progress.

V. Conclusions

All Eastern European countries – except for Poland and Yugoslavia where the collectivisation of agricultural land had been abandoned – were confronted with the same question in the early 1960s: how could the newly created structures in the agricultural sector be made to work? In all these states the collectivisation of agriculture had made major strides; most agriculture was in the socialist sector and the bulk of it was formally in the hands of agricultural cooperatives. In broad

[33] The journal, Pártélet [Party Life], designed to inform party members properly is particularly noteworthy in this regard.

[34] Lajos Fehér (1917-1981) was a secretary of the Central Committee responsible for the agricultural sector. Later on his remit was to supervise agriculture as a vice-Prime Minister, in which capacity he became one of the leaders of the agrarian lobby.

terms, the traditional Stalinist techniques were to persist in Rumania and Albania. Czechoslovakia, the German Democratic Republic and Bulgaria reformed the Stalinist techniques to create a set of practices that might be termed 'neo-Stalinist'. The reforms Hungary adopted went considerably further than any of this and became the basis of a uniquely Hungarian solution to the agrarian problem under state socialism (Swain, 1998: 5-19).

The key to understanding the 'Hungarian model' lies in the analysis of the changing relationship between those who held state power and rural society after the revolution of 1956, especially during the final phase of collectivization in 1958-1961. Here a particular 'dialogue' developed between the political decision-makers and the presidents of the collective farms. The Hungarian party and state leadership initially tolerated and later actively supported a re-organisation of the cooperatives that at once departed from the Soviet *kolkhoz* model while being more in tune with local preferences.

This pragmatic approach was the result not of conscious planning but took shape as a by-product of a number of decisions made in response to conditions and developments in the countryside. The political decision-makers did play a role in what came to transpire, as did the leaders of the cooperatives and their memberships. In this sense the new policy can be said to have come about as the product of an interaction between pressures originating from above and below. As a result of the 'dialogue' between the country's leaders and the representatives of agrarian society we can observe in Hungarian agriculture, from the early 1960s on, a cautious and gradual deviation from the Soviet model (Varga, 2007: 216-218).

It is important to emphasize, however, that the Hungarian leadership never openly drew attention to this departure from the *kolkhoz* pattern. Kádár and his associates did not want to get into an ideological dispute with the leadership of the Soviet Communist Party; they satisfied themselves with implementing procedures that were de facto at variance with Soviet agrarian practices. The result of this pragmatic approach was to lay the foundations of a uniquely 'Hungarian model' of collectivised agriculture.

Archival sources

Hajdú-Bihar Megyei Levéltár [Hajdú-Bihar County Archive].

Magyar Országos Levéltár [Hungarian National Archive], MOL HBML.

Zala Megyei Levéltár [Zala County Archive], ZML.

Bibliography

BAUERKÄMPER, Arnd (2002), *Ländliche Gesellschaft in der kommunistischen Diktatur. Zwangsmodernisierung und Tradition in Brandenburg 1945-1963,* Wien, Böhlau.

BEREND, Iván T. (1996), *Central and Eastern Europe 1944-1993: Detour from the Periphery to the Periphery,* Cambridge, Cambridge University Press.

BESSEL, Martin & JESSEN, Ralph (eds.) (1996), Die Grenzen der Diktatur: Staat und Gesellschaft in der DDR, Göttingen, Vandenhoeck und Ruprecht.

BOKOVOY, Melissa (1998), *Peasants and Communists. Politics and Ideology in the Yugoslav Countryside, 1941-1953*, Pittsburgh, University of Pittsburgh Press.

CRAMPTON, Richard J. (1997), *Eastern Europe in the Twentieth Century and After,* London/ New York, Routledge.

FAZEKAS, Béla (1976), *A mezőgazdasági termelőszövetkezeti mozgalom Magyarországon [The Movement of Agricultural Cooperatives in Hungary],* Budapest, Kossuth.

FEHÉR, Lajos (1961), 'Erősítsük termelőszövetkezeteink szocialista jellegét' [Accentuating the Socialist Character of Our Cooperatives], *Társadalmi Szemle [Social Review], 2,* p. 16-24.

FITZPATRICK, Sheila (1994), *Stalin's Peasants: Resistance and Survival in the Russian Village after Collectivization,* New York, Oxford University Press.

HANN, Chris (1980), *Tázlár: a Village in Hungary,* Cambridge, Cambridge University Press.

IORDACHI, Constantin & DOBRINCU, Dorin (eds) (2009), *Transforming Peasants, Property and Power. The Collectivization of Agriculture in Romania, 1949-1962*, Budapest & New York, CEU Press.

Ö. KOVÁCS, József (2013), 'The Application of Collectivization in Hungarian Society', in Arnd BAUERKÄMPER & Constantin IORDACHI (eds), *The Collectivization of Agriculture in Communist Eastern Europe: Comparison and Entanglements from the 1930s to the 1980s,* Budapest, CEU Press (forthcoming).

KUNSZABÓ, Ferenc (1974), *Elnöktípusok a termelőszövetkezetekben [Types of Presidents in Agricultural Cooperatives],* Budapest, KJK.

LAMPLAND, Martha (1995), *The Object of Labour. Commodification in Socialist Hungary*, Chicago / London, University of Chicago Press.

NAGY, József (1995), 'A paraszti migráció irányai és okai az ötvenes években' [Directions of and reasons for the migration of peasants in the 50s], in Tibor VALUCH (ed.), *Hatalom és társadalom a xx. századi magyar történelemben [Society and Power in Hungary in the 20th Century],* Budapest, Osiris, p. 647-661.

NAGY, József (1999), 'A kulákkérdés és megoldása az 1948-1953-as években' [The 'Kulak' Question and its Solution, 1948-1953], *Múltunk*, 3, p. 41-97.

PETŐ, Iván & SZAKÁCS, Sándor (1985), *A hazai gazdaság négy évtizedének története 1945-1985. I.: Az újjáépítés és a tervutasításos irányítás időszaka 1945-1968. [The History of Four Decades of Domestic Economy, 1945-1985*, Vol. 1: The Period of Rebuilding and Command Economic Planning, 1945-1968], Budapest, KJK.

PITTAWAY, Mark (2004), *Eastern Europe 1939-2000,* London, Arnold.

ROMSICS, Ignác (1999), *Hungary in the Twentieth Century,* Budapest, Osiris.

SCHÖNE, Jens (2005), *Frühling auf dem Lande? Die Kollektivierung der DDR-Landwirtschaft,* Berlin, Ch. Links Verlag.

SIPOS, Levente (1991), 'Reform és megtorpanás. Viták az MSZMP agrárpolitikájáról (1956-1958)' [Reform and Baulking. Debates on the Agrarian Policy of the HSWP, 1956-1958], *Múltunk,* 2-3, p. 188-197.

SWAIN, Nigel (1985), *Collective Farms which Work? Cambridge,* Cambridge University Press.

SWAIN, Nigel (1998), 'A Framework for Comparing Social Change in the Post-socialist Countryside', *Eastern European Countryside,* 4, p. 5-19.

SZAKÁCS, Sándor (1979), 'A Szovjetunió mezőgazdasága (1917-1939)' [Agriculture of the Soviet Union, 1917-1939], *Agrártörténeti Szemle [Agrarian History Review]*, 1-2, p. 95-141.

SZAKÁCS, Sándor (1998), 'From Land Reform to Collectivization (1945-1956)', in Péter GUNST (ed.), *Hungarian Agrarian Society from the Emancipation of Serfs (1848) to the Re-privatization of Land (1998),* New York, Columbia University Press, p. 257-298.

SZAKOLCZAI, Attila & VARGA, László Á. (eds) (2003), *A vidék forradalma, 1956. [Revolution in the Countryside, 1956]*, Vol. I, Budapest, 1956-os Intézet and BFL.

SZAKOLCZAI, Attila (ed.) (2006), *A vidék forradalma, 1956. [Revolution in the Countryside, 1956]*, Vol. II, Budapest, 1956-os Intézet and BFL.

TŐKÉS, Rudolf L. (1996), *Hungary's Negotiated Revolution. Economic Reform, Social Change and Political Succession, 1957-1990*, Cambridge, Cambridge University Press.

VALUCH, Tibor (2001), *Magyarország társadalomtörténete a XX. század második felében [The Social History of Hungary in the Second Half of the Twentieth Century],* Budapest, Osiris.

VARGA, Zsuzsanna (2001), *Politika, paraszti érdekérvényesítés és szövetkezetek Magyarországon 1956-1967. [Politics, the Assertion of Agrarian Interests and Cooperatives in Hungary, 1956-1967]*, Budapest, Napvilág.

VARGA, Zsuzsanna (2004), 'Agrarian Development from 1945 to the Present Day', in János ESTÓK (ed.), *History of Hungarian Agriculture and Rural Life, 1848-2004*, Budapest, Magyar Mezőgazdasági Múzeum, p. 221-294.

VARGA, Zsuzsanna (2007), '"The 'Modernizing" Role of Agriculture in the Hungarian Economic Reforms', in Christoph BOYER (ed.), *Zur Physiognomie sozialistischer Wirtschaftsreformen. Die Sowjetunion, Polen, die Tschechoslowakei, Ungarn, die DDR und Jugoslawien im Vergleich*, Frankfurt/Main, Max-Planck-Institut für Europäische Rechtsgeschichte, p. 201-218.

VIOLA, Lynne (1996), *Peasant Rebels under Stalin: Collectivization and the Culture of Peasant Resistance*, New York, Oxford University Press.

WÄDEKIN, Karl-Eugen (1977), 'The Soviet Kolkhoz: Vehicle of Cooperative Farming or of Control and Transfer of Resources', in Peter DORNER (ed.), *Cooperative and Commune*, Madison, The University of Wisconsin Press, p. 95-116.

WÄDEKIN, Karl-Eugen (1982), *Agrarian Policies in Communist Europe*, The Hague/ London, Allanheld, Osmun-Nijhoff.

10. Comparing agro-structural change in the Alpine regions: developmental trends and driving forces, 1950-2000

Flavio V. Ruffini, Thomas Streifeneder, Christian Hoffmann & Kathrin Renner

I. Introduction

Over recent decades mountain agriculture in the Alps has been affected by profound structural change. Even though this change has meant human displacement and the abandonment of farms on a very large scale, distinct variations are evident in the different regions (Streifeneder & Ruffini, 2007; Streifeneder *et al.*, 2007). Many farms have been abandoned throughout the Alpine arc due to a complex configuration of economic and non-economic driving forces (Giuliani, 2003; Mann, 2003b; Tappeiner *et al.*, 2003). In nearly 90 per cent of the cases, the immediate reason for abandoning a farm has been the retirement of the farmer in a situation where potential successors and heirs had little interest in taking over the business. Behind these individual decisions, powerful national and regional structural forces, as well as forces specific to families and to business concerns (Mann, 2003a: 32), can be identified at work.

Changes in the nature and intensity of land use, as well as the abandonment of farms have had negative consequences for biodiversity, the character of the landscape, the protection of managed areas and the economic and social viability of upland rural communities (MacDonald *et al.*, 2000; Tappeiner *et al.*, 2003). Yet, in spite of these negative consequences, agriculture remains an important sector throughout the Alpine regions of Europe. Its multi-stranded contribution to the cultivation of natural resources, to the persistence of decentralised settlement and to the conservation of distinct cultural heritages is still immense and widely recognised by politicians on a national and European level[1].

Even though agriculture throughout Europe's mountain areas was characterized by a number of similar developments in the second half of the twentieth century, the different regions within the Alpine arc were still significantly different from one another at the beginning of the twenty-first century. In this chapter we will first

[1] See also EU directives 1257/1999/EEC, No. 1698/2005, Section III-220 of the EU Constitution, the protocol on Mountain Farming of the Alpine Convention and minutes from the first forum on mountain areas held in Brussels on 31 March 2009.

describe the pattern of the structural change in Alpine agriculture and then explore some causes (driving forces) of this change. In seeking relevant answers, the following questions will be addressed:

- What regional and spatial differences and similarities can be observed in the agro-structural change that has occurred?

- What are the causes and factors (driving forces) that have led to similar and to different developmental patterns in the Alpine regions?

- What can be considered typical of agro-structural change within mountain regions as compared to non-mountain regions within the Alpine countries?

II. Approach

The area under study is identical with the territory of the Alpine Convention as defined by Eurac[2] (Ruffini *et al.*, 2004) and as applied within the Agralp[3] project. It encompasses a total of 5,954 municipalities (LAU2[4]) in Germany (DE), France (FR), Italy (IT), Austria (AT), Switzerland (CH), Slovenia (SI), Liechtenstein (LI) and Monaco (MC)[5]. In all, this extensive territory covers 190,558 km^2 and was home to 14,1 million inhabitants in 2007 (Ruffini & Streifeneder, 2007). Austria and Italy alone comprise more than half of the area and of the inhabitants of the Convention territory.

The period under study dates from 1950, when the first post-war censuses were conducted, to the surveys completed in 2000[6]. Since the available data do not permit a precise analysis of the trans-Alpine agro-structural changes over the whole 1950-2000 period, the 1950-1980 and the 1980-2000 periods will be analysed separately.

It is important to keep in mind that all farms regardless of size were taken into account in the first period, while only units that comprised at least 1 hectare of utilised agricultural area were considered to be farms (EU standard) in the second period. Consequently, the agro-structural changes identified for the 1950-1980 period

---

[2] European Academy, Bozen.

[3] Agralp was conceived to investigate the development of agricultural structures in the Alps. For further information about the project, see www.eurac.edu/agralp.

[4] Eurostat: Local Administration Units (LAU2 = former NUTS5 level). NUTS-Nomenclature des unités territoriales statistiques. Official territorial classification (Online: http://ec.europa.eu/comm/eurostat/ramon/ nuts/splash_regions.html).

[5] No agricultural statistical data exist for Monaco.

[6] For the sources of the statistical data we have relied on, see the sources accompanying Tables 10.1 and 10.3.

were somewhat different from those of the later period. The patterns uncovered for the first period are nonetheless useful since they provide a historical context for our analysis of the 1980-2000 period.

III. Structural changes between 1950 and 1980

III.1. Farm abandonment

Table 10.1. Farm abandonment rates in the Alpine Regions, 1950-1980

	Change in number of farms (absolute values)					*Change in number of farms (%)*			
	1950	*1955*	*1960*	*1970*	*1980*	*50/60*	*60/70*	*70/80*	*average p.a.*
AT	162,989	---	148,252	138,220	119,837	-9.0	-6.8	-13.3	-1.02
CH	---	85,989	---	61,589	41,363	---	-28.4**	-32.8	-2.88
DE	52,648	---	49,539	37,302	32,653	-5.9	-24.7	-12.5	-1.68
FR	---	114,250	---	66,868	52,647	---	-41.5**	-21.3	-3.05
IT*	---	---	---	371,220	309,146	---	---	-16.7	-1.81
LI	---	1,366	898	783	494	-34.3**	-12.8	-36.9	-3.99
Alps	---	---	---	675,982	555,110	---	---	-17.9	-1.95

* IT: First census 1960 not available; ** values refer to 1955. Sources: AVW (1955, 1960, 1969, 1980); AGRESTE (1998); BFS (1955, 1960, 1969, 1980); INSEE (1955); ISTAT (1961, 1970, 1982); SI-STAT (1960, 1969, 1971, 1981a, 1981b); STATISTIK AUSTRIA (1951, 1960, 1970, 1980); STATISTIK BAYERN (1949, 1950, 1960, 1971, 1979); EUROSTAT (2000: 23).

An analysis of the farm abandonment process in the period between 1950 and 1980 (Table 10.1) reveals the following patterns:

- The average annual farm abandonment rate varies between 1 percent in Austria and 4 percent in Liechtenstein.

- France shows a particularly dramatic change in its annual farm abandonment rate between 1955 and 1970 – a finding that may confirm more the warnings of the statistical office regarding data validity than any material real development. But it is well known, all the same, that the French Alps were the first Alpine Region to be heavily affected by rural exodus in the post-war period (Blanchard, 1956; Veron, 1996).

- Rather surprisingly, the abandonment rates of small farms (less than 5 hectares) in Switzerland are even higher than in France (Figure 10.2).

- The changes occurring in the Italian Alps before 1980 are small in comparison with the abandonment sustained in the years after 1980 (Barberis, 1996, Table 10.3). It would appear that the distinct pattern assumed by agro-structural change in Italy is heavily influenced by the late industrialisation experienced in these areas.

- The greatest changes in the German Alps were recorded in the 1960s and 1970s, and were mainly caused by the industrialisation of the pre-Alpine centres.

III.2. Changes in the utilised agricultural area

Compared to the abandonment rates of farms the changes in the utilised agricultural area between 1950 and 1980 are rather low (see Table 10.2). The main observable pattern here indicates that while many farms were given up in this period, others increased their cultivated area. Less than 0.4 per cent of the cultivated area was given up per year in the whole alpine region. In Switzerland the cultivated area remained stable whereas in France it decreased by more than 1 per cent per year.

Table 10.2. Changes in utilised agricultural area (UAA) in the Alpine regions, 1950-1980

	Change in UAA (hectares)					*Change in UAA (%)*			
	1950	*1955*	*1960*	*1970*	*1980*	*50/60*	*60/70*	*70/80*	*Average 50**/80 p.a.*
AT	2,168,872	-	2,041,697	1,875,402	1,835,369	-5.9	-8.1	-2.1	-0,55
CH	-	308,598	-	303,753	309,964	-	-1.6**	+2.0	+0,02
DE	536,837	-	497,298	484,296	471,501	-7.4	-2.6	-2.6	-0,43
FR	-	1,205,251	-	850,318	849,389	-	-29.5**	-0.1	-1,39
IT*	-	-	-	1,716,385	1,576,365	-	-	-8.2	-0,85
LI		4,334	4,110	3,786	3,634	-5.2**	-7.9	-4.0	-0,70
Alps	-	-	-	5,233,940	5,046,223	-	-	-3.6	-0,36

* IT: First census 1960 not available; ** CH, FR, LI: 1955, IT: 1970.

Sources. see Table 10.1.

III.3. Developmental trends

Since there are no consistent time series data covering changes in the number of farms and the cultivated area in all countries during the 1950-1980 period, agro-structural developmental patterns will have to be inferred from the fragmented available data. Given this proviso, the following trends can be identified:

a) Austria experienced a relatively moderate structural change during the whole 1950–2000 period. This suggests that the specific measures and programmes for mountain farms and mountain areas introduced relatively early did have an immediate effect. Young farmers interested in taking over the family farm were encouraged by state support to do so, even if the individual's economic prospects might have been more promising outside agriculture.

b) Liechtenstein and France stand out as the Alpine regions registering substantial structural change on both indicators, *i.e.* high farm abandonment rates and a significant decrease in the utilised agricultural area.

c) Switzerland displays high farm abandonment rates, but practically no change in utilised agricultural area.

IV. Structural changes between 1980 and 2000

IV.1. Farm abandonment

In 2000 a total of 287,000 farms (as compared to 447,500 in 1980), two-thirds of which were located in Austria and Italy (Table 10.3 and Figure 10.1), were functioning in the territory of the Alpine Convention area. In the two decades between 1980 and 2000 around 160,000 farms (35.8 per cent) were given up or abandoned.

Table 10.3. Farm abandonment rates inside and outside the Alpine regions, 1980-2000

Country	*Number of farms alpine region*				*Change in number of farms 1980-2000*			
	1980		*2000*		*Alpine region*		*Whole country*	
	abs.	*%*	*abs.*	*%*	*abs.*	*%*	*abs.*	*%*
AT	109,554	24.5	96,205	33.5	-13,349	-12.2	-100,577	-31.6
CH	37,256	8.3	24,546	8.5	-12,710	-34.1	-31,786	-31.1
DE	29,041	6.5	22,017	7.7	-7,024	-24.2	-419,000*	-50.1*
FR	52,647	11.8	28,128	9.8	24,519	-46.6	-598,865	-47.4
IT	165,607	37.0	93,046	32.4	-72,561	-43.8	-676,102	-20.7
LI	358	0.1	191	0.1	-167	-46.7	-	-
SI	53,089	11.9	23,149	8.1	-29,940	-56.4	-105,754	-51.4
Alps	447,552	100.0	287,282	100.0	-160,270	-35.8	-	-

* 1980-2000 only the former federal states in West DE. Sources: Avw (1980, 1990, 2000); AGRESTE (1998, 2001); BFS (1980, 1991, 2001); ISTAT (1982, 1990, 2003); SI-STAT (1981 a, b, 1990, 2000); STATISTIK AUSTRIA (1980, 1990, 1999); STATISTIK BAYERN (1979, 1991, 1999).

What is most striking in this extraordinary development within a very short period is the significant difference between the northern (with the notable exception of

Liechtenstein) and the southern Alpine regions, a difference that was less clear in the 1950-1980 period (Table 10.1). The French, Italian, Liechtenstein and Slovenian segments of the Alps are where we find a very high percentage of farms being given up in the 1980-2000 period. In Italy the abandonment rate of 43.8 per cent (72,600 farms) for the Alpine region was considerably higher than that for the farms in the country as a whole (20.7 per cent). The very high abandonment rate in the 1980–2000 period also contrasts with the rather moderate changes occurring in the 1970–1980 period. France and Liechtenstein, on the other hand, recorded similarly high abandonment rates that were more or less in line with the pattern established in the 1950-1980 period (Table 10.1).

Looking at the period from 1970–2000 reveals an accelerating increase of farm abandonment rates. While almost every third farm was abandoned in the 1990s, this happened 'only' to every sixth farm during the 1980s and to every tenth farm in the 1970s. As a general pattern, however, this does not hold true for the Alpine region as a whole. Farm abandonment in Austria was roughly the same in the 1990s as it was in the 1970s. The rate in Switzerland in the 1980-1990s even fell below that of the 1970s. But Austria and Switzerland were the 'exceptions'; in all other countries the farm abandonment rate increased extra-proportionally after 1980 (see Table 10.1 and 10.3 as well as Figure 10.2).

IV.2. Development in disadvantaged and non-disadvantaged mountain areas

'Non-disadvantaged areas' are officially designated spatial blocs that include the areas classified under EU Regulation Nr. 1257/1999 as non-disadvantaged in the Alpine arc. Such areas include, for example, the broad valley bottoms in the autonomous provinces of Bolzano-South, Tyrol and Trento that are classified as only partially deprived areas but officially designated as disadvantaged communities. Accordingly, 31.9 per cent of all communities in the Alps are not classified as disadvantaged.

Within the whole Alpine region, the abandonment rate in non-disadvantaged areas is higher than in disadvantaged areas (Figure 10.1). The possible causes of this in Austria and Germany are most likely the diverse official support schemes for mountain areas as well as the relatively good opportunities for part-time working. One particularly striking contrast can be seen in the border areas of Italy and Slovenia. The contrasting pattern here can be explained by the fact that while in Italy potential farm successors prefer to move to the pre-Alpine towns, the still widespread practice of subsistence farming on the Slovenian side of the border leads to a greater persistence of the agricultural population.

Figure 10.1. A comparison of farm abandonment rates in less favoured mountain areas and non-disadvantaged areas in the Alps between 1980 and 2000

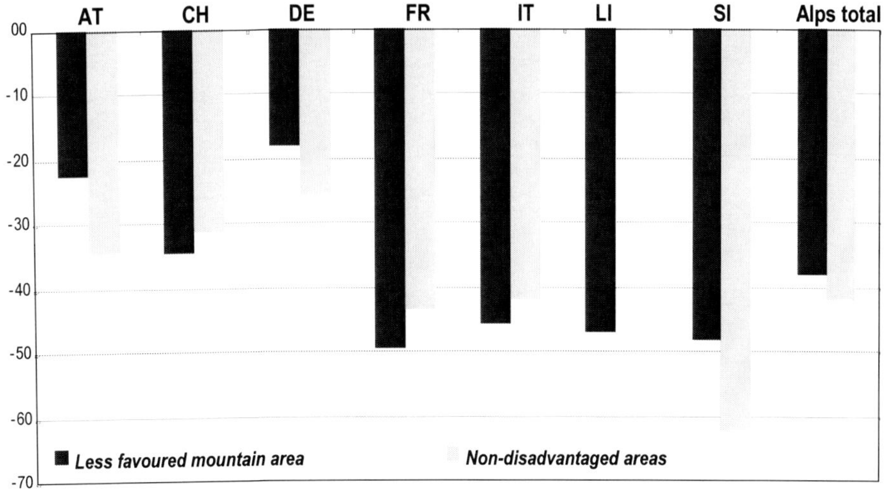

Sources. see Table 10.3.

IV.3. Changes in the utilised agricultural area and in farm size

Changes in the utilized agricultural area in the 1980-1990s were quite unevenly distributed across farms of different sizes. Between 1980 and 2000 the utilized agricultural area of farms more than 1 ha (alpine pastures included) decreased by 7.6 per cent or 0.39 per cent annually (432,925 ha, Table 10.4). The decrease is even higher (9.2 per cent) if the summer grazing areas are not considered. The highest decrease was recorded in Italy (16.5 per cent) and Slovenia (34.7 per cent), where the utilized agricultural area was mainly reduced on small farms of less than 5 hectares. The 'winners' in terms of structural change were larger farms of more than 20 ha (Figure 10.2); these tended to absorb land from smaller farms when they were given up. Even so the agricultural sector within the Alps still operated on a relatively small scale with an average area of 13 ha per farm in 2000 (without alpine pastures), as compared to an average farm size of 18.7 ha in the EU as a whole (Lebensministerium, 2004).

Figure 10.2. Farm size changes in Alpine regions, 1980-2000*

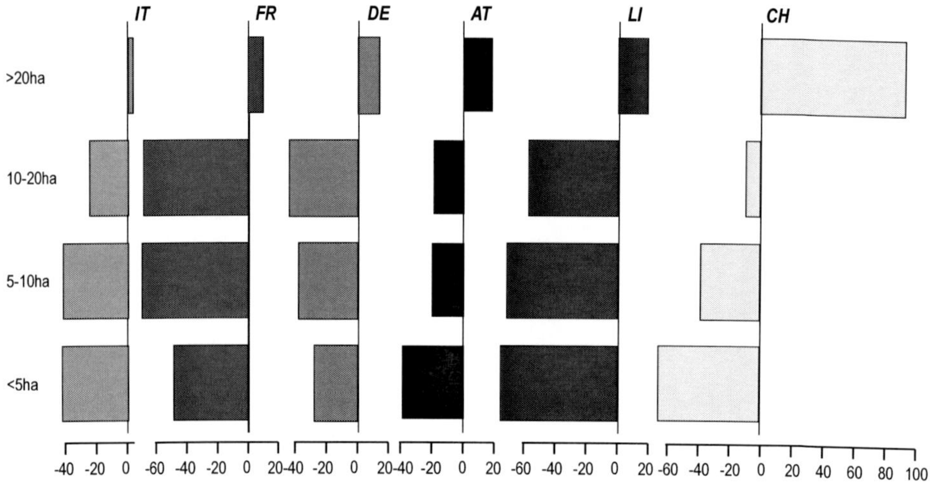

* The Swiss and Liechtenstein data omit summer grazing areas; the German data omit cooperative pastures; and no Slovenian data are available.

Sources. see Table 10.3.

Table 10.4. Utilised agricultural area (UAA), 1980-2000

State	*UAA in alpine regions*				*UAA in the alpine areas and on the national level*					
	1980		*2000*		*alpine regions 1980-2000*			*national level 1980-2000*		
	hectares	*%*	*hectares*	*%*	*hectares*	*%*	*annually %*	*hectares*	*%*	*annually %*
Austria	1,835,369	32.1	1,734,369	32.8	-101,000	-5.5	-0.28	-120,082	-3.4	-0.17
Switzerland*	805,360	14.1	791,938	15.0	-13,422	-1.7	-0.08	-563	-0.1	0.00
Germany**	511,996	9.0	505,433	9.6	-6,563	-1.3	-0.06	-751,000	-6.2	-0.32
France	849,389	14.9	858,650	16.2	9,261	1.1	0.05	-1,890,000	-6.0	-0.31
Italy***	1,502,027	26.3	1,254,044	23.7	-247,983	-16.5	-0.90	-2,789,000	-17.6	-0.96
Liechtenstein	3,634	0.1	3,593	0.1	-41	-1.1	-0.06	-41	-1.1	-0.06
Slovenia	210,751	3.7	137,566	2.6	-73,185	-34.7	-2.11	-15,925	-3.2	-0.16
Alps	5,718,526	100.0	5,285,593	100.0	-432,933	-7.6	-0.39	-5,566,611	-8.6	-0.45

* Inclusive of alpine pastures; ** inclusive of cooperative alpine pastures; the national figures for UAA refer to the old federal states; *** UAA only considers farms > 1 ha.

Source. see Table 3.

France, on the other hand, with an average utilized agricultural area of 30 ha per farm even recorded a slight increase of 1.1 per cent in utilized agricultural area (although the overall rate in the country decreased by 6.0 per cent). The reason for this extraordinary result lies in the inclusion of cooperatively owned Alpine pastures which were not recorded in previous surveys (Figure 10.3).

The demarcation of new land, due to subsidies for extensive animal husbandry, also contributed to this increase. In Switzerland, on the other hand, the utilized agricultural area hardly changed at all (1.7 per cent in the mountain area compared to the national average of 0.1 per cent, Table 10.4). One reason for this was that the land of farms given up was mainly taken over by other farms. The same phenomenon can be observed for the 1950-1980 period (Table 10.2).

IV.4. Types of agro-structural change and their driving forces

Figure 10.3. Comparison of the relative change in farm numbers and in utilized agricultural area (UAA), 1980-2000

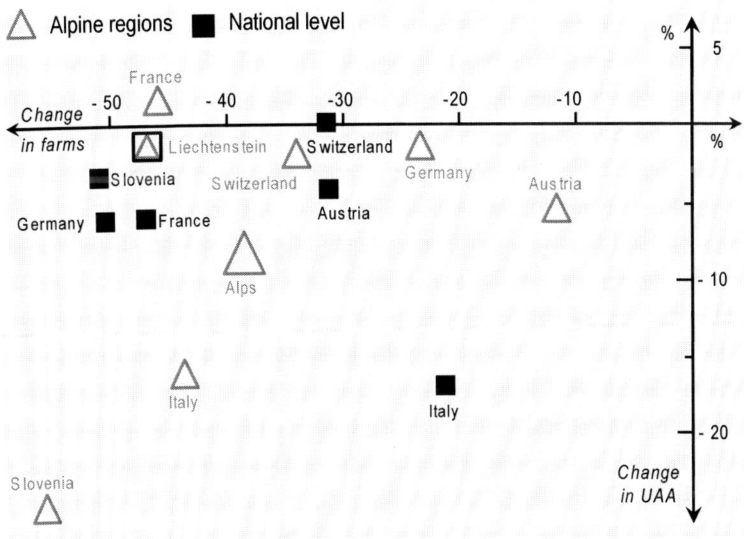

Sources. see Tables 10.1; 10.2; 10.3 and 10.4.

Combining the indicators 'farm change' and 'change of utilized agricultural area' we can distinguish converging and diverging patterns at the national level. In all, four key trends can be identified (Figure 10.3):

a) The Alpine regions with the most moderate structural changes are found in Germany, Austria and Switzerland, where both the average decrease of utilized agricultural area and the farm abandonment rates were relatively low. Compared with the previous period, Germany and Switzerland in fact even 'improved' in the most recent period. The German region, it seems, underwent an earlier agro-structural change; and, during recent decades, special agro-environmental measures and regulations for the preservation of pastures have affected positively farmers' decisions to stay on the land. In Switzerland, throughout the whole 1950-2000 period, the remaining farms expanded to absorb the cultivated area that was becoming available during the process of farm abandonment.

b) Despite a still high decline in the number of farms, Liechtenstein registered only an average level of structural change. Throughout the two periods this highly industrialised country has displayed the same pattern, with high farm abandonment rates and a relatively moderate decrease in utilized agricultural area. The various measures and support schemes, which were introduced rather late on, have only contributed partially to preserving the utilized agricultural area.

c) Italy and Slovenia have displayed considerable structural changes with high farm abandonment rates and a loss of the utilized agricultural area that was greater than the average in the Alpine region. The Italian pattern presents a sharp contrast between the two periods, highlighting the manner agro-structural change was delayed until the 1980s.

d) The French Alpine structural change is somewhat exceptional: the utilized agricultural area increased in spite of high farm abandonment rates in the 1950–1980 period.

V. Driving forces

Agro-structural change can be viewed as the product of a complex configuration of economic, social (including familial), cultural and ecological driving forces (Zanetti, 1999: 46; Baur et al., 1999: 234). Within the economic sphere the balance between full-time and part-time farming has changed significantly over recent decades. In 1980, full-time farming still outweighed part-time farming across the entire Alpine region. But this had changed by 2000 when the two were equally balanced in the Alpine arc as compared to the entire EU-15 where part-time farming was practiced by less than 25 per cent of cases (European Commission, 2003). What we can also observe is that the agrarian sector itself affects all parts of the wider socioeconomic system of rural areas. Individual farmers, or farmers in a certain region, may therefore react rather differently to a change of conditions compared to farmers in general (Mann, 2003b). Since several studies suggest that agro-structural changes may significantly be hindered or accelerated by state policies (Mann, 2003a; Weiss, 2006; Baur, 1999 and Dax, 2008), it makes sense to analyse national and EU policy measures directed towards the mountain farms of the Alpine territories.

V.1. EU policy for the development of rural areas

Many regional or spatial measures are brought together in the EU's current rural development policy. Of particular relevance for our purposes are the following:

- compensatory allowances for farmers in less-favoured areas;
- initiatives designed to regulate the development of rural areas, especially agro-environmental measures;
- measures within the framework of cohesion and regional structural policy to support the economic revitalisation and diversification of rural areas;
- the Common Agricultural Policy (CAP).

Since the mid-1990s, EU rural development and agriculture policies have been closely linked with rural development becoming the so-called 'second pillar' of the CAP (Grabski-Kieron & Krajewski, 2007). This alignment represents an important step for both the rural areas and the agricultural sector. While agricultural development was long regarded as the 'engine' of rural development, a new holistic approach towards rural areas, including mountain regions, has now become accepted and established.

Of major significance to the development of agriculture in mountain areas are the EU's compensatory allowances as well as measures that acknowledge and seek to counter less favourable production conditions in Alpine countries. Certain mountain areas in the Alps have also been classified as Objective 2 and Objective 5b areas and so have become eligible for specific EU subsidies. The EU's embrace of territorial measures has followed in the footsteps of policies already introduced much earlier by nation-states. In all the Alpine countries subsidies came to be regarded as the most important measure to maintain mountain farming and to safeguard agricultural incomes (Hovorka, 2006)[7]. The introduction of agro-environmental measures in the EU in 1992[8], by allowing for agriculture to be supported for environmental reasons, represented a milestone in the development of a specific type of policy intervention.

---

7 In the Austrian Alps these subsidies represent an average of 62.3 % of farming incomes (Hovorka, 2006).

8 Regulation (EEC) No. 2078/92.

These environmental subsidies have come to be accepted as supporting a wide range of activities within the agricultural sector.

V.2. Agro-political instruments for mountain regions in the Alpine countries

Among the national policy initiatives worth highlighting are the German Cultural Landscape Programme (KULAP, since 1993), the Bavarian Nature Protection Programme (VNP), the Austrian Programme for the Development of an Environmentally Compatible, Extensive and Natural Habitat-protecting Agriculture (ÖPUL, since 1995) and the Swiss ecologically motivated direct payments as well as their predecessors dating to the 1950s. All these programmes are of key importance for underpinning the incomes of Alpine mountain farmers (Wessely & Güthler, 2004; Broggi *et al.*, 1997). There can be little doubt but that participation in them has reduced the likelihood of farm abandonment (Weiss, 2006).

In the German Alps no other economic sector received higher subsidies than agriculture (Wessely & Güthler, 2004: 29). Financial aid was provided by the EU, the national federal government, the Bavarian state as well as by rural districts and municipalities. The legal basis for supplying financial aid to Bavaria's agricultural sector between 1974 and 2006 has been the act for supporting agriculture (*Landwirtschaftsförderungsgesetz*). The most important instruments have been the EU's livestock subsidies along with various direct and indirect benefits that have their origins in the commitment to support mountain regions. Of some importance as well has been the federal rural development programme (StMLU, 2003) that has favoured the maintenance of mountain regions (B IV 2.2).

State support of mountain farming has a long tradition in Austria where an act for supporting mountain farmers (*Bergbauernhilfsfondsgesetz*) was introduced as early as 1937 (Penz, 1996). And the mountain farming cadastre (*Berghöfekataster*) of 1953 identified a number of obstacles (*Erschwernispunkte*) to be taken into account when deciding the sum of the financial aid to be paid to the individual farmer under the grant for mountain farmers (*Verharrungsrente*). Remunerating farmers for maintaining the Alpine heritage has contributed to the survival of mountain farming in Austria. The *Berghöfekataster* was revised in 2001 when the national classification of farms was adapted to a regional system in accordance with European guidelines.

In Switzerland article 2 of the agricultural act (*Landwirtschaftsgesetz*) of 1951 required that all measures should take into consideration the difficult production and living conditions of the mountain areas (Baumann & Moser, 1999: 340-351). The official recognition that most of the mountain farms would have been unable to survive without specific subsidies has functioned as a counterbalance against the market forces accelerating the agro-structural change in these disadvantaged areas (Flury, 2005; Lauber, 2006).

Instruments for supporting mountain farming have also been introduced in France. A programme for modernising mountain farming was first devised in 1947 and again in 1952 in the guise of an economic development plan. Direct payments and various other subsidies have continuously contributed to supporting agricultural incomes in the mountain area. In the context of the significant structural adjustments taking place in the French Alpine regions (Blanchard, 1956), grants for investment and for the provision of infrastructure services, as well as premiums for newly established farmers have played a major role (Véron, 1996). The *prime à l'herbe*, a premium for grassland, is of particular importance for the income of permanent grassland farms in mountain regions (Berthelot *et al.*, 1999). A combination of all these measures has helped to slow down emigration and to limit the abandonment of farms in the French Alps.

Italian mountain regions (and farming areas in particular) have also received financial aid. But some of the relevant measures here were relatively late in appearing, and then only as a response on the part of the authorities to crisis conditions. Furthermore, such measures comprise only a small fraction of the total agricultural budget and vary significantly in their size and in their means of distribution among the regions (Streifeneder & Ruffini, 2007).

Since the 1970s in Slovenia different subsidies have been available to help promote agriculture and to develop the economies of the mountain regions more generally (Gosar, 1991). Further significant agro-political instruments were introduced in the 1980s and in the first half of the 1990s. However, one assessment suggests that these agrarian measures for mountain farming were '... mostly not well coordinated and were rarely integrated strategically with non-agrarian considerations' (Gosar & Cundar, 1996: 208). There are additional subsidies and compensation payments available at the municipal level. Ever since the political system changed in 1993, an ambitious programme of specific agro-structural measures, adapted according to the European guidelines and to CAP objectives, has commenced (Erjavec *et al.*, 2001; Erjavec *et al.*, 2003).

Similarly to Switzerland direct payments for farmers who provided ecological services or specific modes of production, were introduced in Liechtenstein in the 1990s (LI, 2006).

In conclusion we can say that two apparently inter-related tendencies are observable in Germany, Austria and Switzerland: firstly, direct payments, agro-environment measures and other policies from the second pillar of the CAP have been introduced or continued; and, secondly, the rates of farm abandonment in the 1980/90s have risen less dramatically than in areas where such policies were largely absent (Table 10.3 and Figure 10.3). The pattern in Liechtenstein, where substantially more medium and large farms were given up than in Switzerland (where similar policy measures were introduced), can be read either as the exception that proves the rule – or as a case where the absolute numbers are too small to amount to anything significant when expressed in relative terms.

V.3. Regional policies for mountain areas

Among the numerous forces that influence regional policies are cultural norms and values as well as the character of the political-administrative system and the type and duration of public policy initiatives. But only rarely, as in the recent Italian mountain legislation, have integrated policy programmes been developed for mountain areas. In most cases measures designed specifically for mountain areas were integrated into other policies (Dax & Hovorka, 2005). In Austria and Switzerland, the policies for rural Alpine areas represent an 'aid for an appropriate development of sub-areas' and on that basis they can be equated with a policy for mountain areas (Dax, 2000; SAB, 2002).

Approaching the subject more analytically, the mountain policies of individual states not entirely defined as mountainous (like Liechtenstein) can be grouped under the following three broad headings (Nordregio, 2004):

- States that lack specific mountain policies; in these cases (as in Slovenia), the welfare of mountain areas becomes part of general (economic) policy;
- States with mountain policies that seek to pursue a multi-sectoral developmental approach (such as Germany and Austria);
- States with a mountain policy focused on the promotion of a holistic development in areas not covered by specific mountain laws (France, Italy and Switzerland). Such an approach consists of a legal definition of mountain areas, the introduction of particular laws for these areas and the provision of institutions and programmes that focus on training and research.

In Austria the upkeep of mountain farming is regarded as essential not only for maintaining the cultural landscape as a tourist attraction but also for safeguarding economic and social viability in the mountains. To this end a special national regional policy for mountain farmers (*Bergbauernsonderprogramm* 1972-1990) was inaugurated; and in the 1990s this was devolved to an independent regional level in order to support project initiatives (Dax, 2000). While legal regulations in the field of spatial planning and regional development exist only at the sub-national level, municipalities possess more room for manoeuvre. The Austrian Conference on Spatial Planning (ÖROK) has a co-ordinating function, but it does not possess any legally binding powers. Generally speaking, cultural conceptions (*kulturelle Leitbilder*) appear to play a key role within Austrian regional policy (Borsdorf, 2007).

Mountain areas represent a focal point within Swiss federal and regional policies as well. The origin of Switzerland's current regional policy is based on a national law for investment support in mountain areas (Bundesgesetz über Investitionshilfe für Berggebiete) that dates from 1974 (Messerli, 1991). Under this law the policy of providing financial aid was replaced by a policy of integrative economic support. The key elements of the Swiss regional policy since the 1970s have been the existence of regional administrative structures, an integrative developmental approach, the encouragement of decentralised settlements, ecological protection and maintaining the cultural landscape as well as supporting the combination of agriculture and tourism (Messerli, 1991). But despite many successes, the Swiss Working Group for Mountain Areas (SAB, 2002) believes 'that the actual instruments for regional policies are not flexible enough to be able to react in an adequate manner to the new challenges'. The SAB therefore has called for a reconfiguration and a revitalization of Swiss regional policy.

Germany is characterized by a hierarchical regional planning system in which the top ranking bodies are the state regional planning departments and the Ministerial Conference on regional planning. The individual 'states' (Bundesländer) are responsible for implementing the national state's policy that aims to reduce regional disparities and to create more equal living standards. It is at the county and municipal levels where the national objectives are given specificity and effect in the Alpine countryside. At the heart of regional planning is an Alpenplan that has provided the basis of Bavaria's regional development programme since 1972 (Recreation Landscape Alps).

The mountain areas of France were given legal recognition during the 1960s, most notably in 1961 with the official demarcation of the mountain area and again in 1967 with the mountain area commission and report as well as with legislation for rural development (Véron, 1996). Upon the enactment of the acts of decentralisation of 1982 and 1983, the state pulled back from regional planning in favour of granting local authorities greater room to manoeuvre. Since 1985 previous measures (such as compensatory payments) have been adjusted (Bazin, 1999).

Italy's mountain territory was deemed in the 1980s to be 'a structurally weak and nationally insignificant peripheral region' (Martinengo, 1991: 216). In view of the high rates of farm abandonment and of population decrease in the 1990s, this may still be the case today despite a recognition of the particular significance of mountain regions contained in Article 44 of the Constitution and the earlier mountain act of 1952. The creation of so-called mountain communities in 1971 may have marked a politically progressive step and brought the Alps closer to the centre of political power, but the specific content of mountain policy today differs from region to region so much that one observer has concluded that the situation 'has become very complex' (Martinengo, 1991). Although the number of these mountain collectives has changed significantly, this aspect of the situation has been brought into line with the new Finance Act of 2007 and in particular by the earlier Mountain Act of 1994 (Barberis, 1996). Public contribution to the scope of the 'National Fund for Mountain Areas' has been insignificant, and cannot be regarded as a specific policy or strategy in any real sense.

After 1945 the Slovenian Alps were affected by a Leninist-Stalinist policy of industrialisation that pushed urbanisation and influenced migration and settlement patterns (Gosar, 1991). In the 1970s and 1980s, the centrally derived concept of an 'equal, decentralised polycentric development policy/industrial development' (Gosar & Cundar, 1996) was extended to the Alpine areas on the periphery. Besides, tourism was not considered as important. Only in 1969 and 1971 were the mountain areas legally recognised as 'less developed areas' eligible for infrastructure schemes, investment facilitation, tax exemption and so on. But there was no specific state policy for mountain areas and none is currently being devised. The measures for mountain areas that do exist amount to a traditional approach that focuses on direct support for agriculture and the development of physical infrastructure, human capital and other investments as well as the improvement of industrial facilities in the valleys.

V.4. Non-economic driving factors

The forces driving agro-structural change can transcend economics and politics. In spite of low incomes in agriculture, some people remain in the sector for social and cultural reasons (Groier, 2004). The situation on full-time farms in Austria demonstrates, for instance, how traditional family farm values are deeply rooted among potential farm successors (Vogel, 2005). Statements such as 'the farm should remain in the family' and 'due to the family tradition I am a farmer' reveal a deeply embedded commitment to family farming values. Birkenhauer (2001) points to the different mentalities that emerged in individual territories over the centuries and to the different strategies employed to confront feelings of desolation in the community when land is abandoned.

Generally speaking, the commitment of the young to take over a farm can be deemed to be much more strongly embedded in the northern than in the southern Alpine regions. Mountain farmers in many (but not all) German-speaking areas have tended to identify themselves more strongly with their role and they have enjoyed a higher social standing than their colleagues in southern areas, all of which is reflected in the relatively low abandonment rates compared to the national and regional averages (Wessely & Güthler, 2004). The same holds true for the Austrian and Swiss Alpine areas[9]. Ruth Rossier (2005), who has explored the different values systems of agricultural trainees from mountain and valley regions, finds that the former are much more attached to their farms than the latter. Stefan Vogel (2005) also detects a higher rate of committed successors in the Austrian mountain areas. All these studies, however, clearly show that the willingness to take over the farm is not separated from perceptions of the long-term prospects of the farm's survival and its economic potential to generate an adequate income. Hence, it is not surprising that full-time farmers were generally more likely to have identified a successor than part-time farmers. In Italy by contrast, where farmers increasingly diminish in number, farms were mostly abandoned after the farmer retired. This Italian pattern may also be the result of the association of very small farms with the system of partible inheritance that persisted until 1994.

VI. Conclusion

That farms in mountain regions have been abandoned since the late nineteenth century is a well-known development. But the fact that more than one in every three Alpine farm units was given up in the last two decades of the twentieth century indicates that a critical new level of abandonment has now been reached. What is equally important to note is that the alpine regions have displayed different patterns of structural development and change in farming due to a wide range of natural, economic, social, cultural and political conditions.

We can conclude from the statistical evidence presented above that agricultural development is very much influenced on a regional level by the presence of economic, policy and socio-cultural driving forces. But how exactly these policies influence the decisions of individuals to continue with or to give up farming remains a largely unanswered question. What can be said from our findings is that regions with sound

9 This is so despite different agricultural structures. While part-time farming continues to dominate in Austria (national average: 60 per cent; Lebensministerium, 2004), full-time farming predominates in Switzerland (approximately 60 per cent; national average 69.8 per cent; Bfs, 2004). Some 74 per cent of all Swiss farm businesses, however, operate an ancillary or para-agricultural activity, especially in mountain areas (66 per cent of farms; Bfs, 2007).

economies (high salaries and income, working opportunities outside agriculture), favourable demography among farmers, relatively high subsidies and the persistence of supportive socio-cultural values tend to register lower rates of farm abandonment. Off-farm job opportunities, especially those related to tourism, generally make a substantial contribution to a lowering of the farm abandonment rate. That said, it is also clear that in many cases exogenous factors are less important for an individual who has to decide whether or not to continue working the farm he or she inherits than endogenous factors within the family and the community. The development of Alpine agriculture in the foreseeable future, if we are to go on the experience of the recent past, will most likely depend more on the interplay of a multitude of factors than on a single one.

Bibliography

AGRESTE (2001), *Recensement agricole 2000. La fiche comparative Rhône-Alpes et Provence-Alpes-Côte d'Azur,* CD-Rom, Paris.

AGRESTE (1998), *Recensements généraux de l'agriculture 1970 – 1979 – 1988 – Fiches comparatives,* CD-Rom, Paris.

AVW (Amt für Volkswirtschaft Liechtenstein) (1955/1960/1969/1980/1990/2000), *Liechtensteinische Landwirtschaftszählung 1955,* 1960, 1969, 1980, 1990, 2000, Vaduz.

BARBERIS, Corrado (1996), 'Die Landwirtschaft im italienischen Alpenraum', in EUROPÄISCHE AKADEMIE BOZEN (ed.), *Landwirtschaft im Alpenraum – unverzichtbar, aber zukunftslos? Eine alpenweite Bilanz der aktuellen Probleme und der möglichen Lösungen*, Berlin, Blackwell Verlag.

BAUMANN, Werner & MOSER, Peter (1999), *Bauern im Industriestaat. Agrarpolitische Konzeptionen und bäuerliche Bewegungen in der Schweiz 1918-1968*, Zürich, Orell Füssli.

BAUR, Priska (1999), *Agrarstrukturwandel in der Schweiz. Dissertation ETH,* Nr. 13240, Zürich.

BAUR, Priska, PEZZATTI, Mauro, RIEDER, Peter & SCHLUEP, Isabelle (1999), *Langfristige Entwicklung der Agrarstrukturen in Südtirol, Bozen*, Europäische Akademie.

BAYERISCHES STAATSMINISTERIUM FÜR LANDWIRTSCHAFT UND FORSTEN (StMLU) (2003), *Bayerischer Agrarbericht 2003. München,* Mayer & Söhne.

BAZIN, Gilles (1999), *L'Évaluation de la politique de la montagne,* Vol. 1, Rapport de l'instance d'évaluation, Paris, La Documentation francaise.

BERTHELOT, Philippe, CHATELLIER, Vincent & COLSON, François (1999), 'L'Impact des mesures agri-environnementales sur le revenu des exploitations agricoles francaises', *Économie Rurale,* 249/Janvier-Février, 1999.

Bfs (Bundesamt für Statistik) (1955/1960/1969/1980/1991/2001), *Landwirtschaftliche Betriebszählung 1955, 1960, 1969, 1980, 1991, 2001 (und Betriebsstrukturerhebung)*, Bern.

Birkenhauer, Josef (2001), 'Alpen 2000: Eine realistische Bestandsaufnahme', *Mitteilungen der Geographischen Gesellschaft in München*, Bd. 85, p. 19-43.

Blanchard, Raoul (1956), *Les Alpes Occidentales*, Tome 7: *Essai d'une synthèse*, Arthaud, Grenoble / Paris.

Borsdorf, Falk (2007), 'Kulturelle Einflussfaktoren auf Raumordnung und Regionalpolitik – Österreichs Sonderstellung', in Axel Borsdorf & Sigrun Lange (eds), *Politische, kulturelle und wissenschaftliche Perspektiven der nachhaltigen Raumentwicklung in den Alpen. Erste Ergebnisse des an der LFU Innsbruck koordinierten EU-Projektes DIAMONT*, Sigrun, Innsbruck, Universität Innsbruck, p. 53-88.

Broggi, Mario F., Kusstatscher, Kurt & Sutter, Roman (1997), *Ökologisch motivierte Direktzahlungen in der Berglandwirtschaft des Alpenbogens*, Berlin, Blackwell.

Bundesamt für Statistik (Bfs) (2004), *Einblicke in die schweizerische Landwirtschaft. Ausgabe 2004*, Neuchâtel, BFS.

Bundesamt für Statistik (Bfs) (2007), 'Nebentätigkeiten von bäuerlichen Familien in der Schweiz 2005', *BFS Aktuell*, Neuchâtel, BFS.

Dax, Thomas (2000), 'Berggebietspolitik in Europa – Start einer umfassenden Bewertung', *Zolltexte Nr. 36*, Juni 2000, p. 36-39.

Dax, Thomas & Hovorka, Gerhard (2005), 'Trends der Regionalentwicklung der Berggebiete in Europa', *Ländlicher Raum*.

Erjavec, Emil, Kavici, Stane, Mergos, Georges & Stoforos, Chrysostomos (2001), 'Agricultural Policy Options for Slovenia in the Prospect of EU Accession', *Eastern European Economies* 39/1, p. 36-53.

Erjavec, Emil, Rednak, Miroslav, Volk, Tina & Turk, Jernej (2003), 'The Transition from "Socialist" Agriculture to the Common Agricultural Policy: the Case of Slovenia', *Post-Communist Economies*, 15/4, p. 557-569.

Europäische Akademie Bozen (1996), *Landwirtschaft im Alpenraum – unverzichtbar, aber zukunftslos? Eine alpenweite Bilanz der aktuellen Probleme und der möglichen Lösungen*, Berlin, Blackwell Verlag.

European Commission (2003), *Farm Structure, Survey 1999/2000*, Luxembourg. European Communities.

Eurostat (2000), *Farm Structure. Historical Results – Surveys from 1966/67 to 1997. Final Version: October 2000*, Luxembourg.

Flury, Christian (2005), *Bericht Agrarökologie und Tierwohl*, Bern, Bundesamt für Landwirtschaft.

Giuliani, Gianluca (2003), 'Das schweizerische Berggebiet: aktuelle Probleme, *erwartete Entwicklungen, Lösungsansätze'*, *Atlas*, 25, p. 29-38.

GOSAR, Anton (1991), 'Die Schlüsselprobleme in den Slowenischen Alpen und die Zukunft der Alpen aus jugoslawischer Sicht', in Werner BÄTZING & Paul MESSERLI (eds), *Die Alpen im Europa der neunziger Jahre. Ein ökologisch gefährdeter Raum im Zentrum Europas zwischen Eigenständigkeit und Abhängigkeit,* (Bernensia 22), Bern, Geographisches Institut der Universität Bern, p. 92-119.

GOSAR, Anton & CUNDER, Tomaz (1996), 'Die Landwirtschaft im slowenischen Alpenraum', in EUROPÄISCHE AKADEMIE BOZEN (ed.), *Landwirtschaft im Alpenraum – unverzichtbar, aber zukunftslos? Eine alpenweite Bilanz der aktuellen Probleme und der möglichen Lösungen*, Berlin, Blackwell Verlag.

GRABSKI-KIERON, Ulrike & KRAJEWSKI, Christian (2007), 'Ländliche Raumentwicklung in der erweiterten EU', *Geographische Rundschau,* 59/3, p. 12-19.

GROIER, Michael (2004), *Wachsen und Weichen. Rahmenbedingungen, Motivationen und Konsequenzen von Betriebsaufgaben in der österreichischen Landwirtschaft. (Forschungsbericht Nr. 51)*, Wien, BA für Bergbauernfragen.

HOVORKA, Gerhard (2006), 'The Influence of Agricultural Policy on the Structure of Mountain Farms in Austria', *Ländlicher Raum,* p. 1-30.

INSEE (Institut National de la Statistique et des Études Économiques de la France) (1955), *Recensements Généraux de l'Agriculture 1955,* Paris.

ISTAT (Istituto Nazionale di Statistica Italia) (1961/1970/1982/1990/2003), *Censimenti generali dell'agricoltura 1961, 1970, 1982, 1990, 2001,* Rome.

LAUBER, Stefan (2006), 'Veränderungen der Betriebsstrukturen im Alpenraum. Modellergebnisse aus dem Projekt SULAPS', *Agrarwirtschaft und Agrarsoziologie,* 2, p. 55-67.

LEBENSMINISTERIUM (Bundesministerium für Land- und Forstwirtschaft) (2004), *Grüner Bericht 2004. Wien,* AV & Astoria.

LEL SCHWÄBISCH GMÜND (Infodienst Landwirtschaft - Ernährung - Ländlicher Raum) (2009): Flächennutzung, Online: http://www.landwirtschaft-mlr.baden-wuert-temberg.de/servlet/PB/menu/1041998l2/ index.html [Januar 2009], Schwäbisch Gmünd.

LI (Fürstentum Liechtenstein) (2006): Liechtensteins Landwirtschaft im Reformprozess. Thema der Woche. http://www.liechtenstein.li/fl-portal-ak¬tuell?newsid=13730, [April 2006], Vaduz.

MACDONALD, Daisy *et al.* (2000), 'Agricultural Abandonment in Mountain Areas of Europe: Environmental Consequences and Policy Response', *Journal of Environmental Management,* (59), p. 47-69.

MANN, Stefan (2003a), 'Bestimmungsgründe des landwirtschaftlichen Strukturwandels', *AgrarForschung,* 10/1, p. 32-36.

MANN, Stefan (2003b), 'Theorie und Empirie agrarstrukturellen Wandels?', *Agrarwirtschaft,* 52/3), p. 140-148.

MARTINENGO, Edoardo (1991), 'Le Alpi per l'Europa - una proposta politica', in Werner BÄTZING & Paul MESSERLI (eds), *Die Alpen im Europa der neunziger Jahre. Ein ökologisch gefährdeter Raum im Zentrum Europas zwischen Eigenständigkeit und Abhängigkeit*, ed. Geographica Bernensia, 22, Bern, Geographisches Institut der Universität Bern, p. 205-229.

MESSERLI, Paul (1991), 'Herausforderungen und Bedrohungen des schweizerischen Berggebietes durch Europa an der Wende zum 21. Jahrhundert', in Werner BÄTZING & Paul MESSERLI (eds), *Die Alpen im Europa der neunziger Jahre. Ein ökologisch gefährdeter Raum im Zentrum Europas zwischen Eigenständigkeit und Abhängigkeit*, Geographica Bernensia, 22, Bern, Geographisches Institut der Universität Bern, p. 142-176.

NORDREGIO (2004), *Mountain Areas in Europe: Analysis of Mountain Areas in EU Member States, Acceding and Other European Countries*, Stockholm, Europäische Kommission-DG Regionalpolitik.

PENZ, Hugo (1996), 'Die Landwirtschaft in den österreichischen Alpen', in EUROPÄISCHE AKADEMIE BOZEN (ed.), *Landwirtschaft im Alpenraum – unverzichtbar, aber zukunftslos?, Eine alpenweite Bilanz der aktuellen Probleme und der möglichen Lösungen*, Berlin, Blackwell Verlag.

ROSSIER, Ruth (2005), 'Projekt Hofnachfolge'. Protokoll der Sitzung vom 25. November 2005 der Arbeitsgemeinschaft ländliche Sozialforschung. Bundesanstalt für Bergbauernfragen, Wien, p. 1-20.

RUFFINI, Flavio V., STREIFENEDER, Thomas & EISELT, Beatrice (2004), 'Definition des Perimeters der Alpenkonvention', in UMWELTBUNDESAMT DEUTSCHLAND (ed.), *Die Veränderungen des Lebensraumes Alpen dokumentieren. Indikatorensystem und Konzept für einen Alpenzustandsbericht*, Anhang III, Berlin, Umweltbundesamt, p. 1-15.

RUFFINI, Flavio V. & STREIFENEDER, Thomas (2007), 'The Alps: One region – many realities', *Geographische Rundschau*, International Edition, Jg. 3 (4), p. 70-73.

SCHWEIZERISCHE ARBEITSGEMEINSCHAFT FÜR BERGGEBIETE (SAB) (2002), 'Positionspapier Regionalpolitik.' 8. Februar 2002, Bern. Online: http://www.sab.ch/fileadmin/user_upload/ Politik/Positionspapiere/O-200202dRegionalpol.pdf [Accessed 15 July 2008].

SI-STAT (STATISTICAL OFFICE OF THE REPUBLIC OF SLOVENIA) (1981a/1990), Agricultural statistics 1981, 1990, Ljubljana.

SI-STAT (STATISTICAL OFFICE OF THE REPUBLIC OF SLOVENIA) (1971/1981b), Agricultural results within the population censuses of 1971, 1981, Ljubljana.

SI-STAT (STATISTICAL OFFICE OF THE REPUBLIC OF SLOVENIA) (1960/2000), Census of agriculture 1960, 2000, Ljubljana.

SI-STAT (STATISTICAL OFFICE OF THE REPUBLIC OF SLOVENIA) (1969), Sample census of agricultural holdings 1969, Ljubljana.

STATISTIK AUSTRIA (1951/1960/1970/1980/1990/1999), Land- und forstwirtschaftliche Betriebszählung 1951, 1960, 1970, 1980, 1990, 1999 (Agrarstrukturerhebung), Wien.

STATISTIK BAYERN (BAYERISCHES LANDESAMT FÜR STATISTIK UND DATENVERARBEITUNG) (1949/1960/1971/1979/1991/1999), Landwirtschaftliche Betriebszählung 1949, Landwirtschaftszählung 1960, 1971, 1979, 1991, 1999, München.

STREIFENEDER, Thomas & RUFFINI, Flavio V. (2007), 'Ausgewählte Aspekte des Agrarstrukturwandels in den Alpen - Ein Vergleich harmonisierter Agrarstrukturindikatoren auf Gemeindeebene im Alpenkonventionsgebiet. Berichte über Landwirtschaft', *Zeitschrift für Agrarpolitik und Landwirtschaft*, Bd. 85 H. 3, p. 406-440.

STREIFENEDER, Thomas, TAPPEINER, Ulrike, RUFFINI, Flavio V., TAPPEINER, Gottfried & HOFFMANN, Christian (2007), 'Selected Aspects of Agro-structural Change within the Alps - A Comparison of Harmonised Agro-structural Indicators on a Municipal Level in the Alpine Convention Area', *La revue de géographie alpine, Journal of alpine research*, Bd. 95 (3) 2007, p. 27-52.

TAPPEINER, Ulrike, TAPPEINER, Gottfried, HILBERT, Andreas & MATTANOVICH, Ernst (eds) (2003), *The EU Agricultural Policy and the Environment. Evaluation of the Alpine Region*, Berlin, Blackwell.

VÉRON, François (1996), 'Die Landwirtschaft im französischen Alpenraum', in EUROPÄISCHE AKADEMIE BOZEN (ed.), *Landwirtschaft im Alpenraum – unverzichtbar, aber zukunftslos? Eine alpenweite Bilanz der aktuellen Probleme und der möglichen Lösungen*, Berlin, Blackwell Verlag, p. 73-116.

VOGEL, Stefan (2005), 'Hofnachfolge in Norddeutschland und Österreich im Vergleich', in *60. Sitzung der Arbeitsgemeinschaft ländliche Sozialforschung*, Wien, Bundesanstalt für Bergbauernfragen (BABF), p. 20-27.

WEISS, Franz (2006), *Bestimmungsgründe für die Aufgabe/Weiterführung landwirtschaftlicher Betriebe in Österreich*, Diskussionspapier DP-14-2006, Universität für Bodenkultur, Institut für nachhaltige Wirtschaftsentwicklung, Wien.

WESSELY, Helga & GÜTHLER, Andreas (2004), *Alpenpolitik in Deutschland- Anspruch und Realität. Politische Zielaussagen, Situation, Handlungsbedarf, Nürnberg*, Bund Naturschutz in Bayern.

ZANETTI, Gerhard (1999), *Strukturwandel und Flächennutzungsänderungen in der österreichischen Land- und Forstwirtschaft. Gutachten des Österreichischen Instituts für Raumplanung (ÖIR)*, Wien, Schriftenreihe der Österreichischen Raumordnungskonferenz (ÖROK) Nr. 145.

11. Why did nineteenth-century states establish agricultural research institutions? The origins of the south German plant-breeding stations c. 1900

Jonathan HARWOOD

I. Introduction

Almost all studies of German agricultural policy between the late nineteenth century and the First World War have been narrowly focused upon trade policy. In so doing, they have paid much less attention to a wide range of other kinds of intervention which were just as important, though less visible and thus rarely the subject of public debate: for example, support for cooperatives, improved credit facilities, promotion of 'internal colonisation', or the establishment of institutions for education, extension or research. Of course the late nineteenth century establishment of these latter institutions is too obvious to be overlooked entirely, but even when historians have attempted to account for their emergence, they have tended to regard them as simply a solution to an economic problem. As international competition forced down prices from the 1870s, farmers were thought to need help in adapting to changing markets. And this meant that state-funded institutions providing testing, experiment and extension had a role to play. While German agricultural officials clearly did hope that creating such institutions would further technical modernisation - in order to lower production costs and increase competitiveness - it seems to me that too little attention has been given to the political context in which such institutions were established.

Elsewhere I argued that one of the reasons underlying Prussia's creation of the chambers of agriculture - an institution responsible for extension, secondary education and research - during the 1890s was that, faced with growing political turbulence in the countryside from the 1890s, state officials introduced a variety of measures intended to curb rural unrest, of which the chambers were one (Harwood, 2008). In this chapter I pursue this issue further, turning to the decision in several Central European states around 1900 to set up plant-breeding stations. Like earlier public institutions charged with quality control (experiment stations, seed-control stations), the breeding stations conducted testing so as to provide farmers with better information on the choice of crop variety, but they were also designed to assist regional farmers to breed new varieties better suited to their locale. What made them distinctive among experiment stations, moreover, is that they were explicitly designed to assist *peasant*-farmers.

Since the movement for what I call 'peasant-friendly' plant-breeding, though especially strong in southern Germany, was quite widespread, I begin by outlining developments in Germany as a whole from the 1880s, then turning to the southern German political context in the 1890s, and finally focussing more briefly on the establishment of one particular station: that in Bavaria (1902). One major point which emerges is that state involvement came relatively late. During the 1880s those who took the initiative in improving local varieties (*landraces*) *via* breeding were either the occasional estate-owner or agricultural scientist. From the early 1890s a few cooperatives undertook such work (with technical assistance from local agricultural scientists). At this time, however, quasi-official organisations representing farmers such as the agricultural associations (*landwirtschaftliche Vereine*) or the State Agricultural Commission (*Bayerischer Landwirtschaftsrat*) played a largely passive role, and only from the late 1890s did state officials begin to explore the possibility of promoting varietal-testing or the development of plant-breeding. The other main point of this story is that the state's motives were not merely economic. To be sure, there was growing awareness among state officials that, despite the stereotype of 'peasant conservatism', small farmers were not averse to changing their cultivation practices and had in fact been relatively quick to adapt to changing late nineteenth century markets. The modernisation of peasant-farming thus looked to be not just desireable in principle but also viable in practice. In addition, however, governments were nervous about the rise in peasant radicalism - and their declining support for mainstream parties - following the reduction of protective tariffs in 1891. During the 1890s, therefore, agricultural officials took every opportunity to emphasise the wide range of measures - among them the creation of experiment stations - which the state had introduced in order to shore up the position of peasant-farmers.

Thus the reasons behind the establishment of the Bavarian station were both economic *and* political.

II. The growth of commercial plant-breeding and its consequences

The earliest German successes in commercial breeding were in the sugar industry. Attempts to increase the sugar content of sugar-beet began there in the 1820s, and by the 1870s one company (Rabbethge and Giesecke) had succeeded in increasing sugar content from 6 per cent to about 14 per cent. By 1908 this firm had further increased sugar content to 23 per cent and was producing one-third of the world market for sugar-beet seeds[1]. During this period, moreover, German industrialisation increased urban demand for food, making higher yields in cereals attractive to farmers. In the

1 Shull, G.H., 'Notes on a European Tour undertaken by George H. Shull...', (1908: 87) (Shull Papers, American Philosophical Society).

1870s and 1880s a small number of German estate-owners - among them Ferdinand Heine, Gustav Bestehorn, O. Beseler, Friedrich Strube, Otto Cimbal and Ferdinand von Lochow – began to try to improve the best of the existing cereal (and, to a lesser extent, potato) varieties. By 1900 they had increased the yield of cereals by 30-40 per cent and of potatoes by over 100 per cent (Raum, 1972: 7; Lochow, 1913; Hillmann, 1910: 347; Keune, 1952: 171-174). Late nineteenth century demand for the new varieties in Germany was buoyant. The number of sugar-beet breeders, for example, increased from ten in 1867 to over forty by 1895 while the number of cereals breeders increased about ten-fold between the 1880s and the 1910s (Rümker, 1895; Rümker, 1928; Kühle, 1926).

Since the growth of commercial breeding brought a large number of new crop varieties onto the market, innovation-minded farmers were confronted with a problem. Which commercial breeders could be relied upon to produce seed which was not only pure but which would also germinate in high proportion? When similar questions concerning the quality of commercial fertilisers had arisen a generation earlier, farmers had turned to the voluntary sector for help: the first experiment stations were established by the regional Agricultural Associations (albeit with modest state support). By the 1880s state governments were more prepared to intervene, and 'seed control stations' (*Samenkontrollstationen*) were established in many parts of German-speaking Europe to assess seed-quality. But the state offered farmers little help in choosing which of the available *varieties* were actually suitable for their particular growing conditions. Throughout the nineteenth century various individuals had conducted comparative tests of this kind, but these had been isolated and uncoordinated efforts. The first systematic and standardised nationwide varietal-testing programme grew out of the establishment in 1885 of the German Agricultural Society. Founded by a group of estate-owners who were keen to promote the use of new cultivation practices and technologies, one of the Society's first steps the following year was to create a Seed-Breeding Division which brought together breeders and academics in order to obtain and distribute information about new varieties (Hillmann, 1910; Hansen & Fischer, 1936).

Over the following decade or so, the Division introduced a variety of measures designed to help farmers make informed choices about the varieties they planted. In 1888, for example, it began a programme of varietal-testing in nine regions across the country, based on seed samples supplied by those breeders willing to take part. The initial focus was on cereals, though from the mid-1890s other crops were included and the results were published in the Society's journals. These tests provided useful information for any farmer who was willing to wade through tables of data before choosing the most appropriate variety. But in 1897 the Division also became the first organisation in Germany to 'certify' varieties whose production met its standards of excellence (Hillmann, 1910; Hansen and Fischer, 1936; Fruwirth, 1911). In the certification process members of the Division visited the farm where a variety was bred (and/or multiplied) in order to establish that what was growing in the breeder's fields was in fact the same variety as in the sample the Division had received, that it was free from disease, that it was growing in visually uniform stands and that proper breeding procedures and facilities were in place. Once certified, the variety was placed on the Society's 'Seedlist'. Finally, in 1905 the Division established a small and selective list of the most outstanding of its certified varieties, the 'Hybred Varieties Register' (*Hochzuchtregister*).

While the Seed-Breeding Division's activities were undoubtedly an important step in promoting the use of improved varieties among farmers, they did not serve some sectors of the farming community very well. For one thing, testing was largely focused upon the hybred varieties then emerging from the commercial sector; only a limited attempt was made to test the large number of traditionally planted local varieties. Thus the test results were of little use to those farmers who could not afford to purchase commercial varieties but who might have benefitted from planting a good but cheaper local variety. For another, the fact that the tests compared varieties when grown under intensive cultivation meant that such tests were of little relevance for small farms where such growing conditions seldom obtained (Kryzymowski, 1913: 466-467). In addition, as various experts pointed out, the Division's varietal-testing programme was just too small to do justice to the enormous diversity of growing conditions across the country (von Lochow, 1904; Rümker, 1898; Saatzuchtstelle, 1905; Böhmer, 1914; Rümker, 1899). This was a serious problem, for by the 1880s it was becoming clear that hybred varieties often did not fare very well in those regions - among them southern Germany - where soil types and climatic conditions were very different from those in northern Germany where commercial breeding was concentrated. Moreover, this neglect of certain regions' needs was aggravated by the fact that the programme relied upon farmers volunteering to plant the test-varieties. The only farms which could take part were those which possessed sufficient staff who could be allocated to such work, and the result was that only large farmers could afford to participate, few of whom were to be found in regions like southern Germany[2]. Even where the results were potentially relevant for a region's farms, it did not help that for the first twenty years or so test results were published only in journals of the Society (and occasionally in the weekly magazines published by regional Agricultural Associations) which were readily accessible to members of the Society or to those with access to good libraries, but neither of these outlets was seen

2 Minutes of the meeting of the Seed-Breeding Division, 18 Feb 08, Jahrbuch der DLG, 23 (1908: 119-167.

regularly by the average farmer. Lastly and most fundamentally, in several respects the Division's procedures were designed more to protect German breeders than to aid German farmers[3].

That the Seed-Breeding Division's activities should have been of limited utility to the smaller farmer was predictable. For the German Agricultural Society was established by a group of estate-owners keen on technical innovation in order to serve a limited constituency of the like-minded (the Society had only 12,000 members in 1896). The Division, for example, was mainly the brainchild of the breeder, Wilhelm Rimpau, who had also been one of the three founding members of the Society, and it was dominated for the first fifteen years or so by commercial breeders, estate-owners and sympathetic academics. Representatives of the peasant-farmer or of south German interests only began to appear in small numbers around 1900 and never amounted to more than a small minority of the main committee before the war. How, then, were peasant-farmers supposed to identify the most suitable varieties for their growing conditions, and who was going to produce new and better ones for them?

III. The emergence of 'peasant-friendly' approaches to plant-breeding

Historians have recognised for some time that the view of peasant-farmers so common in the nineteenth century - that they were technically backward and unresponsive to the market - is wrong (Moeller, 1981; Mooser, 1992; Dipper, 1993; Kopsidis, 2006). But the issue important here is when *contemporaries* began to reach this conclusion. And by the 1890s there were signs in Germany that experts as well as officials were also questioning the traditional view (Koning, 1994). Agricultural economists observed that, unlike many estate-owners, peasant-farmers were intimately familiar with the problems of running a farm and less likely to get into debt. Moreover, they were quicker to respond to the market by moving to intensive cultivation methods with the result that yields were substantially higher on small farms than on estates (von der Goltz, 1903; Werner & Albert, 1900; Ballwanz, 1978). Thus for officials looking for solutions to the late nineteenth century agricultural

[3] For example inclusion of a variety on the Register entitled the breeder to use a protected trademark in promoting it (Anon., 1907). In addition the Division was not prepared to admit all varieties to its certification procedure; applications from foreign breeders were considered only where those countries were open to the import of German varieties (Hansen & Fischer, 1936). The same bias is evident in the Division's definition of what was to qualify as a '*hybred*' variety and thus of those varieties which were to be singled out for inclusion in the Hybred Varieties Register (The definition was originally proposed in Rümker, 1907). The decisive criterion was not to be a variety's outstanding performance in comparative tests, but rather the type of method used to breed it: only varieties bred using the most demanding methods (in von Rümker's view, pedigree breeding) were to qualify for the Register. This meant that local varieties which had been improved by mass selection, no matter how well they performed, were not eligible for inclusion.

crisis, peasants were important, not just because they farmed a substantial majority of the acreage, but also because they looked like a promising target for assistance and investment.

More specifically, as far as promoting high-quality seed among small farmers was concerned, given the limitations of the German Agricultural Society's testing programme, during the 1880s and 1890s various other organisations began to fill the gap, among them agricultural schools and colleges but also seed-growing cooperatives (*Saatbauvereine*). These were groups of (generally small) farmers who bought 'original' (i.e., freshly bred) hybred varieties from commercial breeders, tested them alongside local varieties for their performance in local growing conditions, grew up the best ones in large quantities, and sold them at substantially lower prices than the original hybred (Rümker, 1911a; Fruwirth, 1911). In some regions the local varieties had such good qualities that the cooperative concentrated on their sale. The earliest of these were formed during the 1880s. In the Probstei region of Schleswig-Holstein, for example, local varieties of oats and rye had been grown for seed and sold in other German states as well as in Eastern Europe since the mid-nineteenth century (Kochendoerfer, 1996; Hillmann, 1910: 581-582). In the Pirna region of Saxony (near Dresden), similarly, a local variety of rye possessed properties which had made it attractive as seed-rye in various other parts of Germany for many years. In the Fichtel Mountains region of Bavaria, too, a local variety of seed-oats, grown and sold by several cooperatives formed during the 1880s, was widely planted in southern Germany and in highland areas elsewhere and was awarded medals at World Fairs in 1900 and 1904 (Raum, 1912; Kraus, 1899a; Fruwirth, 1907: 19)[4].

In addition, faced with most commercial varieties' limited utility for small farmers, some agronomists were beginning to suggest that *local* varieties might also be worth improving through breeding. In a paper to the International Congress of Agriculture in 1890, for example, Emmanuel von Proskowetz and Franz Schindler argued that in view of local varieties' valuable properties (*e.g.*, quality, winter-hardiness), it would be worth trying to increase their yield through breeding at specialised experiment stations (Pribyl et al., 1890: 348). Proskowetz (1849-1944) was a Moravian estate-owner who, noticing the decline of cereal yields there in the early 1870s, began to collect local barley varieties from the Hanna region and to test them against the commercial hybreds then favoured. Finding that one of them was consistently superior to the hybreds in yield, quality and early ripening, he began to improve it by mass selection, and in 1884 it was introduced to the market as the 'Hanna pedigree

[4] Similar cooperatives emerged in east and west Prussia around the turn of the century, with support from their regional chambers of agriculture, and for the same reasons: growing conditions there were quite different from the rest of Germany so that German hybred varieties often did quite poorly there (Gisevius, 1901; Krech, 1904; Modrow, 1904).

barley'. It was very popular with brewers, and by 1900 commercial breeders had begun to use the Hanna as their starting material (Proskowetz, 1937; Tschermak, 1958: 81-88). During the 1890s, however, the suggestion that local varieties should be improved was fairly controversial. Many agricultural experts and innovation-minded farmers believed that the only realistic way to increase production was *via* intensive cultivation. And since local varieties did not yield as highly as hybred ones, nor did they respond well to mineral fertiliser (tending to fall over or 'lodge'), to launch a breeding programme based upon them was perceived as a step backward (Kiessling, 1912; Broili, 1899: 218).

Nevertheless, Proskowetz and Schindler's recommendation may have fallen upon receptive ears. For by the 1890s even the most sought-after local varieties were beginning to face competition from higher yielding commercial varieties. In Pirna, for example, a botanist at the nearby experiment station in Dresden, Bruno Steglich, concluded that only if the Pirna variety were improved through breeding would it be able to continue to compete. The best way to do this, he believed, was for a group of interested local farmers to organise themselves into a cooperative for whom he would serve as technical advisor[5]. In 1896 a breeding- and sales-cooperative was duly set up, and the improved Pirna rye's rapid success – including a gold medal at the Paris world's fair in 1900 – attracted attention from state agricultural officials elsewhere who saw in it a potential model of how small farmers could secure high quality seed at moderate cost through self-help[6]. The idea was also attractive in southern Germany where there were scarcely any commercial breeding firms around 1900[7]. A case in point was the Bavarian seed-growing cooperative in the Fichtel Mountains region whose members also embarked in 1900 upon a breeding programme to improve their local oats variety. The programme enjoyed financial support from the Interior Ministry, and as at Pirna and the Probstei, the cooperative relied upon technical advice from an agronomist (Prof. Carl Kraus of the Agricultural College at Weihenstephan in Bavaria)[8].

---

5 The idea that plant-breeding could be organised on a cooperative basis was not a radical departure. Animal-breeding cooperatives were extremely common in Bavaria by this time (Jahresbericht des Bayrischen Landwirtschaftsrats für das Jahr 1899: 10-11).

6 STEGLICH, 1910. On official interest see 25 July 98 - [Interior Ministry?] to governments of Oberpfalz, Regensburg and Oberfranken, ML 309, BayHStA.

7 This was probably because the enormous variation in soil and climate over quite short distances in the south meant that a variety bred in any one locality was unlikely to flourish across the region - or elsewhere in Germany - such that the market for it would be quite small.

8 5 Mar 1900 - Georg Heim (of the Fichtelgebirgsverkaufsgenossenschaft) to Min. des Innern/ Landwirtschaft; ML 309, BayHStA. Although these three breeding-cooperatives were the best known, by the turn of the century a handful of similar organisations had been set up in other German-speaking states (KRAUS, 1899a; KRAUS, 1900).

By the late 1890s the possibility of improving local varieties through systematic plant-breeding was starting to get discussed in agricultural organisations and publications (Kiessling, 1922; Anon, 1900). In addition, several experts were suggesting that since existing arrangements for meeting the needs of small farmers were insufficient, some new kind of institution might be required. In 1898, a high profile proposal of this kind was made in a lecture to the German Agricultural Society by the professor of agronomy at the University of Breslau, Kurt von Rümker (Rümker, 1898; in a similar vein his 1900 and 1911b)[9]. Acknowledging that commercial varieties were probably too expensive for the small farmer, he suggested that the way forward could be through the proliferation of breeding-cooperatives such as those recently established in the Probstei and Pirna regions. There was no reason, he insisted, why small farmers could not be encouraged to undertake basic plant-breeding themselves. While some crops were unsuited for technical reasons and some breeding-methods were too complicated for non-experts, cereals-breeding using simple methods (mass- and individual-selection) would certainly be feasible on small farms. A specialised state institution, however, was not necessary, he believed, since technical support could be provided by teachers at local agricultural secondary schools or the regional chamber of agriculture.

A year later a Bavarian estate-owner, Josef Broili, made a similar appeal, calling upon the German Agricultural Society to make a greater effort to address the interests of small farmers. His lecture to the Seed-Breeding Division ended with a formal proposal that the Society should encourage the improvement of southern Germany's local varieties and that the state should lend a helping hand (Broili, 1899). One of the experts who spoke up in support of Broili in the ensuing discussion was Carl Kraus of the College at Weihenstephan who had been making many of the same points in papers of his own that year (as well as starting to work on the improvement of local cereals varieties) (Kiessling, 1910: 156). Despite some individuals' scepticism, he argued, local varieties could be successfully improved *via* selection, and it was important that more work of this kind be undertaken. As to how such proposals could be implemented, Kraus also endorsed breeding-cooperatives as a useful way to meet the needs of small farmers, at least in some circumstances. But unlike von Rümker, he suggested that state plant-breeding stations could be valuable in other cases. The latter could be useful in overseeing a state-wide programme of varietal-testing which would include monitoring the quality of the available commercial varieties as well as studying the range of local varieties in order to identify which might be worth improving. And while such a station could not attempt to breed varieties for an entire

[9] Rümker's sensitivity to the widely varying circumstances in which farmers worked is also evident in his argument that intensive agriculture was not necessarily the most sensible strategy; in some conditions extensive cultivation methods would yield greater profits (see summary in Pommer, 1898: 485-486).

state, it could at least provide the technical support for small farmers who wished to take up breeding (Kraus, 1899a, 1899b, 1900; Liebscher, 1896).

Only from the late 1890s did the state begin to take an interest in these matters. The proposals by Rümker et al., for example, seem to have been noted by the Bavarian Interior Ministry which in 1897 and 1898 had asked the State Agricultural Commission to consider whether a more systematic programme of state funding for varietal-testing might make sense. And by the start of 1900 the minister was exploring whether the state should also initiate a programme of plant-breeding[10]. By that time there were several models available as to how the state might support such work. One was to support the formation of a breeding-cooperative which would be advised by a technical expert from a local experiment station, as had been done by Steglich at Pirna. Alternatively, the expertise could be provided by a local chamber of agriculture or by creating a section for plant-breeding within an existing experiment station[11]. But the institutional model most widely cited in the German-language literature around the turn of the century was probably the state plant-breeding station established in 1886 at Svalöf in southern Sweden[12]. For as one German observer pointed out, although Svalöf was a public-sector institution, its varieties were competing successfully against those of older and more famous European private breeders[13].

Thus by 1900 the economic rationale for promoting the use of quality seed among peasant-farmers was clear. And although neither the existing arrangements for varietal-testing nor commercial varieties were well-suited to the needs of small farmers, alternative approaches based on local varieties were beginning to be pursued with some success, and institutional models as to how the state might support such activity were available. Nevertheless the states' relatively late decision to establish plant-breeding stations was not simply an economic one.

10 On varietal-testing, see Staatsministerium des Innern, Massnahmen 1905, 109. On breeding, see 9 Mar 00 - Min. des Innern/Landwirtschaft to Bayerischer Landwirtschaftsrat, ML 309, BayHStA.

11 In 1899 the Posen chamber's experiment station was providing farmers with instruction on individual selection (Steinbeck, 1997: 130). And from around 1900, at the experiment station in Lausanne, Gustave Martinet had recruited local farmers, not just as multipliers but as collaborators in his cereals-breeding programme (Moser, 2003: 28). I thank Peter Moser for sharing additional unpublished results with me).

12 For example, in a lecture in 1900 Kraus had cited Svalöf (Kraus, 1900), and the station was extensively discussed by Bruno Steglich in a 1904 lecture to the Oekonomische Gesellschaft in Dresden (Steglich, 1905).

13 Arnim-Schlagenthin, 1901. This success is presumably why the association of German chambers of agriculture had recommended that varietal tests should routinely include a Svalöf variety among its standard test varieties (Ulander, 1906).

IV. The political context in southern Germany

In response to declining cereals prices on the international market during the 1870s, the Reich government chose in 1879 to introduce tariffs. But prices continued to fall, and during the 1880s there were constant references in Germany to an 'agricultural crisis'. The Reich's decision to reduce tariff protection in 1891, therefore, unleashed rural political unrest throughout the country. Disgusted by the apparent lack of an organisation capable of lobbying government on their behalf, farmers - mainly in northern and Protestant areas - flocked to join a newly formed militant organisation, the Agrarian League (*Bund der Landwirte*, est. 1893) which campaigned for the restoration of tariffs. In Bavaria 1893 was also a pivotal year since wheat prices dropped for the first time in many years. And while the League managed to attract a few large Bavarian farmers, the small and medium peasants who dominated Bavarian agriculture evidently perceived the League as a lobby-group for estate-owners and kept their distance (Schlögl, 1954: 567; Tirrell, 1951:195).

But who was to represent them? The (Catholic) Centre Party, the largest party in the Bavarian parliament, no longer enjoyed their confidence since it had endorsed tariff-reduction and was seen as having done nothing for agriculture. Nor did the Bavarian Agricultural Association (est. 1809), a voluntary organisation recognised by the state as the quasi-official representative of the farming community. The Association's membership was in fact dominated by estate-owners, civil servants and professionals; only a small proportion of the farming community were members. As a result, when the founder of a radical movement called for a new farmers' organisation which had 'no nobles, no priests, no doctors, no professors but just peasants', it was clear whom he was attacking (Haushofer, 1963: 215)[14]. In addition, the facts that the Association's statutes largely forbade 'political' activity, and that its leading members were also state officials, undermined the Association's credibility as an independent organisation which would fight on farmers' behalf (Schlögl, 1954: 567; Haushofer, 1986; Hundhammer, 1926: 103-104). That the Association might be in need of reform had been raised by a Bavarian farmers' organisation in the 1880s but to no avail. The issue of representation became acute, however, with the emergence of the Agrarian League, and in 1894 Prussia attempted to solve the problem by creating chambers of agriculture. This move found support in Bavaria among radical critics of the Agricultural Association (Hundhammer, 1926: 106; Harwood, 2008; Bergmann, 1986: 80-88), prompting the Bavarian Agriculture Ministry to ask the State Agricultural Commission to consider whether existing arrangements

[14] For data on membership, see Bayerisches Staatsministerium des Innern, 1890: 785. At a rough guess, in the 1890s at most ten per cent of Bavarian farmers were members of the Association; another estimate reckons the proportion at its peak was barely twenty per cent (Hundhammer, 1926: 103).

for the representation of farmers' interests were satisfactory. But the Commission recommended retaining the Agricultural Association while making a few reforms in order to meet critics' objections. Having conceded so little, the Association was thenceforth under pressure to defend its record, as well as that of the government, in looking after the farming community's interests[15].

Though criticism continued, peasants also began to organise themselves. Initially they formed a number of local 'peasant leagues' in order to contest local elections against the Centre Party. This strategy succeeded, and in 1895 they amalgamated to form the statewide Bavarian Peasants' League (*Bayerischer Bauernbund*), the only peasants' organisation in Bavaria to operate as a political party at both state and Reich level (Hochberger, 1991; Bergmann, 1986). In the Bavarian parliament the League called *inter alia* for the restoration of tariffs, the establishment of agricultural advice bureaus and more practically oriented agricultural education. With these policies it regularly secured about 10 per cent of the votes in the Bavarian parliament before 1914. Given this electoral success and its ability to politicise the rural population, the Peasants' League has been considered in some respects the Bavarian equivalent of the Agrarian League in Prussia (Albrecht, 1974; Hochberger, 1991: 173).

The Peasants League's rapid rise certainly alarmed the Centre Party which soon regarded the League as its main opponent. One wing within the Party responded by organising local Party-affiliated peasants' associations which merged in 1898 to form a state-wide Christian Peasants' Association (*Bayerischer Christlicher Bauernverein*). Its aim was to lure peasant-farmers toward the Centre Party and away from the more radical and secular Peasants League[16]. The League's radicalism was also a source of concern for the government, not least because it raised the awkward prospect of alliances among the opposition parties. The League's stance on political issues (*e.g.*, opposing increased military spending), was generally closer to that of left-liberals and Social Democrats (with both of whom it formed a coalition to fight the election of 1912).

Among the other opposition parties, the Social Democratic Party first entered the Bavarian legislature in 1893. It might have posed little threat to the government,

---

15 The Association's powers were widened so that it could henceforth make recommendations to government, not only on technical matters but also on economic and political ones, and an attempt was made to increase peasant representatation within the membership (Bayerisches Staatsministerium des Innern, 1897: 336-339; Jahresbericht des Bayerischen Landwirtschaftsrats für das Jahr 1895, 1-5; Hundhammer, 1926: 107). On the Association's defence, see Jahresbericht des General-Comites des Landwirtschaftlichen Vereins in Bayern für das Jahr 1893, 3-4, as well as the volume for 1894, 1.

16 Unlike the Peasants League which demanded state action on the peasants' behalf, the Christian Peasants' Association stressed self-help *via* the formation of cooperatives (Haushofer, 1986; Hochberger, 1991).

had it taken as strict a line on policy as did the national Party (which decried the backwardness of the peasantry and advocated land nationalisation), and since there were as yet few industrial workers in Bavaria. But the Bavarian wing of the Party was much more pragmatic than its national counterpart and had no hesitation in championing the interests of peasants. Its leader, Georg von Vollmar - a gifted orator who spoke in upper-Bavarian dialect – was especially effective in challenging the Centre Party to say what they had done to help peasant-farmers. Nor was Vollmar averse to election coalitions with the Peasants League or with oppositional elements within the Centre Party (Lehmann, 1970: 64-76). The political gains achieved by all three of these groups in the 1893 elections, therefore, put pressure on the Centre Party to address peasant concerns more seriously (Möckl, 1972).

The government's response to the Peasants League was in part a defensive one. A close eye was kept on its activities *via* informants' reports on its meetings, and in 1897 dossiers on leaders were initiated, watching for signs of libel which could lead to prosecutions. But more important is the fact that the League's success forced the government to reconsider its agricultural policies. Aware of growing rural discontent, the government stepped up its aid to farmers, *e.g. via* the provision of cheap credit and the promotion of cooperatives. Sensitive to the League's criticisms of the Agricultural Association, the government pressured the Association to increase its activity and, according to one historian, there was a burst of government legislation on rural matters after 1893 (Farr, 1978). The State Agricultural Commission's annual reports emphasised during the 1890s how much effort was being expended on the farmers' behalf. As the report for 1898 declared, the government, legislature and agricultural officials had stood by the farming community, and one could be grateful for the funding they had allocated toward the establishment of public institutions, especially of cooperatives. Here was 'cause for optimism'[17]. Indeed, state appropriations for agriculture were increasing rapidly; between 1880 and 1904, for example, the Interior Ministry's expenditure on agriculture increased nearly four-fold[18].

Another indication of the government's concern that its efforts should be properly recognised was its publication of a survey in 1897. At various points during the nineteenth century government surveys on the state of Bavarian agriculture had been published (in 1862, 1872 and 1890). And in 1896, noticing that Prussia, Saxony and Baden had been releasing such reports at regular intervals, the Ministry decided

[17] Jahresbericht des Bayerischen Landwirtschaftsrats für das Jahr 1898, 2; 15; see also Jahresbericht for 1897.

[18] Bayerisches Staatsministerium des Innern, 1905: tab. XI. During the 1890s spending on the breeding of small animals increased five-fold, on poultry-breeding four-fold and on dairying six-fold ('Uebersicht der Aufwendungen aus öffentlichen Mittel zur Förderung der Landwirtschaft in Bayern…', ML 204, BayHStA).

that it would again be appropriate to publish a further volume[19]. The volume was to have two functions. Hoping for a lead by technique (*Vorsprung durch Technik*), one aim was to stimulate technical improvement in those regions thought to be lagging behind by promoting best practice. In the covering letter sent with the volume to all district governments, for example, the Ministry urged that more attention be given to encouraging studies of cultivation methods (especially those concerned with fertiliser-use) and to the organisation of seed markets[20]. Its other aim, however, was to chart the progress in Bavarian agriculture which had been made since 1890, though in a very narrowly defined way. For the volume was supposed to focus only upon those measures favourable to agriculture which had been introduced by government, state legislature and the (quasi-state) Agricultural Association. There is little doubt that measures taken by voluntary organisations such as cooperatives, peasant associations or the German Agricultural Society had also contributed to agricultural progress, but these were evidently not to be included, suggesting that the volume was conceived, less as a survey of progress *per se* than as an exercise in public relations[21].

Significantly, the 1897 volume's focus upon government measures was rather different from that taken in the preceding volume published in 1890. In part this is evident from their respective titles; whereas the earlier volume was entitled simply 'Agriculture in Bavaria', the later one was entitled 'Measures taken by the Bavarian Agricultural Administration'. But it is also noticeable in the fact that when the possibility of a new volume was first raised in 1896, the top civil servant in the Ministry suggested to his minister that simply to continue the older form of presentation - a compilation of statistics and analyses of the current state of various areas comprising agriculture - would be unwieldy as well as expensive and time-consuming to compile (since the government wished to have the volume ready for the next legislative session). Preferable, he argued, was a smaller volume which would not attempt to survey the state of agriculture but would focus primarily upon the measures taken by government bodies since 1890 and their effects upon the state's agriculture. The minister agreed[22].

In the introduction to the 1897 volume the Ministry conceded that while it had not been able to secure an increase in agricultural prices since 1890, success had

19 18 Dec 96 - Min. des Innern/Landwirtschaft to Finanzministeriums, MF 71660, BayHStA.

20 4 Oct 97 - Min, des Innern/Landwirtschaft to Kgl. Regierungen und Kammern des Innern, ML 204, BayHStA.

21 When preparations for a successor volume were being made in 1903, for example, the top civil servant in the Interior Ministry wrote to various other ministries, requesting their cooperation in compiling the necessary evidence and stressing that the volume should include only the new measures introduced by government (7 Aug 03 - Min. des Innern/Landwirtschaft to Finanzministerium, MF 71660, BayHStA).

22 9 Dec 96 - Ministerialrat Brettreich to Excellenz, ML 204, BayHStA.

been achieved through improvements in production and sales of commodities, better organisation of credit arrangements, increased protection of farmers against animal disease and hailstorms and a reduction in the tax-burden. Over the following 350 pages each area of policy relevant to agriculture – animal and plant production, agricultural industry, infrastructure, taxation, labour policy, research and education – was discussed in turn, noting which measures the state had taken, how much they had cost and with what success. In the case of plant production, for example, the volume explained that the Agricultural Association had actively promoted the use of the best seed through funding for seed-markets, encouraging the collective purchase of seed, varietal-testing and better procedures for cleaning and sorting seed. It then listed every region in which such a measure had been introduced, noting how much the state had spent in each case[23].

On publication in 1897 the volume was sent out to all provincial governments in the state as well as to other Bavarian ministries, the State Agricultural Commission, and cooperatives among others[24]. Evidently determined to make sure that the book's message had escaped no one, a government spokesman in the state legislature devoted a lengthy speech to its contents. The legislature had often been criticised recently for having neglected agriculture, he noted, but the newly published volume clearly demonstrated just how much the government had done for the farming community since 1890. Indeed, Bavaria was the envy of other German states in several respects including its support for cooperatives. And not wishing to forego the party-political opportunities which the volume presented, he rounded off his address with an attack upon the behaviour of individual members of the Bavarian Peasants League as well as a dig at the Social Democratic Party[25].

As the new century opened, agricultural officials continued to take every opportunity, when attending meetings of farmers' organisations, to emphasise their support. Following a talk recommending that the state should support plant-breeding, for example, the top agricultural civil servant endorsed the recommendation, noting that although 'much had been done in Bavaria to improve agriculture', further measures could still be introduced to foster the use of better seed. The Ministry, he reassured

---

[23] Bayerisches Staatsministerium des Innern, 1897: 3, 73ff. From 1898, as the Ministry pointed out, freight charges for seed had been reduced by half (Bayerisches Staatsministerium des Innern, 1905: 108).

[24] 4 Oct 97 - Min. des Innern/Landwirtschaft to Kgl. Regierungen u. Kammern des Innern, ML 204, BayHStA.

[25] 378th session of the Kammer der Abgeordneten on 7 Feb 1898, Stenograph. Bericht d. Kammer der Abgeordneten, vol. 11 (1898), 16-22, ML 204, BayHStA. Interestingly, some contemporaries were aware of the book's political implications. As word of its forthcoming publication leaked out, one Munich newspaper declared its suspicion of the draft document which was supposed to demonstrate the government's role in securing 'improvements' for agriculture (18 Feb 96 - cutting from General-Anzeiger d Kgl Haupt- u Residenzstadt, no. 41, ML 204, BayHStA).

his audience, was 'warmly in favour of this development and keen to support it'. Considerable sums had already been spent for this purpose, and 'I can assure you that the government will carefully consider all such applications' (for funding) (Kraus, 1900: 88). But the pressure did not let up; in 1901 even the German Agricultural Commission - hardly a radical voice - asked the Bavarian government to establish a chamber-like organisation that would grant all farmers direct representation, but the request was declined (Hundhammer, 1926: 109-110). Later that year rural unrest reached a crescendo when a large meeting of farmers took place in Regensburg with the aim - for the first time - of unifying the Bavarian farming community. Several speakers called for state intervention to protect farmers against foreign imports while one spokesman for the cooperative movement (Georg Heim) went further. It was 'noteworthy', he said, 'how unresponsive our [Bavarian] government has been to the peasantry's demands' (Schlögl, 1954: 571).

A year later the opening of the new plant-breeding station provided the government with a partial defence. To my knowledge, the Bavarian Peasants League had never demanded peasant-friendly experiment stations (though as we have seen, they were keen on more practically-oriented education), and it is difficult to know what peasant-farmers more generally may have thought about such stations. What is clear, however, is that the plant-breeding station was just one of several Bavarian experiment stations which were all established (or taken over by the state) at about the same time: a research station for moorland agriculture in 1900 (*Bayerische Moorkulturanstalt*), a fisheries research station in 1900 (*Bayerische Biologische Versuchsanstalt für Fischerei*), and an experiment station for agricultural botany (*Agrikulturbotanische Anstalt*) in 1902. Moreover, the government and allied agricultural organisations devoted considerable space to the stations - and to the state expenditure which made them possible - in various publications. In its 1897 survey of agricultural progress, for example, the Bavarian Interior Ministry drew attention to the fact that the number of experiment stations had increased since 1890 and that the stations' testing work had increased substantially in recent years. (The Ministry was also evidently sensitive to the fact that farmers had to pay for such tests, for it pointed out that from January 1897 the test-fees had been halved and were thus 'not very high')[26]. Whatever *farmers'* perception of the stations may have been, therefore, it seems clear that agricultural

[26] Bayerisches Staatsministerium des Innern (1897: 326-327). In its annual report the following year, the State Agriculture Commission pointed to the establishment of the moorland experiment station as well as to state support for varietal-testing as examples of the state's concern for farmers' welfare (Jahresbericht des Bayerischen Landwirtschaftsrats, 1898). In its 1905 survey the Ministry devoted a lengthy section to the state's moorland experiment stations, pointing out that recent visiting experts from Prussia, Austria, Sweden and Russia had been impressed with the Bavarian facilities (Bayerisches Staatsministerium des Innern, 1905: 16-24). Sections were also devoted to the stations for agricultural botany (*ibid.*, 102-105) and plant-breeding (*ibid.*, 105-107), in each case with figures on expenditure. For a general discussion of these and other similar stations, see LUNDGREEN *et al.* (1986: 96-100).

officials themselves were keen to promote the stations as an important contribution to the well-being of the farming community.

V. Conclusion

Although the stations were obviously designed to promote agricultural development, that was probably not their sole function. For it is also clear that the Bavarian state came under considerable pressure from the early 1890s to demonstrate to farmers that government - along with closely associated elite organisations such as the Bavarian Agricultural Association - was taking the worsening agricultural crisis of the 1890s seriously. Agricultural officials took every opportunity to declare their solidarity with farmers and to list the variety of measures which the state had introduced in order to assist the small farmer. While there is no direct archival evidence that political motives underlay the decision to establish the Bavarian plant-breeding station, the indirect evidence makes it highly likely.

But did the factors which were formative in Bavaria apply elsewhere in southern Germany? Although I cannot present the evidence here, the answer seems to be 'yes'. The situation in the southern German state of Württemberg, for example, suggests that similar factors led to the establishment of a station there in 1905. On the other hand, if peasant radicalism, along with the absence of commercial plant-breeding, were the important spurs to the formation of the southern German stations, one ought not to find such peasant-friendly institutions in northern Germany. For in the north the number of estate-owners was much larger, such that rural protest was dominated, not by peasant organisations but by the elite-led Agrarian League. Moreover it was in the north where commercial plant-breeding was well-established. To be sure, public-sector plant-breeding institutions were also established in the north in the decade before the First World War. The interesting thing about them, however, is that none of them was explicitly designed to serve peasant-farmers. Their research and development work was intended either to *support* the northern private-sector or to *complement* its breeding-work rather than providing small farmers with an *alternative* to it (Harwood, 2012).

Moving further afield, what implications might this case have for our understanding of European agricultural policy? It is sometimes suggested that, apart from trade policy, most nineteenth century state agricultural policy was designed to assist farmers to help themselves. Only in the First World War did the state intervene directly in the agricultural economy. The southern German stations, however, do not fit altogether easily into this framework. To be sure, they did foster self-help by providing information on suitable varieties and by teaching the region's small farmers how to do plant-breeding. Moreover, their founding statutes forbade 'commercial activity', no doubt with an eye to the sensitivities of the large private-sector breeding industry which had by then emerged. Accordingly, the stations did not sell varieties which they had bred on the open market. In Bavaria, for example, those varieties bred in-house were turned over to the neighbouring state farm for multiplication and sale at minimal cost. Alternatively, the station carried out the technically demanding early stages of breeding in-house and handed over such 'unfinished varieties' free or at cost to selected cooperatives or individual breeders who finished off the breeding process, multiplied the seed and sold it at moderate cost (since the station's technical assistance had relieved these breeders of having to maintain expensive facilities).

Despite distancing themselves from 'commercial activity', however, the stations nevertheless had a definite impact upon the seed market. They were, after all, intended to free small farmers from dependence upon commercial plant-breeding and to foster the growth of small breeding enterprises in their region. Accordingly, the stations encountered private-sector suspicion from an early stage. Sure enough, following the stations' establishment, the number of individual or cooperative breeders in southern Germany increased markedly, and by the 1920s quite a few of the varieties bred in association with the Bavarian station, at least, had become very successful (Anon., 1935). With the deterioration of the German agricultural economy from the mid-1920s, therefore, older well-established breeding firms began to feel the pressure from the Bavarian station and the new smaller breeding enterprises which it was supporting. Recurrent complaints of 'unfair competition' and improper state intervention were directed at the Bavarian station in particular, and in the early 1930s the older firms vigorously lobbied both the Bavarian Ministry of Agriculture and the state legislature, calling for the station to abandon all work in varietal development. Although the Ministry defended the station's policies, after 1933 older firms were able to secure support for their campaign from the new regime (Harwood, 2012). Thus what had started out in 1900 as a relatively uncontroversial attempt to help 'little players' in the agricultural economy proved so effective that by the 1920s it was threatening the big ones. Though seemingly innocuous at first, the state's promotion of self-help eventually turned out to be market-intervention in disguise. It might, therefore, be worth asking whether other forms of state assistance before 1914 had a similar historical trajectory.

Archival sources

Bayerisches Hauptstaatsarchiv, Munich (BayHStA).

Bibliography

ALBRECHT, Dieter (1974), 'Von der Reichsgründung bis zum Ende des ersten Weltkrieges (1871-1918)', in Max SPINDLER (ed.), *Handbuch der bayerischen Geschichte,* vol 4, 1, Munich, C.H. Beck, p. 283-386.

ANON, (1900), 'Über den Anbauwerth verschiedener Weizensorten', supplement to *Allgemeine Norddeutsche Zeitung*, 70 (24 March).

ANON, (1907), 'Die Einrichtung der DLG zur Förderung der Pflanzenzüchtung und des Pflanzenbaues', *Badisches Wochenblatt*, p. 220-222.

ANON, (ed.) (1935), *Reichssortenliste der zugelassenen Kulturarten. Nach dem Stand vom 1. Juni 1935*, Berlin, Reichsnährstand.

ARNIM-SCHLAGENTHIN, Graf von (1901), 'Die schwedische Saatzuchtanstalt Svalöf', *Deutsche Landwirtschaftliche Presse*, 28, p. 138-140.

BALLWANZ, Ilona (1978), 'Der Zusammenhang zwischen der Produktionsentwicklung und der Betriebsgrösse in der deutschen Landwirtschaft von 1871 bis 1914', *Jahrbuch für Wirtschaftsgeschichte*, 3, p. 77-99.

BAYERISCHES STAATSMINISTERIUM DES INNERN (1890), *Die Landwirtschaft in Bayern. Denkschrift*, Munich, R. Oldenbourg.

BAYERISCHES STAATSMINISTERIUM DES INNERN (1897), *Die Massnahmen auf dem Gebiete der landwirtschaftlichen Verwaltung in Bayern, 1890-1897. Denkschrift*, Munich, R. Oldenbourg.

BAYERISCHES STAATSMINISTERIUM DES INNERN (1905), *Die Massnahmen auf dem Gebiete der landwirtschaftlichen Verwaltung in Bayern, 1897-1903. Denkschrift*, Munich, R. Oldenbourg.

BERGMANN, Hannsjoerg (1986), *Der Bayerische Bauernbund und der Bayerische Christliche Bauernverein, 1919-1928*, Munich, C.H. Beck.

BÖHMER, G. (1914), 'Die Entwicklung der Sortenfrage und ihre Lösung durch Sortenprüfung', *Kühn-Archiv*, 5, p. 191-206.

BROILI, Josef (1899), 'Welche Sorten von unseren Getreidearten werden jetzt in Süddeutschland angebaut...?', *Jahrbuch der DLG*, 14, p. 214-219.

DIPPER, Christoph (1993), 'Landwirtschaft im Wandel: Neue Perspektiven der preussisch-deutschen Agrargeschichte im 19. Jahrhundert', *Neue politische Literatur*, 38, p. 29-42.

FARR, Ian (1978), 'Populism in the Countryside: the Peasant Leagues in Bavaria in the 1890s', in R. J. EVANS (ed.), *Society and politics in Wilhelminian Germany*, London, Croom Helm, p. 136-159.

FRUWIRTH, Carl (1907), *Sorten, Saatfruchtbau und Pflanzenzüchtung in Württemberg*, Plieningen, Friedrich Find.

FRUWIRTH, Carl (1911), *Die Beziehungen der Pflanzenzüchtung zu den Wirtschaftsbetrieben*, Leipzig, Reichenbachische Verlagsbuchhandlung.

GISEVIUS, Paul (1901), *Die Sortenfrage in den Nordostprovinzen, ein Führer für die Auswahl passender Getreide- und kartoffel-Sorten...*, Berlin, Parey.

GOLTZ, Theodor von der (1903), *Geschichte der Deutschen Landwirtschaft,* vol 2, Stuttgart / Berlin, J.G. Cotta.

HANSEN, Johannes & FISCHER, Gustav (1936), *Geschichte der Deutschen Landwirtschafts-Gesellschaft*, Berlin, DLG.

HARWOOD, Jonathan (2008), 'Research and Extension in Political Context: Rural Unrest and the Origins of the Prussian Chambers of Agriculture', in Nadine VIVIER (ed.), *The State and Rural Societies: Policy and Education in Europe, 1750-2000*, Turnhout, Brepols, Rural History in Europe, 4, p. 135-157.

HARWOOD, Jonathan (2012), *Europe's Green Revolution and Others Since: the Rise and Fall of Peasant-friendly Plant Breeding*, London, Routledge.

HAUSHOFER, Heinz (1963), *Die deutsche Landwirtschaft im technischen Zeitalter*, Stuttgart, Ulmer.

HAUSHOFER, Heinz (1986), 'Bäuerliche Führungsschichten in Bayern im 19.-20. Jahrhundert', in Friedrich PANKRAZ & Wolfgang ZORN (eds), *Aus der Bayerischen Agrargeschichte 1525-1978. Gesammelte Beiträge zur Bayerischen Agrargeschichte von Heinz Haushofer*, Munich etc., BLV Verlagsgesellschaft, p. 39-48.

HILLMANN, Paul (1910), *Die deutsche landwirtschaftliche Pflanzenzucht*, Berlin, DLG-Verlag.

HOCHBERGER, Anton (1991), *Der bayerische Bauernbund, 1893-1914*, Munich, C.H. Beck.

HUNDHAMMER, Alois (1926), *Die landwirtschaftliche Berufsvertretung in Bayern*, Munich, doctoral dissertation, University of Munich.

KEUNE, Otto (ed.) (1952), *Männer die Nahrung schufen*, Hannover, Landbuch Verlag.

KIESSLING, Ludwig (1910), 'Die Königliche Bayerische Landessaatzuchtanstalt', in P. HILLMANN (ed.), *Die deutsche landwirtschaftliche Pflanzenzucht*, Berlin, DLG, p. 155-168.

KIESSLING, Ludwig (1912), 'Die züchterische Bearbeitung der Landsorten in Bayern', *Beiträge zur Pflanzenzucht*, 2, p. 74-96.

KIESSLING, Ludwig (1922), 'Einige Gegenwartsfragen der Pflanzenzuchtung und des Saatbaues', *Mitteilungen der DLG*, p. 454-460.

KOCHENDOERFER, Martin (1996), 'Landeshoheitliche und genossenschaftliche Getreidezüchtungsbestrebungen zur Jahrhundertwende', *Zeitschrift für Agrargeschichte und Agrarsoziologie*, 44, p. 174-193.

KONING, Niek (1994), *The Failure of Agrarian Capitalism: Agrarian Politics in the United Kingdom, Germany, the Netherlands and the USA, 1846-1919*, London/New York, Routledge.

KOPSIDIS, Michael (2006), *Agrarentwicklung: Historische Agrarrevolutionen und Entwicklungsökonomie*, Stuttgart, Franz Steiner.

KRAUS, Carl (1899a), 'Förderung der Landwirtschaft in Bayern durch praktische Anbau- und Düngungsversuche', *Vierteljahresschrift des Bayerischen Landwirtschaftrathes*, 4, supplem. vol, p. 332-353.

KRAUS, Carl (1899b), 'Wie erzielt man die beste Braugerste', *Jahrbuch der DLG*, 14, p. 220-231.

KRAUS, Carl (1900), (untitled talk), in *Bericht über die XXXVI. Wanderversammlung bayerischer Landwirte in Nördlingen vom 28. und 29. Mai 1900*, Augsburg, Ph. J. Pfeiffer, p. 68-88.

KRECH (1904), 'Die Einrichtung von Saatbauvereinen im nordöstlichen Deutschland, insbes. des westpreussischen Saatbauvereins Althausen', *Jahrbuch der DLG*, 19, p. 240-243.

KRYZYMOWSKI, Richard (1913), 'Beziehungen zwischen der Betriebsintensitat und der Sortenfrage', *Jahrbuch der DLG*, 28, p. 456-467.

KÜHLE, Ludwig (1926), 'Der Stand und die Lage der deutschen Pflanzenzucht', *Mitteilungen der DLG*, 41, p. 856-865.

LEHMANN, Hans Georg (1970), *Die Agrarfrage in der Theorie und Praxis der deutschen und internationalen Sozialdemokratie*, Tübingen, J.C.B. Mohr/ Paul Siebeck.

LIEBSCHER, Georg (1896), 'Die Getreidezüchtung, ein Mittel von grosser Bedeutung für die Rentabilität des Getreidebaues', *Würtembergisches Wochenblatt für Landwirtschaft*, p. 217-218.

LOCHOW, Ferdinand von (1904), 'Die Aufgaben der Saatzucht-Abteilung', *Jahrbuch der DLG*, 19, p. 25-31.

LOCHOW, Ferdinand von (1913), '25 Jahre deutsche Pflanzenzucht', *Illustrierte Landwirtschaftliche Zeitung*, p. 448-450.

LUNDGREEN, Peter *et al.* (1986), *Staatliche Forschung in Deutschland, 1870-1980*, Frankfurt a.M./New York, Campus.

MÖCKL, Karl (1972), *Die Prinzregentenzeit. Gesellschaft und Politik während der Ära des Prinzregenten Luitpold in Bayern*, Munich/Vienna, R. Oldenbourg.

MODROW, Wilhelm (1904), 'Saatzucht im nordöstlichen Deutschland', *Jahrbuch der DLG*, Vol 19, p. 243-255.

MOELLER, Robert (1981), 'Peasants and Tariffs in the *Kaiserreich*: How Backward were the *Bauern*?', *Agricultural History*, 55, p. 370-384.

MOOSER, Josef (1992), 'Preussische Agrarreformen, Bauern und Kapitalismus. Bemerkungen zu Hartmut Harnischs Buch "Kapitalistische Agrarreform und industrielle Revolution"', *Geschichte und Gesellschaft*, 18, p. 533-554.

MOSER, Peter (2003), *Züchten, säen, ernten: Agrarpolitik, Pflanzenzucht und Saatgutwesen in der Schweiz, 1860-2002*, Baden, Switzerland, Hier und Jetzt Verlag.

POMMER, E. (ed.) (1898), *Jahresbericht über die Erfahrungen und Fortschritte auf dem Gesammtgebiete der Landwirtschaft,* 13, Braunschweig, Vieweg.

PRIBYL, Leo *et al.* (ed.) (1890), *Das Goldene Buch der Land- und Forstwirtschaft in Oesterreich-Ungarn*, Vienna.

PROSKOWETZ, Emanuel (1937), *Memorabilien*, Prague, Tschechische Akademie der Landwirtschaft.

RAUM, Hans (1912), 'Züchtung und Saatbau des Fichtelgebirgshafers', *Landwirtschaftliches Jahrbuch für Bayern*, 2, p. 841-940.

RAUM, Hans (1972), 'Neuere Zeit: Geschichtliches, persönliche Erinnerungen', *Beiträge zur Geschichte von Weihenstephan*, p. 1-19.

REMY, Theodor (1907), 'Einige Gedanken über die Gefahren und Nachteile des modernen Pflanzenzuchtbetriebes', *Deutsche Landwirtschaftliche Presse*, 34, p. 687-688.

RÜMKER, Kurt von (1895), 'Die Rassenzüchtung landwirtschaftlicher Kulturpflanzen als Forschungsgebiet und Lehrgegenstand', in *Festschrift zum 70. Geburtstag J. Kühns*, Berlin, Parey, p. 51-77.

RÜMKER, Kurt von (1898), 'Der wirtschaftliche Mehrwert guter Kulturvarietäten und auserlesenen Saatgutes', *Arbeiten der DLG*, 36, p. 127-153.

RÜMKER, Kurt von (1899), 'Kann Deutschland seinen Getreidebedarf noch selbst decken?', *Mitteilungen der landwirtschaftlichen Institute der königlichen Universität Breslau*, 1, p. 152-195.

RÜMKER, Kurt von (1900), 'Üeber die Bedeutung der landwirt. Versuchsstationen und speciell ueber den Wert und die Organisation von Versuchswirtschaften in historischer Beleuchtung', *Journal für Landwirtschaft*, 48, p. 179-203.

RÜMKER, Kurt von (1907), 'Die systematische Einteilung und Benennung der Getreidesorten für praktische Zwecke', *Fühlings Landwirtschaftliche Zeitung*, 56, p. 545-566.

RÜMKER, Kurt von (1911a), 'Über Bedeutung und Methoden der Saatgutzucht', *Mentzel und Lengerkes Schreibkalender*, p. 64-90.

RÜMKER, Kurt von (1911b), *Das landwirtschaftliche Versuchs- und Unterrichtswesen in den USA*, Berlin, Parey.

RÜMKER, Kurt von (1912-13), 'Die Entwicklung der landwirtschaftlichen Pflanzenzucht in Deutschland und ihre betriebswirtschaftliche Aufgaben', *Zeitschrift für Pflanzenzüchtung*, 1, p. 329-346.

RÜMKER, Kurt von (1928), '20 Jahre Gesellschaft zur Förderung deutscher Pflanzenzucht', *Mitteilungen der DLG*, 43, p. 248-250.

SAATZUCHTSTELLE DER DLG (1905), 'Einrichtungen der DLG zur Förderung der Pflanzenzüchtung und des Pflanzenbaues', *Mitteilungen der DLG*, 20, p. 104-108.

SCHLÖGL, Alois (1954), *Bayerische Agrargeschichte: die Entwicklung der Land- und Forstwirtschaft seit Beginn des 19. Jahrhunderts*, Munich, Bayerischer Landwirtschafts-Verlag.

STEGLICH, Bruno (1905), *Einrichtung und Tätigkeit der Saatzuchtanstalt Svalöf in Schweden und Fortschritte auf dem Gebiet der Pflanzenzüchtung*, Leipzig, Reichenbach.

STEGLICH, Bruno (1910), 'Dresden: Landwirtschaftliche Abteilung d. Königlich Sächsischen Pflanzenphysiologischen Versuchsstation', in P. HILLMANN (ed.), *Die deutsche landwirtschaftliche Pflanzenzucht*, Berlin, DLG, p. 47-59.

STEINBECK, Norbert (1997), *Bürokratisierung und betriebliche Rationalisierung: die preussischen Landwirtschaftskammern und die Entwicklung der Agrarverfassung im wilhelminischen Kaiserreich*, Bern etc., Peter Lang.

TIRRELL, Sarah R. (1951), *German Agrarian Politics after Bismarck's Fall: the Formation of the Farmers' League*, New York, Columbia University Press.

TSCHERMAK, Erich von (1958), *Leben und Wirken eines österreichischen Pflanzenzüchters*, Berlin / Hamburg, Parey.

ULANDER, Axel (1906), 'Die schwedische Pflanzenzucht zu Svalöf. Eine kurze Darstellung', *Journal für Landwirtschaft*, 54, p. 105-124.

WERNER, Hugo & ALBERT, Friedrich (1900), *Der Betrieb der Deutschen Landwirtschaft am Schluss des 19. Jahrhunderts (= Arbeiten der DLG, No. 51)*, Berlin, DLG.

12. Mediators between the industrial state and agriculture: the social profile and professional activities of agronomists in Switzerland, 1871-2007

Daniel FLÜCKIGER

I. Introduction

As the number of people working in the agricultural sector decreased drastically in Europe during the second half of the twentieth century, the number of employees in closely related areas of the economy continued to multiply. Fewer and fewer farmers, farmers' wives and farm labourers faced more and more advisors, salesmen, teachers, clerks and researchers – many of whom were trained agronomists. For a long time historical research neglected the rise of these new occupations, even though their increasing prominence and interaction with the farming population is one of the more significant features of agricultural change in the twentieth century, as recently published studies have maintained (Auderset, Bächi & Moser, 2012; Moser, 2003: 25; Uekötter, 2006a: 148).

Historians have often argued that improvements in agriculture were initiated 'from above' and failed because of the 'stubbornness' of farmers. Frank Uekötter (2006a: 160) has pointed out that this interpretation, however popular it may be among political elites, certainly does not hold good in the sphere of environmental stewardship. By downplaying the relevance of practice in the development of theories and policies, the 'from above' perspective might even be said to be more part of the problem than the solution.

Éric Mollard (2002: 10-12) and Deborah Fitzgerald (2003: 5-6), for instance, suggest that the ongoing and intimate contact farmers have with the soil, plants and animals leads to a knowledge that enables them to use these resources in ways conducive to sustainable development. They describe the modernization of agriculture in the twentieth century, with its negative social and environmental effects, as a project that was forced upon agricultural communities from outside and as one that has mainly reflected the ideals of political elites. Agronomists, according to this view, were aligning themselves with the interests of the state and the political elites and represented their interests in the sphere of agricultural practice. Under such circumstances the social distance between the farming population and the new occupations was bound to widen.

While this perspective is convincing, the question remains: how well is the suggested social distance between the farming population and the agronomists supported empirically? Until recently, a major obstacle to answering this question was the absence of sources produced by the farming population themselves. Uekötter (2006a: 148) has suggested that the technical literature of agricultural advisors might be a fruitful proxy source of evidence. But such documents lamentably provide little information on the social background of agronomists and their relationships with farmers.

Fortunately, in the last ten years a great variety of sources produced by farmers, farmers wives and farm labourers have been unearthed and catalogued by the Archives of Rural History (ARH) in Bern. Since its foundation in 2002 the ARH have made accessible the papers not only of more than a hundred farmers and farmers organizations at the local, regional and national levels, but also a substantial body of new source material from agricultural colleges and research stations of the state. Many of these collections contain texts, minutes and copious correspondence written by young farmers who were educated in agricultural colleges by agronomists or by farmers and farmers wives who worked closely with scientists in plant or cattle breeding projects (Brodbeck, Flückiger & Moser, 2007; Brodbeck, Ineichen & Schibli, 2012). Switzerland, therefore, is a particularly suitable case to conduct an empirically based analysis of the relationship between farmers and agronomists and the process of knowledge creation and transfer between the new occupations and the farming population[1].

For more than a century here the agronomists (*Ingenieur-Agronomen*) were exclusively trained at the Swiss Institute of Technology (ETH) in Zurich. This training course was established in 1871 by the Swiss Federal State with the aim of forming a national elite in the agricultural sector. And, indeed, many graduates soon began to occupy important positions in the civil service, in agricultural associations and in private companies within the agro-industrial complex. This chapter aspires to provide a clearer picture of the Swiss agronomists as a distinctive and well-organised professional group. It aims, in addition, to shed light on the relationship between the agronomists and the farming population in the process of knowledge production.

To this end a social profile of agronomists is provided by first looking at their social origins, training and occupational outlets between 1871 and 2006. Then their professional self-organisation, and their means of advancing and defending their

[1] This article is based on a research project of the Archives of Rural History on agronomists which has been funded by the State Secretariat for Education and Research (SER) in the course of the COST action A35.

occupational interests will be discussed. Thirdly, some of the day-to-day activities of agronomists, in particular in relation to the topic of feed conservation techniques from the 1920s to the 1950s, will be examined more closely. The influence that social and institutional circumstances have had on the professional activities of agronomists will be assessed here as well as the claimed development of agriculture towards a science-based industry. Finally, the question of whether it is appropriate to see the modernization of agriculture as something imposed from outside and 'from above' will be discussed.

II. Social profile

The term 'agronomist' dates back to the middle of the eighteenth century (Mollard, 2002: 6) when it came to be associated with a European movement of agricultural reforms and improvements organized by 'economic societies'. Prominent examples of such societies are found in Edinburgh, Dublin, London, Rennes, Zurich and Bern (Holenstein, Stuber & Gerber-Visser, 2007: 13). In the Swiss context, the Economic Society in Bern (*Oekonomische Gesellschaft Bern*) has become the subject of extensive research during the last few years (Stuber *et al.*, 2009).

The agronomists of the eighteenth century operated in significantly different surroundings compared to those of the late nineteenth and twentieth century: they lived in a pre-industrial economy[2]. Agriculture was the leading sector of the economy and their main aim was to achieve an optimal use of biotic resources. Society was viewed as depending on agriculture and as flourishing only when agriculture and forestry provided them with the necessary resources (Pfister, 1990: 37-49; Sieferle et al., 2006: 49-50). Furthermore, the eighteenth century agronomists were not professionals in the modern sense but rather, in most cases, landowners who experimented on their own estates or farms with new methods of crop and animal husbandry and discussed the results in learned journals. Most of these individuals belonged themselves to the political elites (Mollard, 2002: 9). They offered their followers opportunities beyond individual economic profit and their authority was often independent of their practical success or failure (Stuber *et al.*, 2009: 26). In the second half of the nineteenth century, agriculture lost its leading role in societies that had become, or were in the process of becoming, industrial. No longer did 'modernization' primarily mean optimizing the use of biotic resources; it now came to be associated with a more extensive consumption of fossil resources. In this situation, 'old' agronomists such as Albert von Fellenberg-Ziegler in Bern expected agriculture to become an industry

2 The agronomists of the eighteenth century were familiar with the word 'industrious', but used it with a different meaning from what later came to be understood as 'industrial'. For them, 'industrious' people were essentially diligent labourers (Holenstein, 2007: 17; de Vries, 1994).

in the same rank as other industries (Flückiger, 2009a: 171-172). They regarded the creation of an academically trained profession as crucial for implementing their vision of a 'new' agriculture and used the newly established agricultural societies in the 1860s to further this demand.

For over a century the Agricultural Department at the Swiss Institute of Technology (ETH), founded in Zürich in 1871, remained the main institution for the education of agronomists in Switzerland. In contrast to their often self-educated predecessors, the 'new' agronomists were all formally trained and earned their living from pursuing their vocation. In the beginning it was difficult for the Agricultural Department to attract students from Switzerland; in fact in the first two decades most students were foreign and all the professors were German (Kraemer, 1896; Düggeli, 1929). The small number of Swiss graduates soon gave rise to criticism, and in 1883 the Federal government responded by supporting the cantons in establishing or expanding agricultural colleges (Schweizerischer Bundesrat, 1883: 871). Before long these colleges had become the most important single source of students for the Agricultural Department at the ETH.

Figure 12.1 illustrates how, in the aftermath of these reforms, an increasing number of students completed their degrees in agriculture. It also shows how the number of agronomists rose particularly strongly after World War I, World War II and again from the 1960s to the 1980s. These three waves, associated with the two World Wars and with the boom in the second half of the twentieth century, are equally evident across the entire range of disciplines taught at the ETH. But since the 1990s, as expansion gave way to contraction, the Agricultural Department has suffered the sharpest decline in student enrolment.

Since the 1970s the ETH has no longer been the sole institution educating agronomists in Switzerland. Several other institutions began to confer a similar, but not identical title on their graduates. The most important of these is the Hochschule für Agrar-, Forst- und Lebensmittelwissenschaft (HAFL) in Zollikofen. In contrast to the ETH, the numbers of graduates at the HAFL has remained stable (Figure 12.2). However, education at the HAFL also has been subject to fundamental changes during the last 30 years. Agriculture, compared to rural-based leisure activities such as horse keeping, has declined in importance at the HAFL as well.

Figure 12.1. Graduates in agricultural sciences at the Swiss Institute of Technology (ETH Zürich), 1872-2006

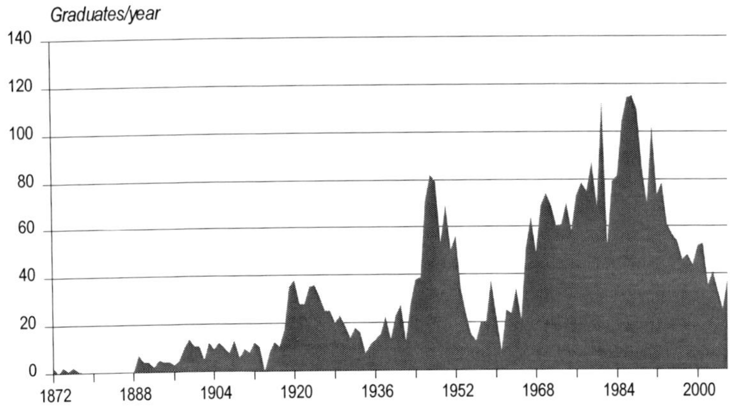

Sources. Erteilte Diplome an der ETH Zürich 1855-2002, url: http://www.ethistory.ethz.ch/materialien/statistiken (accessed: 02.01.2007); Diplome für Bachelor-, Master- und Diplomstudium 1990-2006, url: http://www.fc.ethz.ch/facts/studierende/zeitreihen/ (accessed: 15.12.2006).

Figure 12.2. Graduates in agricultural sciences at the Swiss Agricultural College (SHL Zollikofen), 1969–2007

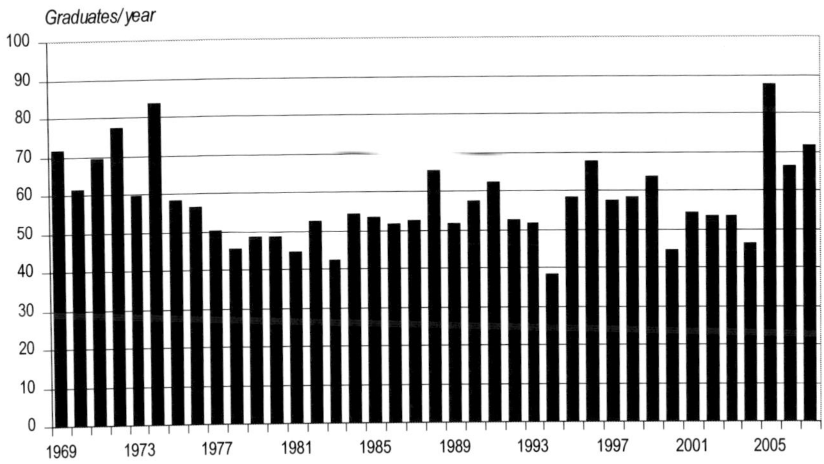

Sources. Schweizerische Hochschule für Landwirtschaft, 1969-2006; Pressemitteilung, 17.01.2008, url: http://www.shl.bfh.ch/, accessed: 29.01.2008; Pressemitteilung, 25.04.2008, url: http://www.shl.bfh.ch/, accessed: 04.05.2008.

Beside the statistical information, extensive information on agronomists has been amassed by the ARH from obituaries in the *Bulletin of the Swiss Association of Agronomists and Agricultural Teachers*. The analysis of these obituaries makes it obvious that the social composition of those attending the Agricultural Department at the ETH has changed since the turn of the new century. In the twentieth century, most students were Swiss, predominantly from the German-speaking area; they came from both religious denominations, Catholic and Protestant, and represented all four language groups. (Zollikofer, 1954). In Zürich, as in Wageningen (Burg & Bos-Boers, 2003: 38), women began to study agricultural sciences later than men. The first female student to graduate in agronomy in Zürich was Sophie Sawicka who was awarded her diploma in 1904[3]. But it was only in the late twentieth century that the number of women studying agricultural sciences increased significantly. Between 1970 and 1998 the proportion of female students rose from five to forty-six per cent[4]. For most of the period under survey the ETH offered graduates of the agricultural colleges a special entrance examination. Since many of them were from a farming background, they were in a relatively privileged position when studying agronomy at the ETH where all students at the Department of Agriculture needed at least twelve months work experience in agriculture before they could graduate (Zollikofer, 1954).

Our sample consists of 131 agronomists, among them most of the renowned members of the profession in the twentieth century[5]. Overall, at least a third of those who graduated before 1971 had formerly attended an agricultural college or completed an agricultural apprenticeship. Some two-thirds were graduates from grammar schools (Gymnasien) or had completed a non-agricultural apprenticeship before studying at the ETH. The proportion of sons of farmers among all agronomists was most likely even higher. Of the 131 examined obituaries relating to agronomists who died between 1933 and 2005, two-thirds (87) were the sons of farmers. Although only a few per cent were the sons of agronomists, a small number of highly prominent dynasties of agronomists (such as the Keller family) did exist. Sometimes the sons of agronomists even became farmers themselves. Farming was, at least until the 1970s, not an inferior activity for agronomists. For Switzerland, therefore, it is hardly appropriate to assume a big social distance between agronomists and farmers or to claim that this 'new class' forced an alien modernization project upon farmers. The results for the agronomists fit fairly well with the general pattern of the professions

3 See the biographical database Personen der ländlichen Gesellschaft im 19. und 20. Jahrhundert (www.agrararchiv.ch).

4 Weibliche Studierende an der ETH Zürich 1917-2002, url: www.ethistory.ethz.ch/statistik/, accessed: 15.12.2006.

5 All the biographical information relating to Swiss agronomists used in this article is to be found in the biographical database Personen der ländlichen Gesellschaft im 19. und 20. Jahrhundert (www.agrararchiv.ch).

(lawyers, medical doctors, and engineers) in Switzerland which seem to depend less on the central state than elsewhere in continental Europe (Siegrist, 1988: 94; Siegrist, 1996: 353). On the other hand, state institutions were crucial in Switzerland as well for creating and promoting the agronomists as a profession.

III. Professional organization and activities

Professions can be viewed as occupational groups who obtain a jurisdictional authority over specific fields on the basis of their monopoly of specialised skills. They usually form their own organizations, accredit the skills of their members, justify their claims with reference to public purposes and functional necessities and generate considerable individual incomes on the basis of eliminating the competition of laymen (Ridder-Symoens, 1996; Brassley, 2005; Siegrist, 2005: 70). Although Harold Perkins (1989: 83) uses the notion of 'professional classes', more recent contributions emphasize that professions were more open than closed in a way that allowed for a certain social mobility (Ackroyd et al., 2006: 17-19). In this context, the Swiss agronomists are rather a profession than a 'new class'.

The vast bulk of the 'new' agronomists spent their working lives in three areas (frequently changing from one to another in the course of their careers): education and extension, central administration and the agricultural organizations. These fields of activity were especially important, but of course many agronomists were also to be found right across the food chain: from the field of production to the processing industries and food retailing.

The Agricultural Department at the ETH had also been established to educate agricultural teachers for the agricultural colleges. When the 'new' agronomists founded their own organization in 1901, teaching was their main concern, the name of the new institution was: Swiss Association of Agricultural Teachers (*Schweizerischer Verband der Lehrer an Landwirtschaftsschulen* SVIAL). The association published textbooks, organized further vocational training, established contacts with colleagues in other countries and kept its members informed of job opportunities. In this institutional context agronomists also reflected on their collective identity and on the defining tenets of their profession (*Schweizerischer Verband der Lehrer an landwirtschaftlichen Schulen und Ingenieur-Agronomen,* 1951: 9–13).

Table 12.1 shows the occupational activities of agronomists during the twentieth century. The columns from 1900 to 1972a are based on the obituaries mentioned above. The columns from 1972b on are based on membership surveys conducted by what was now called the SVIAL Association. Education was by far the most important occupation open to agronomists in 1900; after a slow relative decrease it still accounted for one-fifth of their known employment in 1999. It should be noted that the percentages shown for extension are most likely too low because, initially, advising farmers in the fields was a summer occupation for agronomists who taught winter classes at the agricultural colleges. It was during the long summer break that they were able to visit farms, work for agricultural organizations or conduct experiments. Sometimes these part-time activities developed into new full-time occupations. The connection between education and extension was especially strong in Switzerland where a separate advisory service was established only in 1958 (Blum, 1993; Baumann & Moser, 1999: 122-127, 380-382).

Table 12.1. Professional activities of members of the *Schweizerischer Verband der Ingenieur-Agronomen und der Lebensmittel-Ingenieure (SVIAL)*, 1900-1999

Professional Activity (%)	*1900*	*1925*	*1950*	*1972 (1)*	*1972 (2)*	*1985*	*1989*	*1999*
Public Administration	17	10	11	12	13	16	16	11
Education	44	37	28	22	22	24	20	21
Agr. Unions/ Associations	8	15	20	20	na	na	na	Na
Marketing/Management/Sales	na	na	na	na	15	11	18	10
Research and Development	14	17	20	23	35	20	30	16
Extension	0	0	3	5	8	22	11	30
Development Cooperation	0	1	2	3	4	2	5	na
Divers	17	18	16	17	3	5	0	11
Total	100	100	100	100	100	100	100	100
Number of Persons	20	113	188	150	na	na	1549	352

Sources. Archiv für Agrargeschichte: Personen der ländlichen Gesellschaft im 19. und 20. Jahrhundert; Archiv SVIAL (1971(2)-1999). With the exception of the column for 1999 more than one answer per person was possible.

The expansion of education and extension was crucial for both the survival and the later success of the Agricultural Department at the ETH. At the beginning of the twentieth century a majority of the Swiss students came from agricultural colleges. While these colleges provided the most important reservoir for prospective students of agronomy, they also offered the most important job opportunities for agronomists. Typically agronomists went back to teach in agricultural colleges after graduating. The circle was completed when some of their own students decided – frequently on the advice of their teachers – to become agronomists themselves.

Agronomists played an active role in the development of the agricultural colleges. Adolf Kraemer, one of the first professors at the ETH in Zürich, supported the introduction of winter classes (*Winterschulen*), a new type of agricultural education which he knew from the south-west of Germany. From the 1880s onwards agronomists such as Hans Moos, Josef Käppeli and Carl Moser set up dozens of new winter classes. This led to a big increase in the number of students. Especially the sons of farmers, whose presence on the home farm was indispensable during the summer months, now attended the agricultural colleges in significant numbers (Käppeli, 1911; Wahlen, 1943).

Figure 12.3 shows how the number of students at agricultural colleges increased between the 1890s and the 1920s, the period when winter courses were introduced. The winter schools were essential for the developments of the agronomists as a group, although one should not overestimate the institutionalization of formal education in agriculture before the 1950s when, in addition to the attendance at the agricultural colleges, the number of apprentices increased rapidly (Stuber et al., 2009: 39-41).

Figure 12.3. Students of agricultural colleges in Switzerland, 1860-1989

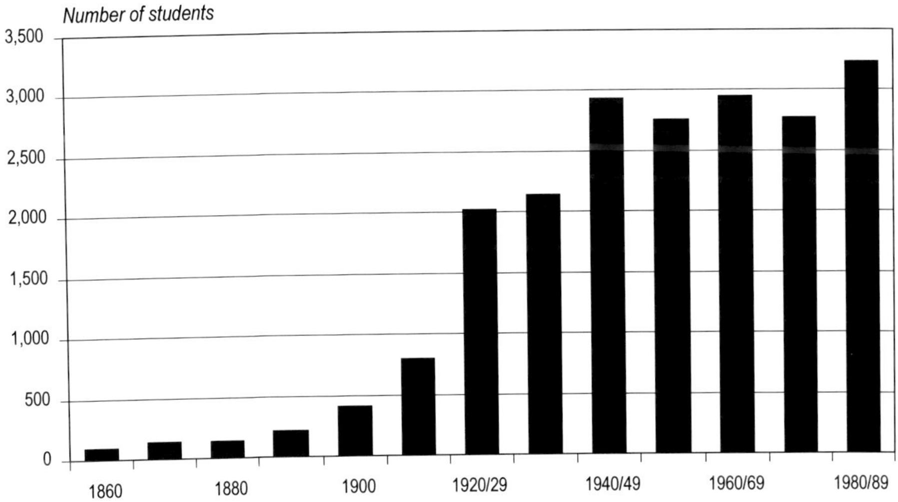

Sources. Brugger & Landwirtschaft (1850-1914: 334); Brugger & Landwirtschaft (1914-1980: 17, 20-29).

Agronomists used the typical strategies of the professions to consolidate their position by relying on a sympathetic state (Siegrist, 2005: 78-81). Agricultural colleges were operated by the cantons, but from 1884 the Federal government paid part of the costs of the textbooks and the teachers' salaries as a means of supporting education and gaining influence. Significantly, the support of the Federal government was granted only if the teachers were graduates of the ETH (Schweizerischer Bundesrat, 1951: 156).

Furthermore, in the first half of the twentieth century, agronomists achieved a monopoly of teaching in agricultural subjects at schools for further training (*Fortbildungsschulen*) in many cantons outside the mountain areas. At the beginning of the twentieth century in many cantons these *Fortbildungsschulen* had become compulsory for rural youth who left school at the age of 16 without taking up formal vocational training; other cantons, mainly in the mountain areas followed suit in the middle of the century (Wegmann, 1960: 56).

Already in the 1930s SVIAL members discussed the desirability of making specialised vocational training compulsory for all farmers who were running a farm – either in the form of an apprenticeship or attendance at an agricultural college (Günthart, 1933: 92; Lichtenhahn, 1942). But it took several decades before this aim was realized. Only the new instrument of direct payments in the 1990s gave the Federal government the power to make farmers conform to whatever the state required them to do or not to do. Since 2007 farmers only receive direct payments for environmental stewardship if they had passed a course of formal vocational training[6]. But the compulsory formal education of farmers is only one part of the ever more encompassing regulation of their economic activities which has been compared to a cage of steel (*stählerner Käfig*) within which European farmers must operate today (Uekötter, 2006b: 123).

State regulation of the agricultural sector contributed substantially to the professional advancement of agronomists. Agronomists were members of the government in the 26 cantons as well as on the national level. And as civil servants they were not only working in the departments of agriculture. Even though the absolute number of politicians and civil servants recruited among the agronomists was never especially high (Table 12.1), those who entered the civil service or made political careers were crucial for the growth of the profession. Many of them were teachers at agricultural colleges before becoming ministers where they often were directly responsible for the expansion of agricultural education. Moreover, since 1913 the director of the Federal Office of Agriculture (BLW) has always been an agronomist. The BLW was crucial in formulating and implementing the agricultural policy that was centralized on the national level since the last quarter of the nineteenth century. During World War I, the director of the BLW, Josef Käppeli, outlined a new conceptual scheme that

6 Weisungen und Erläuterungen vom 31. Januar 2007 zur Verordnung über die Direktzahlungen an die Landwirtschaft vom 7. Dezember 1998, Art. 2.1.c, url: http://www.blw.admin.ch, accessed: 04.02.2008.

determined the shape of agricultural policy for the next few decades (Moser, 2005). In 1959/60 and again in the early 1990s it was again the BLW that formulated (and implemented) yet another new conceptualisation of agricultural policy (Baumann & Moser, 1999: 366-393; Moser & Brodbeck, 2007: 93-109).

One of the consequences of World War I was that the state intervened to a far greater extent in the agricultural sector than before. During the war the BLW established new producer associations as tools to co-ordinate production and provide food supplies for the domestic population. For most farmers membership in producers' associations became compulsory. For instance, farmers no longer were allowed to retail their milk directly to consumers, but had to deliver it to the local cooperative which then sold or transformed it into cheese in accordance with needs at the national level. As a result, the membership of agricultural associations grew to an unprecedented extent up to the 1950s (Moser & Brodbeck, 2007: 15-21).

The growth of the agricultural institutional matrix (Schuurman, above) created more and more job opportunities for agronomists who often became the heads of these new institutions. However, the new policy developed by agronomists also provoked conflicts between farmers and the state. In the 1930s and again in the 1960/70s dissatisfied farmers repeatedly established new associations that opposed the official agricultural policy. Agronomists often tried to appease the members of these movements and pleaded for co-operation with the existing authorities. Overall, agronomists were very loyal to the state and hardly ever played an important role in dissident organizations (Moser, 1994: 393; Baumann & Moser, 1999: 150-186). Renowned agronomists such as Oskar Howald strongly influenced agricultural policy when in office, but tended to become outsiders when they had retired and began to oppose the (new) mainstream consensus around Swiss agricultural policy (Baumann & Moser, 1999: 198-202, 375-380).

After World War I, as we have seen, agricultural associations employed more and more agronomists. Consequently, the Swiss Association of Agricultural Teachers changed its name to the Swiss Association of Agricultural Teachers and Agronomists (SVIAL) in 1933. Agronomists increasingly worked now as advisors, for agricultural associations, as salesmen or as civil servants (Schweizerischer Verband der Lehrer an landwirtschaftlichen Schulen und Ingenieur-Agronomen, 1951: 23–25).

Table 12.1 shows that after 1972 organizational work in agricultural associations disappears as a distinct category. The SVIAL, in its surveys, no longer considered this as a distinguishable form of activity. At the same time, the importance of employment in industry and commerce rose. However, the original questionnaires in the archives of the SVIAL reveal that agronomists themselves designated their work in associations as 'Marketing/Management/Sales', 'Research and Development' or 'Extension'. For the agronomists themselves the difference between working in agricultural associations and 'industry and commerce' was in reality rather fluid.

The fact that representatives of agricultural associations increasingly saw themselves as doing basically the same work as employees of industrial or commercial companies explains to a certain degree the tendency of agronomists to distance themselves from the farming world they often came from. At the ETH itself, professors of the Agricultural Department for the first time questioned their position at the intersection between science and agricultural practice – seeing themselves more and more exclusively as scientists. Already in 1955, Hans Deuel and Edzard Zollikofer demanded the abolition of the special entrance examination (created with farmers' sons in mind). They failed in 1955, but in 1969 a further attempt was successful. Moreover, the Department extended the length of the course of studies from four to five years (Zollikofer, 1954). At the same time the HAFL was established in Zollikofen. The SVIAL was one of the main promoters of the new institution. It planned to leave the 'lower' and more technical and practical occupations to graduates of the HAFL, while the graduates of the ETH – their own members – would compete for the more theoretical (and better salaried) jobs. Teaching, of course, was intended to remain an exclusive domain of graduates of the ETH and the title of 'agronomist' (Agronom) continued to be reserved for ETH graduates[7]. However, the HAFL successfully emphasized the value of practical experience and invested in teaching didactical skills. It didn't take long before their graduates were successfully applying for the same positions as those of their ETH counterparts. In the 1990s they even gained access to teaching posts in the agricultural colleges (*Schweizerische Hochschule für Landwirtschaft,* 1969/70: 10; *Schweizerische Hochschule für Landwirtschaft,* 1970/71: 9; Moser, 2009: 242). Today, HAFL graduates are regarded as much more suitable for teaching jobs in agricultural colleges than their few ETH rivals.

IV. Towards a science-based industry?

In the third part of this paper the promotion of new feed conservation techniques is analysed in order to look at the agronomists' activities in promoting new technologies. Such activities became a crucial element in the frequently described development of agriculture towards a science-based industry. The very notion of a science-based industry (*Verwissenschaftlichung*) implies the increasing dominance of science in a specific field (Wieland, 2004: 8; Straumann, 2005: 313). The success of individual

[7] The lobbying that accompanied these changes is well documented in the archives of the SVIAL.

professions such as the agronomists to establish themselves as a body of experts can be interpreted as a part of the broader trend to build science-based industries.

Historians and sociologists have long argued that politics became less important with the development of science-based industries. But recent studies indicate that scientists too were relying on political power in advancing their claims. Indeed, political decisions were a major driving force behind the development of agriculture towards a science-based industry (Wieland, 2004: 233; Harwood, 2005: 224). A good example to study the link between applied science and politics is the promotion of new feed conservation techniques such as silage-making in the interwar period. While the principles of this technique were already fairly well known in the late nineteenth century (Wahlen, 1979; Brassley, 1996), it was not widely applied in practice in Switzerland until after World War I when the agronomists began to promote silage making on farms as a means of advancing the new agricultural policy of the government.

This policy clearly advocated food security for the domestic population over the production of food for the export market. The young generation of agronomists expected that the shift from a specialised, cattle-based agriculture producing cheese for the world market to a more tillage-based, mixed agriculture would result in a decrease of the land available for fodder production. To maintain a high level of milk and meat supplies for domestic consumption, fodder production had to be made more efficient. This became even more important during World War II, when imports of feeding stuff declined and became insecure. The implementation of more efficient conservation techniques such as silage-making and the use of wooden tripods for hay cocks became increasingly seen as the solution to the problem (Grandjean, 1943: 7; Hess & Ryf, 1943: 6-8).

To improve feed conservation techniques had been one of the aims of the Institute of Animal Feeding (*Institut für Haustierernährung*) ever since it was set up at the Agricultural Department at the ETH in 1925. In close collaboration with agricultural colleges, research stations, large farms and industrial companies, agronomists at the institute conducted surveys and experiments relating to silage making, artificially dried grass and the use of wooden tripods for hay cocks (Wiegner, 1935: 184; Landis, 1945: 16-20). The institute also cooperated with Artturi Ilmari Virtanen, the renowned Finnish silage expert, who was to win a Nobel prize in chemistry for his methods of fodder preservation in 1945 (Wiegner, 1935: 220; Grandjean, 1943: 11). In the late 1930s agronomists and farmers also set up silage associations on a cantonal level. In 1942, the regional associations established the Swiss Silage Association (SSA). The records of the Bernese Silage Association (BSA) illustrate the interaction of farmers and agronomists in producing and transferring knowledge of the new techniques.

The interaction and cooperation between research stations, industrial companies, silage associations and farmers who practised the new method of fodder conservation was facilitated by close personal relations amongst many of the relevant agronomists. Hans Kellerhals, the president of the BSA, for example, had already used Virtanen's methods on the big farm he managed in the 1920s. He cooperated closely with the federal research station in Zürich-Oerlikon whose director, Friedrich T. Wahlen (later to become a minister in the federal government), had been his classmate at the ETH. Hermann Gutknecht, the secretary of both the Bernese and the Swiss Silage Associations, had been a student with Kellerhals and Wahlen and, as director of the federal research station in Liebefeld-Bern, was now a colleague of the latter. And Samuel Grandjean, who conducted an impressive range of experiments on fodder conservation, was research assistant at the Institute of Animal Feeding while working his own farm and acting as a sales representative for industrial companies (Wiegner, 1935: 223; Grandjean, 1943: 11). And: all of them were members of the agronomists' association (SVIAL).

In their oral and written promotion of silage-making, Grandjean and his colleagues constantly emphasized their own practical experience, mentioned the positive feedback received from other farmers, depicted the apparently higher efficiency of silage with bar charts, and calculated the difference of feed value in monetary terms (Grandjean, 1943: Hess & Ryf, 1943; Bickel, 1957). To 'prove' technical advantages in monetary terms was, of course, a widespread practice among engineers in the nineteenth and twentieth century (Smith, 1990: 660; Gudermann, 2005: 37).

Uekötter (2006a: 157) interprets the tactic of highlighting the advantages of specific techniques in monetary terms as a response to farmers' expectations. This is an interesting hypothesis, but does it find support in our data? Until recently, the lack of adequate sources had made it impossible to answer this question. But the newly catalogued papers by the Archives of Rural History (AHR) enable us to analyse farmers' expectations and reactions more closely. Especially valuable are the minutes of students' papers and discussions at the agricultural college of Schwand-Münsingen; they document the interaction between agronomists and young farmers meticulously. A close analysis of this unique source shows that the students used the same semi-scientific language as their teachers and that they indeed also calculated in monetary terms the apparent feed value lost by sticking to traditional conservation techniques. If agronomists estimated the advantages of new techniques in monetary terms they, one can argue, essentially used the same methods as the farmers. And the agronomists used a language that the young farmers were able and ready to understand, as well as to use, themselves.

What the minutes of the college further reveal is that young farmers gave the use of new techniques a meaning that connected their social identity with their preferences for modern and 'progressive' farming techniques. According to the students, young and 'progressive' farmers used silage or wooden tripods for haycocks, while old and 'conservative' ones did not. Thus, the often enthusiastic attitude of young towards the new feed conservation techniques was a question of sober calculation and helped to create a group identity of 'young progressive farmers' prepared to do what they were convinced society expected them to do. When they talked about feed conservation, these young farmers tried to present themselves as a group that was able to provide food for the people – and as a group that at once expected and deserved a bright future.

Interestingly, the teachers, exclusively agronomists for agricultural subjects, did not support the enthusiasm of their students unconditionally. On the contrary, they constantly drew attention to potential problems with the new feed conservation techniques in specific situations. They even demonstrated their own tacit knowledge of agricultural practice. Werner Daepp for example, the director of the agricultural college in Schwand-Münsingen, pointed out the importance of motivated, well-trained and experienced labour if wooden tripods for haycocks were to be used efficiently. He distinguished between more and less motivated and skilled labour exactly in the same way as his students did. But while the students used the argument in favour of the wooden tripods, Daepp used it to justify the practice of making hay without them on the college farm.

Whether used for or against the wooden tripods, the argument concerning the importance of individual skills and motivation went far beyond what could be measured in a scientific experiment at the time. It was only in the 1950s that the experiments of research stations showed that the use of wooden tripods required twice as much labour as the ordinary, long practiced hay harvest methods (Bickel, 1957: 364). Wooden tripods are one, but by no means the only, example of agronomists going just as far as farmers beyond what they could measure when making decisions about what to do and how to do it. Even at the beginning of the present century, students at agricultural colleges were being taught to judge the quality of fodder not only with the help of chemical analyses but also with their own noses, hands and eyes (Flückiger *et al.*, 2008: 219).

It is therefore questionable if an epistemic difference between farmers and scientific experts existed in Switzerland to the extent that has been claimed for France, Germany and the USA. In Switzerland the use of human senses and tacit knowledge was important in the daily work of farmers and agronomists alike – and it remains so even today. Patricia Fry, based on field research at the end of the twentieth century, emphasizes the convergence between farmers and scientists as regards their perceptions of the soil (Fry, 2001). Two of the main reasons for this relative closeness of farmers and agronomists in Switzerland are the similar social background of both groups and the relatively high theoretical standard of teaching at the agricultural colleges from the late nineteenth century. Farming backgrounds and practical experiences gave the agronomists the necessary tacit knowledge to understand the feedback responses of farmers to technical problems. And farmers who graduated from agricultural colleges generally had more than a basic scientific knowledge. They shared, as in other countries, a certain collective identity (Uekötter, 2006a: 158; Fitzgerald, 2003: 175)

In Switzerland agronomists were well aware of the enthusiasm of many young farmers for new technologies. The farmers' openness to new technology was important for their cooperation with the agronomists. However, agronomists also acknowledged difficulties with the new techniques and emphasized the need for 'practical solutions' to these problems. Sometimes scientific methods were used to explain and to provide objections to specific novelties – a practice that can also be observed in the history of organic farming (Vogt, 2000: 302-306). In the case of feed conservation techniques for example, the Institute of Animal Feeding at the ETH studied the methods of Arnold Messmer, an agronomist in Zurich, who sold silage facilities for the production of 'sweet green fodder' (*Süssgrünfutter*) in the interwar period. With scientific methods, the institute proved that Messmer's 'sweet green fodder' actually was acid and had a much lower feed value than ordinary hay. After the publication of the results of these tests, Messmer's method of producing silage was discredited and disappeared (Wiegner, 1935: 219-220; Grandjean, 1943: 9-10).

Other methods were also important in illuminating technical problems. For a long time the breeding of plants and animals has produced undesired effects, such as diseases or exhausted soils, alongside the desired ones. Farmers generally coped well with environmental challenges before the twentieth century by using their intelligence and experience (Sieferle et al., 2006: 17-20). The farmers' perceptions and their generally pragmatic approach to unintended and undesirable effects created by old or new techniques have yet to be studied comprehensively. In the case of feed conservation techniques, the Swiss sources support the suggestion that the foundation of associations for feed production and feed conservation was also prompted by quality problems encountered in cheese making. These problems were mainly detected by the human senses. Werner Daepp's successor as director of the agricultural college told his students that he himself had eaten 'Swiss' cheese from

Denmark produced with milk from cows that were fed silage. But this cheese, he argued, had a completely different taste to the 'real' Swiss cheese made with raw, unpasteurised milk[8]. Because it was impossible to alter the taste of cheese made from silage milk using scientific methods, the use of silage fodder on farms delivering milk to cheese-making creameries was prohibited altogether (Moser & Brodbeck, 2007: 174-175). In order to remain competitive on the world market with a distinctively tasting Swiss cheese, the dissemination of a method of fodder conservation that saved labour and improved food value had to be restricted to certain areas and farms producing milk for purposes other than cheese making.

The story of feed conservation techniques from the 1920s to the 1950s illustrates how the often assumed tendency of agriculture to become a science-based industry was a complex and often contradictory one. Its success depended not only on scientific but also on non-scientific knowledge in the production, marketing and political spheres. Many of the agronomists themselves realised that a systematic communication of farmers' and tacit knowledge to the scientific community supported scientific progress – as well as the funding of scientific research (Wiegner, 1935: 186).

V. Conclusion

Since World War I Swiss farmers have been embedded in a close network of agricultural institutions. They were obliged to attend formal vocational education and operated within regulated markets. And in order to improve their production they received financial grants and had to comply with innumerable rules and regulations. Yet the perception that this modernization has been forced upon the agricultural community by outside forces, such as the agronomists, would be misleading. The national agrarian elite, including agronomists and many farmers, aligned itself with the expectations of an industrial society represented by powerful groups outside agriculture such as state functionaries, consumers, industrialists and scientists. To get individual farmers to conform, agronomists resorted to a variety of means, just as many other professions in western societies did as well. Modernizing initiatives were hardly ever unilaterally forced upon agriculture from the outside; usually they were introduced, negotiated and re-shaped by actors within the sphere of agricultural practice. It is probably exactly this process of negotiation, transformation and re-shaping that made the process of agricultural modernisation so effective. The close interaction of agronomists and farmers was strongly influenced by the social background of the former and the relatively high formal education of the latter. At least until the 1970s, agronomists were mainly recruited from an agricultural

[8] Archives of the Landwirtschafts- und Haushaltungsschule Schwand-Münsingen (AfA Nr. 141), Dossier 215-05)

background and were familiar with the everyday life of farmers from their own personal experience. Indeed, many of them considered this practical knowledge to be a precondition of their professional activity. Acknowledging farming to be a worthy occupation, they functioned in an industrial society and often perceived themselves more as mediators between the state and agriculture than merely one-sided promoters of a specific modernization project.

Many farmers held similar values and attitudes to those of the agronomists. The evidence of the manuscripts and minutes of students' papers and presentations in the agricultural colleges indicates that the young farmers were often as enthusiastic about new feed conservation and other techniques as their scientifically trained teachers. And agronomists as well as the young farmers referred to scientific and to practical and tacit knowledge in conducting a sophisticated discussion of the advantages and disadvantages of new techniques.

The sober, sometimes even critical, attitude of agronomists towards new scientific findings did not impede the implementation of an agricultural policy that aimed at industrialising the agricultural sector. Quite the contrary, it made these efforts even more effective. The Bernese silage association, for example, did not hinder the spread of silage pits when it warned its readers of 'non-reliable' suppliers of silage facilities, because at the same time it advertised the names of 'trustworthy' ones. Receptiveness to the difficulties created by new techniques helped agronomists to create the trust which in turn facilitated close contact between farmers and the agricultural sector within the industrial world. In short, one can say that the enforcement of the state's agricultural policy was probably so successful because the relevant actors – in the civil service, the agricultural associations and private companies – were closely related to the farming population, coped pragmatically with individual problems and cooperated with farmers who were themselves relatively well educated scientifically.

Agronomists played a prominent role in this process of integrating the agricultural sector into the industrial society. Only very seldom did qualified agronomists associate themselves with fundamental opposition to state agricultural policies. Typically they referred to the needs of industrial society (whether current or anticipated) as they helped to change agricultural policies.

Archival sources

Archives of Rural History, Bern (www.agrararchiv.ch)

Bibliography

ACKROYD, Marcus *et al.* (2006), *Advancing with the Army. Medicine, the Professions, and Social Mobility in the British Isles, 1790-1850*, Oxford, Oxford University Press.

AUDERSET, J., BÄCHI, B. & MOSER, P. (2012), *Die agrarisch-industrielle Wissensgesellschaft. Akteure, Diskurse, Praktiken*, in B. BRODBECK, M. INEICHEN & T. SCHIBLI (eds), *Geschichte im virtuellen Archiv. Das Archiv für Agrargeschichte als Zentrum der Geschichtsschreibung zur ländlichen Gesellschaft,* Studien und Quellen zur Agrargeschichte, Band III, Baden, Hier und Jetzt.

BAUMANN, Werner & MOSER, Peter (1999), *Bauern im Industriestaat. Agrarpolitische Konzeptionen und bäuerliche Bewegungen in der Schweiz 1918-1968,* Zürich, Orell Füssli.

BICKEL, Hans (1957), 'Probleme der Heubelüftung mit Kaltluft', *Schweizerische Landwirtschaftliche Monatshefte*, 35, p. 349-367.

BLUM, Abraham (1993), *Wissensentwicklung als System in unserer Landwirtschaft. Wissen erweitern, austauschen, umsetzen, neu gewinnen*, Lindau, Landwirtschaftliche Beratungszentrale.

BRASSLEY, Paul (1996), 'Silage in Britain 1880-1990. The Delayed Adoption of an Innovation', *Agricultural History Review*, 44/1, p.46-62.

BRASSLEY, Paul (2005), 'The Professionalisation of English Agriculture?', *Rural History*, 16/2, p. 235-251.

BRODBECK, Beat, FLÜCKIGER, Daniel & MOSER, Peter (2007), *Quellen zur ländlichen Gesellschaft. Ein Wegweiser zu Archiven und Quellenbeständen der Agrargeschichte im 19. und 20. Jahrhundert*, Baden, Hier und Jetzt.

BRODBECK, Beat, INEICHEN, Martina & SCHIBLI, Thomas (2012), *Geschichte im virtuellen Archiv. Das Archiv für Agrargeschichte als Zentrum der Geschichtsschreibung zur ländlichen Gesellschaft*, Baden, Hier und Jetzt.

BRUGGER, Hans (1978), *Die schweizerische Landwirtschaft 1850-1914*, Frauenfeld, Huber.

BRUGGER, Hans (1985), *Die schweizerische Landwirtschaft 1914-1980. Agrarverfassung, Pflanzenbau, Tierhaltung, Aussenhandel*, Frauenfeld, Huber.

BURG, Margreet van der & BOS-BOERS, Marian (2003), *Vrouwen, Wageningen en de Wereld. Wetenschap, studie en loopbaan, 1918-2003*, Hilversum, Verloren.

DÜGGELI, Max (1929), *Hans Moos 1862-1929*, Bern, Benteli.

FITZGERALD, Deborah (2003), *Every Farm a Factory. The Industrial Ideal in American Agriculture*, New Haven, Yale University Press.

FLÜCKIGER, Daniel (2009a), 'Zeitungen und Schulen für den fortschrittlichen Landwirt. Albert von Fellenberg-Ziegler (1860)', in Martin STUBER *et al.* (ed.), *Kartoffeln, Klee und kluge Köpfe. Die Oekonomische und Gemeinnützige Gesellschaft des Kantons Bern OGG (1759-2009)*, Bern, Haupt Verlag, p. 171-174.

FLÜCKIGER, Ernst *et al.* (2008), *Futterbau und Futterkonservierung*, Zollikofen, Landwirtschaftliche Lehrmittelzentrale.

FRY, Patricia E. (2001), *Bodenfruchtbarkeit. Bauernsicht und Forscherblick*, Weikersheim, Margraf Verlag.

GRANDJEAN, Samuel (1940, repr. 1943), *A. I .V.-Silage und Silobau*, Aarau, Wirz Verlag.

GUDERMANN, Rita (2005), 'Conviction and Constraint. Hydraulic Engineers and Agricultural Amelioration Projects in Nineteenth-century Prussia', in Thomas LEKAN & Thomas ZELLER (eds), *Germany's Nature. Cultural Landscapes and Environmental History*, New Brunswick Rutgers, p. 33-55.

GÜNTHART, Alois (1933), 'Die Erziehung und berufliche Ausbildung der bäuerlichen Jugend in Haus und Schule', *Schweizerische landwirtschaftliche Monatshefte*, 11, p. 82-95.

HARWOOD, Jonathan (2005), *Technology's Dilemma. Agricultural Colleges between Science and Practice in Germany 1860-1934*, Oxford, Peter Lang.

HESS, Hermann & RYF, Ernest (1942, repr. 1943), *ABC der Silowirtschaft*, Bern, Verbandsdruckerei.

HOLENSTEIN, André (2007), 'Industrielle Revolution avant la lettre. Arbeit und Fleiss im Diskurs der Oekonomischen Gesellschaft Bern (2. Hälfte 18. Jahrhundert)', *Cardanus*, 7, p. 17-40.

HOLENSTEIN, André, STUBER, Martin & GERBER-VISSER, Gerrendina (2007), 'Nützliche Wissenschaft und Ökonomie im Ancien Régime. Akteure, Themen, Kommunikationsformen', *Cardanus*, 7, p. 7-16.

KÄPPELI, Josef (1911), *Gründung und Entwicklung der landwirtschaftlichen Schule Rütti. Festschrift zur Feier ihres fünfzigjährigen Bestehens*, Bern, Gustav Grunau.

KRAEMER, Adolf (1896), *Die landwirthschaftliche Schule des Eidgenössischen Polytechnikums in Zürich. Festschrift zur Feier ihres 25 jährigen Bestehens*, Zürich, F. Lohbauer.

LANDIS, Jakob (1933, repr. 1945), *Neuzeitliche Dürrfutterernte und künstliche Grastrocknung*, Bern, Verbandsdruckerei.

LICHTENHAHN, Paul (1942), 'Kann der Besuch der landwirtschaftlichen Schulen obligatorisch erklärt werden?', *Schweizerische landwirtschaftliche Monatshefte*, 20/5, pp. 125-132.

MOLLARD, Éric (2002), 'D'un Malentendu à l'autre, de la jachère à la rationalité paysanne. Pensée agronomique et représentation sociale dans l'histoire de l'agriculture', *Ruralia*, 10/11, pp. 1-15.

MOSER, Peter (1994), *Der Stand der Bauern. Bäuerliche Politik, Wirtschaft und Kultur gestern und heute*, Frauenfeld, Huber.

MOSER, Peter (2003), *Züchten – säen – ernten. Agrarpolitik, Pflanzenzucht und Saatgutwesen in der Schweiz 1860-2002*, Baden, Hier und jetzt.

MOSER, Peter (2005), 'Am Konsum orientiert, über die Produktion thematisiert. Schweizer Agrarpolitik als Ernährungspolitik 1914/18-1960', in Ernst LANGTHALER & Josef REDL (eds), *Reguliertes Land. Agrarpolitik in Deutschland, Österreich und der Schweiz 1930-1960*, Innsbruck, Studienverlag, p. 192-203.

MOSER, Peter (2009), 'Wie Hermann Bieri aus Lehrbuben diplomierte Bauern machte (1973)', in Martin STUBER *et al.* (eds), *Kartoffeln, Klee und kluge Köpfe. Die Ökonomische und Gemeinnützige Gesellschaft des Kantons Bern OGG (1759-2009)*, Bern, Haupt Verlag, p. 239-242.

MOSER, Peter & BRODBECK, Beat (2007), *Milch für alle. Bilder, Dokumente und Analysen zur Milchwirtschaft und Milchpolitik in der Schweiz im 20. Jahrhundert*, Baden, Hier und jetzt.

PERKINS, Harold (1989), *The Rise of Professional Society. England since 1880*, London, Routledge.

PFISTER, Christian (1990), 'The Early Loss of Ecological Stability in an Agrarian Region', in Peter BRIMBLECOMBE & Christian PFISTER (eds), *The Silent Countdown. Essays in European Environmental History*, Berlin, Springer Verlag, p. 37-55.

RIDDER-SYMOENS, Hilde de (1996), 'Training and Professionalization', in Wolfgang REINHARD (ed.), *Power Elites and State Building*, Oxford, Clarendon Press, p. 149-172.

SCHWEIZERISCHE HOCHSCHULE FÜR LANDWIRTSCHAFT (1969-2006), *Tätigkeitsberichte*, Zollikofen, Schweizerische Hochschule für Landwirtschaft.

SCHWEIZERISCHER BUNDESRAT (1883), 'Botschaft des Bundesrathes an die Bundesversammlung, betreffend die Förderung der Landwirtschaft durch den Bund vom 4. Dezember 1883', *Schweizerisches Bundesblatt*, 15 December 1883, p. 859-969.

SCHWEIZERISCHER BUNDESRAT (1951), 'Botschaft des Bundesrates an die Bundesversammlung zum Entwurf eines Bundesgesetzes über die Förderung der Landwirtschaft und die Erhaltung des Bauernstandes vom 19. Januar 1951', *Schweizerisches Bundesblatt*, 1 February 1951, pp. 133-290.

SCHWEIZERISCHER VERBAND DER LEHRER AN LANDWIRTSCHAFTLICHEN SCHULEN UND INGENIEUR-AGRONOMEN (1951), *Schweizerischer Verband der Lehrer an landwirtschaftlichen Schulen und Ingenieur-Agronomen 1901-1951*, Frauenfeld, Verlag Huber.

SIEFERLE, Rolf P. *et al.* (2006), *Das Ende der Fläche. Zum gesellschaftlichen Stoffwechsel der Industrialisierung*, Wien, Böhlau.

SIEGRIST, Hannes (1988), 'Die Rechtsanwälte und das Bürgertum. Deutschland, die Schweiz und Italien im 19. Jahrhundert', in Jürgen KOCKA (ed.), *Bürgertum im 19. Jahrhundert. Deutschland im europäischen Vergleich*, München, Deutscher Taschenbuchverlag, p. 92-123.

SIEGRIST, Hannes (1996), *Advokat, Bürger und Staat. Sozialgeschichte der Rechtsanwälte in Deutschland, Italien und der Schweiz, 18.-20. Jh.*, Frankfurt a. M., Vittorio Klostermann.

SIEGRIST, Hannes (2005), 'The Professions in Nineteenth-century Europe', in Hartmut KAELBLE (ed.), *The European way. European Societies during the Nineteenth and Twentieth Centuries*, New York, Berghahn Books, p. 68-88.

SMITH, Cecil O. (1990), 'The Longest Run: Public Engineers and Planning in France', *American Historical Review*, 95/3, p. 657-692.

STRAUMANN, Lukas (2005), *Nützliche Schädlinge. Angewandte Entomologie, chemische Industrie und Landwirtschaftspolitik in der Schweiz 1874-1952*, Zürich, Chronos.

STUBER, Martin *et al.* (2009), 'Einleitung. Von der Reformsozietät zur bäuerlichen Bildungsinstitution', in Martin STUBER et al. (eds), *Kartoffeln, Klee und kluge Köpfe. Die Oekonomische und Gemeinnützige Gesellschaft des Kantons Bern OGG (1759-2009)*, Bern, Haupt Verlag, p. 13-60.

UEKÖTTER, Frank (2006a), 'Did They Know What they were Doing? An Argument for a Knowledge-based Approach to the Environmental History of Twentieth-century Agriculture', *Bulletin of the German Historical Institute, Washington DC*, 3, p. 145-166.

UEKÖTTER, Frank (2006b), 'Die Chemie, der Humus und das Wissen der Bauern. Das frühe 20. Jahrhundert als Sattelzeit einer Umweltgeschichte der Landwirtschaft', in Andreas DIX & Ernst LANGTHALER (eds), *Grüne Revolutionen. Agrarsysteme und Umwelt im 19. und 20. Jahrhundert*, Innsbruck, Studien-Verlag, p. 102-128.

VOGT, Gunter (2000), *Entstehung und Entwicklung des ökologischen Landbaus im deutschsprachigen Raum*, Bad Dürkheim, Stiftung Ökologie und Landbau.

DE VRIES, Jan (1994), *The Industrious Revolution: Consumer Behaviour and the Household Economy, 1650 to the Present*, Cambridge, University Press.

WAHLEN, Hermann (1943), *Der landwirtschaftliche Bildungsgedanke in der Schweiz. Ein Beitrag zur geschichtlichen Entwicklung des beruflichen landwirtschaftlichen Bildungswesens*, Bern, Francke.

WAHLEN, Hermann (1979), *Rudolf Schatzmann 1822-1886. Ein Bahnbrecher der schweizerischen Land-, Alp- und Milchwirtschaft und ihres Bildungswesens*, Münsingen, Buchverlag Fischer Druck.

WEGMANN, Karl (1960), *Die nichtberuflichen Fortbildungsschulen in der Schweiz*, Winterthur, Verlag Keller.

WIEGNER, Georg (1935), '10 Jahre Institut für Haustierernährung', *Schweizerische Landwirtschaftliche Monatshefte* 13/6-8, p. 182-191, 212-224, 240-247.

WIELAND, Thomas (2004), *'Wir beherrschen den pflanzlichen Organismus besser.' Wissenschaftliche Pflanzenzüchtung in Deutschland 1889-1945*, München, Deutsches Museum.

ZOLLIKOFER, Edzard (1954), 'Das landwirtschaftliche Hochschulstudium', in *Bildungsgang des Bauern und der Bäuerin in der Schweiz*, Urdorf-Zürich, Fritz Lindner, p. 48-56.

13. The prohibition of dogcarts in Belgium: a hidden agricultural policy?

Serge SCHMITZ

I. Introduction

This chapter analyses the context and the consequences of the prohibition of dogcarts in Belgium. It considers in particular the impact of the legislation prohibiting dogcarts on the Belgian peasantry. In the early nineteenth century dogcarts were widely used in agriculture in different European countries (Denmark, the Netherlands, Belgium, France, England, Switzerland, Italy, Germany, Czechoslovakia are examples), especially when the size of the farm was too small to invest in a horse. By the beginning of the twentieth century, however, the use of dogcarts had either been banned (as in England in 1854 and in France in 1925) or strictly regulated (as in the Netherlands in 1910 and in Czechoslovakia in 1926). In Belgium, however, it was only in 1926 that legislators discussed the question of banning the dogcarts seriously. And it was not until 1975 that the dogcarts were formally banned at the national level.

My argument here is that the prohibition of dogcarts in Belgium was meant to contribute to the modernization of agriculture as much as it was concerned with the wellbeing of the animals. Dogcarts had to disappear in order to turn the ordinary farm into a professionally run commercial unit. Of course, the regulations regarding dogcarts were also based on concern for the wellbeing of animals. Nonetheless, the agricultural dimension of the regulation and its impact on the peasantry should not be ignored. This dimension, I will argue, should be taken much more seriously by historians since it could have been part of a hidden agenda for modernizing agriculture.

Three general issues are raised in my discussion of the Belgian dogcarts. The first concerns the contribution dogcarts made to the peasant economy. Secondly, there is the issue of the politics surrounding the state's efforts to ban dogcarts – the failed attempt in the 1920s, a considerable amount of regulation in the municipalities and provinces in the pre-1960 period and finally a prohibition at the level of the central state in 1975. A third concern highlights the consequences of prohibiting the use of dogcarts, both in everyday life and at the level of the law. Through analyses of parliamentary debates and interviews with older farmers, my discussion underlines the gap between the parliamentary reform of agriculture and the realities of agricultural life as experienced by the farming population. It also argues against the tendency to view only policies coming from the state, primarily the department of agriculture as relevant to the agricultural sector.

II. The dogcart in Belgium

Until the nineteenth century, it seems that the dogcart was quite common in Europe. In the *Grande Encyclopédie*, Albert Larbaletrier (1890) wrote that: 'the use of a dog as a beast of burden was usual for many poor people or for less civilized countries; the practice showed that a dog carefully harnessed to a low cart could produce a sustained effort'. Larbaletrier mentioned that dogcarts were not only common in Belgium but also in the north of France. In fact, people used different animals (*e.g.* cows, goats, dogs) to pull a great diversity of vehicles (Bries & Schippers, 1989). Horses were reserved for rich people and big farms. The dog had the advantages of being docile and tough as well as of being able to live on leftover meals. It also suited the practice of children, who often laboured with their parents on farms, working with animals. The spatial distribution of dogcarts in Europe makes clear that they were particularly used in the dairying districts and in highly populated areas. People relied on them as a solution to the lack of horses. Apparently Belgian agriculture was short of horses in the nineteenth century, particularly because of the high demand for them in industry, the mines, transport, and the army.

At the turn of the twentieth century, pictures of dogcarts were very common on postcards. Several collectors have published books and there are internet websites specifically dedicated to dogcarts (Braems, 1977; Rigaud *et al.*, 1996). These documents give an idea of the work done by the dogs and the geographical distribution of dogcarts. They also inform us that artists painted dogcarts because they were such a characteristic element of both the Belgian cities and the countryside. The places shown on the postcards are mainly places in Flanders, especially near the cities, but also in parts of Campine and the Ardennes. Postcards depicting dogcarts on the low loess plateaus or in the Lorraine region are rare.

The mistreatment of cart dogs was also acknowledged as a Belgian habit and several travellers mentioned abuses (Kalof, 2007). The books published by the Society for the Protection of Animals tell us more about the use of dogcarts and about the main criticisms made of dogcarts by the Society. In the Belgian edition of a schoolbook (Ruhl, 1899) – largely translated from the English book *Kindness to Animals* that contains 115 lessons – eight new lessons about the proper use of dogcarts were included. They present the use of dogcarts as a typical Belgian custom but state that the dog is not intended to be harnessed. Despite this assertion, these lessons explain the least harmful way to harness the dog.

The dogcart as a means of transport disappeared in Belgium around the late 1950s. Up to this dogcarts were mainly used to go to the market and to bring in the harvest from the fields (Bries & Schippers, 1989). In 1920 around 150,000 cart dogs[1] and 100,000 dogcarts remained in Belgium. Cesar Daumerie acknowledged that the dog was not a real beast of burden. Nevertheless, the use of dogcarts was so common[2], that their prohibition would have annihilated many industries[3]. Several Belgian citizens, among them Jules Felix (1877), actually proposed an increased use of dogcarts and the improvement of the Belgian mastiff pedigree to that end. At the beginning of the twentieth century, a society even organized several dog competitions in order to improve the dogcart pedigree. And in 1914, Lieutenant Vandeputte recommended the use of this 'important resource for the country' in the army (Vandeputte, 1914).

The testimony of older people is another source of information – but this presents many difficulties because dogcarts remain for them but a vague memory. Usually they can recall only scant concrete information and often have a child's perspective of the dogcarts. In 2008, however, eighteen persons who still remembered the use of dogcarts were interviewed as part of this study[4]. These interviews provide at least partial information about the use of dogcarts in the countryside, about their prohibition, and about the consequences the legal ban had on small farms.

---

1 Both Étienne Olivier (1992), '*Elles avaient du chien, les charrettes à chiens*', (*Le journal du collectionneur*, 19, p. 19-22) and Julien Semninckx (2009), '*L'attelage canin, carriole du pauvre*', *Dimanche du 31 Mai 2009*, p. 10) have suggested such a number.

2 Statistical data relating to dogcarts have not been used in this study. What statistics are available typically arise when dogcarts become the subject of taxation, as in the Netherlands where dogcart users had to obtain a licence from 1910. Here it is possible to find out how many dogcart drivers there were at the national and municipal levels. For instance, 71 licences were recorded in Harderwijk (a city of 8,661 inhabitants in 1930) in 1929. Such a licensing system did not exist in Belgium. The Belgian agricultural data give contextual information about the size of farms and the presence of horses or tractors, but nothing is mentioned about dogs.

3 Daumerie, Cesar (n.d. [c. 1884]), *Rapport de la commission chargée de rechercher les moyens les plus efficaces de prévenir ou de réprimer les mauvais traitements commis sur la voie publique envers les animaux.*

4 I asked the researchers (19) and the master students (25) of the geography department of the University of Liege, professors of rural geography from the other Belgian universities (4), the administrators of the non-profit association 'Qualité village wallonie' (9), and friends and professionals from both sides of the linguistic divide that I met during the research period (+/- 40), if they (or their parents), knew anything about dogcarts. Using a snowball sampling technique, I asked then if they might also ask old farmers in their places of residence about their memories of dogcarts. We got positive information relating to Aalter, Antwerp, Brussels, Chaudfontaine, Couvin, Esneux, Hasselt, Mouscrons, Tongeren, and Vielsalm but there were no memories of dogcarts in Attert, Ciney, Blegny, and Gembloux.

When older people in Belgium are asked whether they remember dogcarts, often childhood memories of riding on a cart or of dogs helping the milk-woman or the baker come flooding back. Frequently people mention with nostalgia that dogcarts are now prohibited. In 1997, Mr Hermann (65 years old at the time), a farmer from Burtonville, explained that the prohibition of dogcarts marked the end of the peasantry (Schmitz, 1999; 2001). He clearly associated the law with the big shift towards the modernization of agriculture in Belgium, with the beginning of carbon-based mineral resources as an energy source and the beginning of farmers getting into debt. Mrs Maillet (66 years old in 2008) from Vaux-sous-Chêvremont remembered the milk-woman with her dogcart when she was a child. She remembered the changing position of the dog: first in front of the cart, later beside the woman and then, finally, the disappearance of the dog altogether. Then the milk-woman was alone. She used to give five cents to the child when it helped her to go up the street. Mrs Briquet (78 years old in 2008) from Couvin explained that her parents owned a dogcart when she was a child. However, they also had two horses. The horses were assigned to ploughing; and the dog and the dogcart were used for going to milk the cows in the meadows situated two kilometres away from the farmhouse. She did not remember the prohibition but she related how the family's dogcart was abandoned around 1954 when her parents bought a tractor.

A gap exists between the testimonies of the farmers we interviewed and the national legislation concerning the Belgian dogcarts. Several of the farmers we interviewed referred to a law apparently prohibiting dogcarts in the 1950s. However, there was no such law in force or introduced at the national level at that time. But, in the 1950s, several rules did actually regulate the use of dogcarts at the communal and provincial levels; and debates regarding the protection of animals at the national level considered the question of dogcarts. These testimonies particularly underline the importance and usefulness of dogcarts in the completion of daily tasks. Some of those interviewed explicitly mentioned the difficulties small farms got into when dogcarts were prohibited. This information, however, is only the beginning; it needs to be reinforced by other sources in order to reveal more fully the reasons for the prohibition and its impact on rural life.

The legislative process that finally led to the prohibition of dogcarts at the national level took five decades in Belgium. Campaigns conducted by the Society for the Protection of Animals, as well as the pressure applied by the proponents of agricultural mechanisation and motorisation played an important role in moving the process forward.

III. The evolution of the Belgian dogcart legislation

Legislation, parliamentary documents, and the transcripts of debates in public are the major sources of information used in this study. At the end of the nineteenth and at the beginning of the twentieth century several countries introduced laws with the purpose of preventing the abuse of animals. In some countries, the law specifically addressed the use of dogcarts. Dogcarts were thus prohibited in England in 1854 and in France in 1925. They were strictly regulated in the Netherlands in 1910 before being prohibited entirely in 1962; and their use was regulated in Czechoslovakia in 1926 (Bocquet, 1934). In Belgium, on the other hand, the prohibition of dogcarts at the level of the central state was introduced only as late as 1975. Prior to this date, however, several municipal and provincial authorities had strictly regulated or even prohibited the use of dogcarts. Therefore, by the late 1950s dogcarts had in practice been prohibited almost everywhere in Belgium. The use of dogcarts in Belgium was first regulated in towns and cities due to concerns raised relating both to the swift flow of traffic and to the protection of animals. In Antwerp, for instance, the provincial rule of 23 August 1867 not only ordered the muzzling of dogs, but it stipulated the length of the leash and the use of a light when the dogcart travelled at night. Furthermore, several local rules made it compulsory to use a plank to protect the dog from the cold when stopping as well as the provision of a dish of drinking water.

The intention to regulate or even prohibit dogcarts was justified by the growing concern about the wellbeing of animals, 'our inferior brothers', all over Europe. Societies for the protection of animals began to combine their lobbying with an education campaign to heighten people's awareness of animal welfare and even the rights of animals. The new distance between people and animals created in the industrial society seems to have been important in giving force to this new awareness. However, the protection of animals was problematic from the legal point of view because the law did not then recognise animals as capable of possessing interests or rights. In many European countries the idea that an animal might obtain rights and become the subject of rights was rejected. It was regarded as reasonable to protect the animal as the property of a person, but acknowledged as difficult to protect the animal from its master or owner. In most countries only the public maltreatment of animals was an offence, based on a belief that such behaviour shocked and offended other people (Bocquet, 1934).

In Belgium, the law on animal protection of 22 March 1929 did not even mention dogcarts. However, in the second article, the law did provide for the punishment (by a prison sentence of one to five days or a fine of ten to twenty francs) of those who abusively imposed on an animal a painful task or an effort that obviously exceeded its strength. The new legislation rendered more precisely what had already been regulated in the penal code of 1865 that had prohibited the cruel or public maltreatment of animals. And in 1924 Senator Asou, and in 1926 Senators Seeliger, Rutten and Spaak, proposed to increase the sentence and to broaden the scope of the concept of maltreatment[5].

The details of the debates show that the proposals of Seeliger, Rutten and Spaak's for legislative change included a series of rules for the users of dogcarts. The senators wanted to create, in the first article of the law, the stipulation of an obligatory licence for owning a dogcart. What was envisaged was the punishment for the maltreatment of animals in the form of a fine, a prison sentence of up to six months, the seizure of the animals concerned and the revoking of the offender's driving licence.

On 4 May 1926, Senator Seeliger argued his proposal on the basis of four points: First, he said, Belgium would lag behind other nations from the point of view of animal protection. His second contention was that the penal code was unenforceable without clear definitions of what exactly constituted animal cruelty and excessive maltreatment. Third, he referred to different forms of animal cruelty – cockfights in general and cases of cruelty reported in the newspapers. Of the seven examples that he cited, three concerned horses and four dogs, including two specific dogcart cases. For Senator Seeliger these examples highlighted the inadequacies of the law, projected a poor image of Belgium to foreigners and were a bad example to children. Fourthly, Seeliger cited a report that had concluded that the maltreatment of livestock in agriculture was a huge economic loss for the country. In particular, farms of less than six hectares were regarded in this report as inadequate from the viewpoint of sanitation, animal wellbeing and economic viability.

On 16 November 1926 when the law returned to the plenary session after its reading by the Justice Commission, members of the Agricultural Commission argued that this law would have a huge negative impact on agriculture and that the Agricultural Commission would therefore have to analyse its propositions in depth. In particular, the members of the Agricultural Commission desired to delete the administrative part of the law, which struck them as being at once irritating and difficult to implement. Senator Deswarte claimed that 'the protection of animals is an

5 Asou was a Liberal senator and the Liberal Party (along with the Catholic Party) had a majority in 1924, but this was no longer the case in 1926 when the Catholics formed a majority with the Socialists. Seeliger and Spaak (as was also true of Deswarte) were socialists. Rutten, a co-opted member, was the former secretary general of the Christian trade unions in Belgium. While the Socialists and Liberals were mainly supporters of the prohibition of dogcarts, the main opposition came from the Catholics, especially from the Boerenbond (the Flemish Farmer's Union). Mullie (a Catholic senator), who defended the small farms, was very active in the Boerenbond and became its president in 1936.

important question that has a real social impact and well known economic effects'. In the discussion, the assumed costs of the proposed obligations were an important issue. Another important topic was the question of vivisection and its importance to scientific research. Some senators suspected that the real concern of the members of the Agricultural Commission was to protect hunting practices.

On 30 May 1928, the law returned to the Senate relieved of its administrative provisions, which had been judged as too expensive and too difficult to apply to agriculture. The proposed law sought to redefine the maltreatment of animals from a contravention to an offence. Senator Asou recalled that 'every day, newspapers report on the maltreatment of transported animals, of dogcarts, and of horses used in the coal mines, the rubbish dumps or along the canals'[6]; he argued that all the political parties and the newspapers were in favour of the proposed law. Nevertheless, there was a notable reticence in relation to cockfights and vivisection. Senator Asou concluded that this law would have a moral effect, but also a beneficial economic effect. He cited a report by Senator Deswarte that stressed the important economic gains to be made from the better care of animals.

Essentially the debate focused on two main points: the necessity of listing the different forms of maltreatment on the one hand and of defining what was 'excessive' maltreatment on the other. At the heart of the debate were technical questions concerning 'what is excessive?' and how the excessiveness of a particular form of maltreatment was to be determined. What about forms of maltreatment that were not on the list? The Legislative Council judged as problematic the idea of enumerating all forms of maltreatment, not only because of the difficulty of listing all the various forms but especially because compiling such a list implicitly attributed rights to animals. There was also a difference of emphasis in the Senate between senators who wanted to condemn the excesses and those mainly interested in advancing the topic of animal welfare in society. We also find some debate between those senators who wanted to recognize the rights of animals and those who wanted to protect the freedom (and property rights) of animal owners.

During the debate several examples of maltreatment were cited such as those involving pigeon fighting and the use of horses in the mines. When the senators mentioned dogcarts, it was argued that everyone knew the painful life of many of these dogs and the numerous forms of maltreatment that they endured. Senator Clesse

6 Newspapers constitute a potentially useful source of information, especially about how people perceived dogcarts and the law on prohibition. Several interventions during the parliamentary debates cited the newspapers as giving information on the ill-treatment of dogs. They also reported the position of the recently formed Society for the Protection of Animals. Nevertheless, the examination of the regional newspaper *La Meuse* in the years 1925, 1926 and 1929 produced only a few results.

even spoke of the martyrdom of the dogcart dogs. From the agricultural point of view, Senator Seeliger reiterated his view that agriculture was the country's premier industry, but that small farms of less than 5 ha were too numerous and inefficient for the good of agriculture and that livestock on these farms were being maltreated and the rules of sanitation ignored. He was convinced that these farms had neither the minimum capital nor the level of training required to upgrade their practices. Senator Mullie strongly disagreed with this view. He was convinced that the animals were more cherished on small than on big farms because of the relatively greater importance of each animal to the wellbeing of the family.

The law was carried by 87 out of a total of 89 votes on 20 June 1928 in the Senate, and by 118 out of 119 votes in Parliament on 6 March 1929. The text of the law did not enumerate the forms of maltreatment, and so it did not explicitly mention dogcarts. According to the text, 'excessive' maltreatment of animals would be punished in the future. As things turned out, it was precisely the difficulties of specifying what was 'excessive' that led to the failure of the law and therefore to its amendment in the 1950s. The first proposition for a reformulation of the law was made on 26 January 1960, but the examination of the law was delayed until 1965 when the proposition explicitly prohibited dogcarts. This article was indeed new, but it merely 'repeated the provincial and communal rules that already existed almost everywhere in Belgium' (Vermeylen & Heger, 1965). Not surprisingly, there was now no discussion in relation to the prohibition of dogcarts. The debates focused instead on pigeon fighting and fox hunting. These 'sports' were acknowledged to be an element of traditional life in Belgium and as important economic activities. On 11 March 1975 the senators unanimously voted for the new legislative proposal without these articles.

IV. Conclusion

The absence of a debate on the question of the prohibition of dogcarts during the legislation process from 1965 to 1975 illustrates that people and politicians by now accepted the view that dogcarts were cruel to animals as well as the fact that they were no longer necessary. The arguments brought forward against prohibition in the 1920s – the social impact it would have within agriculture, the lack of alternatives, the financial burdens for people forced to look for alternatives and the difficulty of applying strong regulations – had nearly all become obsolete by the radical transformation of the rural society in the post-war years. And the political context differed fundamentally as well. The Common Agricultural Policy was now encouraging and forcing farmers to invest in the modernization of agriculture (Schmitz, 2002). And the former policy of the Catholic Party to keep people in the countryside in order to protect them from the perils of the city had been abandoned. Taken together, these developments greatly diminished any lingering social concern with maintaining the livelihoods of a declining number of increasingly marginal farmers.

But up to the 1950s dogcarts had been an important means of transport for small farmers in Belgium. This was one of the main reasons why the Agricultural Commission of the Senate insisted on being consulted about the new law when the Justice Commission first discussed it in the late 1920s. After the 1950s dogcarts still remained an agricultural fixture – but now their persistence was seen as hindering the process of modernising agriculture. From what is known to date, it can therefore be suggested that the prohibition of dogcarts in Belgium, in addition to being the result of concern for the wellbeing of dogs, was also an important move to modernize agriculture. In 1926 the proposed prohibition of dogcarts was defeated under the influence of the then influential Agricultural Commission. Agriculture, along with other spheres of the economy, was not yet ready to abandon the use of dogcarts. Though the protection of animals was at once laudable and important for the image of Belgium, a majority of its political representatives was convinced that the country could not afford to abandon the dogcarts. As horses were rare and expensive (Tulippe, 1927), many peasants could not afford them. The only acceptable regulations in relation to dogcarts were police rules that sought to manage the traffic in the towns and to oversee the co-existence of dogs, horses and people on the streets. But by the 1950s, several municipalities and provinces had moved to regulate the use of dogcarts more strictly or to prohibit their use entirely. Not surprisingly therefore, when a specific article prohibiting dogcarts was introduced in 1965, it no longer provoked negative reaction. Other means of transport were now widespread and Belgium could join other European nations in the prohibition of dogcarts without attracting resistance. As Blomme (1993) has observed in relation to other transitions, 'it was a steady mutation in which elements of both traditional and modern agriculture co-existed and complemented each other'.

The disappearance of dogcarts can be read as part of the broader process of radical change in agriculture after World War II. It is interesting to note that while in the 1960s a cultural exemption was claimed for fox hunting and cockfighting, an agricultural practice such as the use of dogcarts was no longer addressed from a cultural point of view in spite of the fact that the dogcarts were a strong symbol of the Belgian countryside. Dogcarts no longer in common use were now banned because they symbolized not only the peasantry but also underdevelopment – both images which had to be eradicated from modern Belgian society.

The story of the prohibition of Belgian dogcarts illustrates the complex interdependence between society, the economy and natural resources. That a law prohibiting the use of dogs as draft animals had consequences in agriculture was clear to all actors engaged in the debate in the 1920s. The fact, that the prohibition of dogcarts in the 1960s was no longer disputed does not mean that the issue had lost all its relevance within the agricultural sector. Several of the interviewed farmers were convinced that the effects of such a law on small farms were well known to the minister who now, in contrast to the interwar years, wanted to get rid of the small farms.

In conclusion, we can say that it is worthwhile to look behind the arguments brought forward by public representatives and not simply take what they are saying at face value. What is also evident from our case study is that the most influential developments may not be reflected at the level of national parliamentary debate and legislation. The history of the prohibition of the Belgian dogcarts reveals how agricultural development was influenced not only by a narrowly conceived agricultural policy but by the rules administered by many other state departments as well. But much more research is needed to elaborate these preliminary research findings. It is reasonable to assume that a close analysis of the municipal and provincial prohibitions, which preceded the national legislation, might substantiate the case more fully.

Bibliography

BLOMME, Jan (1993), *The Economic Development of Belgian Agriculture, 1880-1980*, Leuven, Leuven University Press.

BOCQUET, Edmond (1934), *La Protection pénale des animaux dans les Législations françaises et étrangères*, Paris, Recueil Sirey.

BRAEMS, Pierre (1977), *Melkmeisjes en hondekarren in oude prentkaarten*, Zaltbommel, Europese Bibliotheek.

BRIES, Luc & SCHIPPERS, Marcel Martin (1989), *De Hondekar, een verdwenen straatbeeld*, Wommelgem, Vlaamse Trekhonden Vereniging Met Hond en Kar.

DAUMERIE, Cesar (n.d. [c. 1884]), *Rapport de la commission chargée de rechercher les moyens les plus efficaces de prévenir ou de réprimer les mauvais traitements commis sur la voie publique envers les animaux*.

FÉLIX, Jules (1877), *Considération sur l'attelage du cheval et du chien*, Bruxelles, imp. H. Manceaux.

KALOF, Linda (2007), *Looking at Animals in Human History*, London, Reaktion Books.

LARBALETRIER, Albert (1890), 'Chien, Zootechnie', in *La grande encyclopédie, Inventaire raisonné des sciences, des lettres et des arts*, ed. by Henri LAMIRAULT *et al.*, tome 11, Paris.

MINISTÈRE DE LA JUSTICE ET MINISTÈRE DE L'AGRICULTURE (1929), 'Loi relative à la protection des animaux du 22 mars 1929', *Moniteur Belge*, 29 mars 1929, p. 1481-1482.

MINISTÈRE DE LA JUSTICE ET MINISTÈRE DE L'AGRICULTURE (1975), 'Loi sur la protection des animaux du 2 Juillet 1975', *Moniteur Belge*, 18 Juillet 1975, p. 8967-8969.

RIGAUD, Muguette, CHOLLET, Jacqueline, CHOLLET, Jacques, LEMELIN, Dany, PETIT, Michel & POULLOT, Gaston (1996), *L'Attelage à chien en France, historique, cartoliste*, Saint Denis en Val, Cercle des Cartophiles du Loiret.

RUHL, Jules (1899), *Soyez bons envers les animaux, petit manuel de protection*, Bruxelles, Bureau du 'Bien social', Librairie Lebègue et Cie.

SCHMITZ, Serge (1999), *Les Sensibilités territoriales, Contribution à l'étude des relations homme-environnement*, Thèse de doctorat, Université de Liège.

SCHMITZ, Serge (2001), 'Micro-analyse des mécanismes de la reprise démographique des villages de l'Ardenne du Nord-Est : le cas de Burtonville', *Espace, Populations, Sociétés*, 1/2, p. 171-179.

SCHMITZ, Serge (2002), 'Walloon Region: An Agro-forestry Landscape without Farmers?', *Dela*, 17, p. 174-182.

SEELIGER, Jules (1926), 'Proposition de loi relative à la protection des animaux', *Sénat, Documents*, 113, session de 1925 à 1926, p. 470-476.

TULIPPE, Omer (1927), *Le Cheval de trait belge*, Travaux du Séminaire de Géographie de l'Université de Liège, Fascicule XVIII.

VANDEPUTTE, M.-J. (1914), *Le Chien militaire belge employé à la traction des mitrailleuses 'Maxim', quelques notes…*, Bruxelles, Guyot frères.

VERMEYLEN, Piet & HEGER, Charles (1965), 'Projet de loi sur la protection des animaux', *Session du 13 avril 1965 de la Chambre des Représentants*, Document 1049.

14. Changing technologies and output increases in United Kingdom agriculture, 1945-1985

Paul BRASSLEY

I. Introduction

During the period 1945-1965 agricultural output in the United Kingdom increased at an average annual rate of 2.8 per cent, faster than it had ever increased before. In the following twenty years to 1985 the annual rate of increase fell to 1.4 per cent, but that was more than twice the annual rate of increase in the classic agricultural revolution period of the eighteenth and nineteenth centuries (Overton, 1996: 84-87; Brassley, 2000: 63). Initially, the prices received by farmers were also higher (in constant price terms) than they had been since the late nineteenth century, and certainly higher than they would be in the later twentieth century (Brassley, 2000). Was the rapid output increase therefore simply a supply response to high prices, as predicted by basic economic theory? If so, how did the mechanism work? If there was more to it than can be explained by basic economics, what other factors need to be considered?

The author's initial supposition was that the simple economic explanation would suffice, and the following discussion briefly investigates the relationship between prices and volumes of output. However, his recent work on the construction and development of technical change in this period suggests that the process is more complex. Sociologically based approaches have much to offer in explaining how the links are made between a government decision to promote increased production on a national scale, a scientist's development of a new technique, and an individual farmer's decision to adopt new technology and make investments in order to increase output. This paper will suggest that a novel approach to the analysis of technical change that combines some of the features of Actor Network theory, Social Worlds theory and Conventions theory can be used to reveal how linkages are made between the decisions of scientists, agricultural policy makers, the ancillary industries, agricultural training and development organisations, and farmers. This new approach requires an analysis of the activities of authorities, the discourses they use, and the media through which they are promulgated, and the following discussion shows how it can be applied to specific technical changes in the forty years after World War II.

This post-war period was remarkable not only because farm output rose so much, but because in doing so it formed such a contrast to the period between the First and Second World Wars. In many, if not most, parts of Europe the interwar countryside was dependent upon agriculture to a much greater extent than it is today, and, probably, to a greater extent than it had been in the nineteenth century. In the traditional rural economy blacksmiths, wheelwrights, carpenters, thatchers, and a range of other trades produced and maintained farm carts, horses, and buildings, while the village bakers, tailors, boot makers and butchers provided the necessities for local consumption. By the interwar period, certainly in the United Kingdom, these trades were less and less able to compete with cheaper factory-made products brought to the rural areas by relatively low-cost transport. Consequently the vitality of the rural economy was increasingly equated to the viability of the agricultural economy, and rural society increasingly became agricultural society (Brassley, 2006).

Thus the fortunes of agriculture were crucial to European rural people, and interwar agriculture had to cope with depressed world prices and a demand for food that was restricted by industrial unemployment. Producers in both food-importing countries and traditional exporters, such as Denmark, Ireland and the Netherlands, were affected by low prices. Previously one response to price pressures had been for part of the labour force – usually its younger members – to leave agriculture, but industrial unemployment restricted this option (Tracy, 1982: 195, 219). Despite the glorification of the peasantry in nationalist rhetoric, argued Yates and Warriner (both of whom had conducted extensive surveys of the pre-war peasantry), 'Neither in health, education nor prosperity can the farming population of any European country be compared favourably with the urban population'. Much of rural Europe, especially in the east, but also in the west, was, in their view, over-populated (Yates & Warriner, 1943: 17, 51). But an impression of unrelieved and unchanging gloom would be a misrepresentation. The First World War had taken hundreds of thousands of peasants out of their villages and brought them into contact with their fellows from other districts and their urban contemporaries, so that they became more aware of their economic and social position. They had heard of the land reforms that followed the Russian revolution. There were some educational improvements, and also communications and transport changes. All these had some impact on the mentality of the agricultural population (Tracy, 1982: 185; Schmidt, 1938: 26-27; Cardoza, 1991: 192-193). They were not incapable of effective political activity, and from the agrarian political parties of eastern Europe, through quasi-fascist organisations such as the Front Paysan in France, to mild-mannered pressure groups such as the National Farmers' Union in England and Wales, they had some impact on agricultural and trade policy (Tracy, 1982). There were tariffs and import restrictions to deal with falling prices, and programmes of land reform, reclamation, and reorganisation. Overall, however, (and despite some contrary views trenchantly expressed by Oyler, 1950) the impression derived from contemporary reporters such as Warriner (1939) and Yates (1940) is of an interwar European countryside beset by social and economic problems.

During the Second World War, the experiences of food producers and consumers varied from country to country across Europe. War obviously disrupted agricultural production in combat zones, for longer or shorter periods, but, equally obviously, its effects were far more widespread. It affected the availability of inputs, as land was requisitioned for airfields, fortifications, military camps and training areas, labour was diverted into military service or armaments production, and the cost of acquiring capital increased. Conversely, in some countries, new inputs became available as the employment of women in agriculture increased and mechanization was encouraged. In most European countries, whether combatant or not, the war disrupted existing patterns of trade to a greater or lesser extent, and in some it encouraged the adoption of new technology. The impact on output between 1940 and 1945 was variable. In the UK it rose a little (8 per cent according to Brassley, 2000), in France and Italy it fell by 20 to 30 per cent, and in Germany it was stable until 1944 (Federico, 2005: 196). In Denmark, both home consumption and export levels were maintained (Nissen, 2006: 188). Almost everywhere, therefore, food supplies were under pressure while demand did not decrease and often increased, while in consequence the state took a greater interest in managing or influencing both food production and consumption.

Of course, all this did not change immediately on 8 May 1945. The original objectives of the Common Agricultural Policy, as laid down in the Treaty of Rome negotiated between 1955 and 1957, emphasised the need to produce more food and safeguard farmers' incomes, and world prices were significantly lower than European prices after the Second World War just as they had been after the First. But whereas the rural problems of the interwar period tended to persist, those of the post-war period tended to change. Whereas labour had been locked into interwar agriculture by the absence of industrial employment opportunities, rapid post-war economic growth demanded labour, agriculture was its most easily available source, and by 1955 the level of agricultural employment in OEEC countries was less than 80 per cent of its pre-war level. This, coupled with technical change, meant that labour productivity in agriculture increased. In fact, in all western European countries except Ireland it increased more rapidly than the average for the whole economy between 1949 and 1959 (Tracy, 1982: 233). Gross agricultural output in western Europe grew by 2.55 per cent per year between 1948 and 1960, compared with only 0.56 per cent per year between 1936 and 1952 (Federico, 2005: 20). By the mid-1980s the list of European agricultural problems – food surpluses, the costs of farm support, its impact on international trade, rural out-migration and its effects on traditional rural culture, environmental problems – were all clearly different from the pre-war list. This post-war period is therefore of immense significance for the whole of European agriculture, and the following analysis examines agricultural change in one of Europe's offshore islands in an attempt to shed some light on the reasons for its occurrence.

II. Outputs, prices and economics in the UK

The volume of agricultural output in the UK nearly doubled (a 74 per cent increase) in the two decades between 1946 and 1965, as Table 14.1 demonstrates. It is important to emphasise that the figures in Table 14.1 refer to the volume of output, because the gross output at current prices increased by significantly more than this (122 per cent). The difference was accounted for by changes in the value of money and farm product prices, and calculating the volume of output removes both of these variables. In the subsequent twenty years the rate of increase was less (47 per cent) but still substantial (Brassley, 2000: 80-84). The conventional agricultural economist's theory attributes such supply variations to changes in prices, costs, technology, and the goals or objectives of farmers (Hill & Ray, 1987).

Table 14.1. Volume of output in UK agriculture, 1923-1985

	Volume of output (£million, constant 1986 prices)
1923-1929	3,335.07
1930-1934	3,681.48
1935-1939	4,133.72
1940-1945	4,344.52
1946-1950	4,633.56
1951-1955	5,639.42
1956-1960	6,738.18
1961-1965	8,065.04
1966-1970	9,013.79
1971-1975	10,060.27
1976-1980	10,400.44
1981-1985	11,835.91

Source. Brassley (2000: 84).

Price changes appear to be fairly well correlated with output for the 1945-1955 period, as Table 14.2 (which shows producer prices inclusive of subsidies – see MAFF, 1966: 85) demonstrates. In the interwar period, farm prices in the United Kingdom reached their lowest point in 1933, and thereafter rose steadily but slowly until the outbreak of war. Following a significant increase in 1940, they remained at roughly the same level, in constant price terms, until after the war. In the late 1940s and early 1950s they reached a peak, but after 1955, although maintained in current price terms, they were eroded by inflation, so that in constant price terms they fell[1].

Table 14.2. UK agricultural prices in constant price terms

	Index of agricultural prices (1986 = 100)
1923-1929	169.1
1930-1934	148.5
1935-1939	155.1
1940-1945	186.0
1946-1950	205.6
1951-1955	206.0
1956-1960	183.1
1961-1965	159.6
1966-1970	143.6
1971-1975	141.7
1976-1980	137.0
1981-1985	111.4

Source. BRASSLEY (2000: 84).

Yet the volume of output continued to increase. Without performing any complicated calculations, it appears that product price changes are a good explanation of output changes for the 1945-1955 period, but a poor one thereafter. It is therefore necessary to consider the factors that determined the production decisions of farmers in greater detail.

1 For the economic and political background to these price changes, see SELF & STORING (1971: 28), HOLDERNESS (1985 and 1996) and BRASSLEY (1997).

III. Economic, sociological, and technological approaches to output changes

The simple relationship between prices and output conceals several more complex decision-making processes in the minds of producers. To begin with, consider the price/output relationship itself. Why do higher prices produce higher outputs? For an existing cereal producer, increasing output of, say, wheat, may simply be a matter of reducing the area of land devoted to oats and using it to grow wheat, and there is no doubt that this was one of the processes at work as tractors replaced horses, the principal consumers of oats, in the post-war years. But once this substitution has been made, how can output be increased further? One solution may be to increase fertilizer use. Another may be to adopt a higher-yielding variety, assuming that one exists, or to drain land to improve its fertility, or change to a higher-yielding breed of dairy cow, or erect specialist buildings to house intensive livestock such as pigs and poultry. Some of these measures, such as changing one variety for another, have virtually no cost implications for the farmer (although to produce the new variety may have required considerable investment on the part of the seed producer); others, such as drainage or new buildings or changing breeds, require significant capital inputs. Thus the price/output relationship is revealed as a simplification. As well as responding to price changes, farmers also reacted to changes in the returns to alternative products, to cost changes, and to changes in technology, which may produce higher outputs without cost increases, or maintain outputs at decreased costs, or a combination of both. And in order to make these changes, farmers presumably also had the confidence to change their business objectives. Whereas in the interwar years they were concentrating on survival, in the post-war years they were more concerned with output and profit maximisation. Economic analysis has little to say about this shift, yet it appears to be crucial to the post-war change.

It would take a much more detailed analysis than the space constraints of this chapter permit to go through all the impacts of these possible changes and to decide upon their relative importance. However, technical change is often identified, in the words of Hill and Ray (1987: 270), as the 'root cause' of post-war agricultural change, and the purpose here is to explore a new method of examining it.

There are several alternative approaches that deal directly with technical change, and by implication with the objectives of producers. Hayami and Ruttan identify 'the powerful role of economic forces' and survey several alternative theories before exploring the induced innovation theory in detail (Hayami & Ruttan, 1985: 41-72, 441). Diffusionist models provide a reasonable description of the timing of changes (initially slow, followed by a more rapid phase, followed by a plateau – the standard

S-shaped curve) but not much explanation of why they occurred (Rogers, 1995; Goldman & Eliason, 2003). The various approaches to the social construction of technology are more useful. User co-constructionist theories suggest an examination of the way in which users contribute to the development of the product, although most of Oudshoorn and Pinch's wide range of studies are based on final consumer products, whereas most technical changes in agriculture may be seen as inputs into the production process (Oudshoorn & Pinch, 2003). Another sociologically-based approach, Social Worlds theory, seeks to identify the different social worlds in which, for example, traditional and highly technical agriculture are carried out (Wilmot, 2007), and there is a related approach in conventions theory[2]. From a different perspective, Edgerton's 'warfare state' identifies a society that supports science and technology, and takes an interventionist or directive approach to the economy, for strategic reasons (Edgerton, 2006).

IV. The technique, authorities, discourse and media (TADM) approach

The approach adopted in the remainder of this paper is derived from a combination of these methodologies. For any specific technical change, it seeks to identify the authorities involved, the discourses they used, and the media through which those discourses were transmitted. It is founded on the idea that users, or potential users, of a technology perceive it differently over time. Initially it may not exist at all, and its function will be performed, if it is performed at all, by other techniques. Then some kind of prototype technology will emerge, which may or may not provide advantages over competing techniques. Subsequently the prototype will be developed or refined, and in some cases, but not all, widely adopted. The time period over which these changes occur may be short or long. So far this argument contains nothing new or innovatory, but the important point about the theory is that it suggests that the authorities, discourses and media will change as the technology changes. Whereas initially the existing technology will be dominated by traditional authorities using well-established discourses transmitted through existing media, as the technology changes new authorities will emerge and use different discourses in new media. The approach is best explained through using a specific technology as an example, and

2 This emerges from a recognition that the formal or informal contracts involved in commodity exchange are often incomplete, and so require rules, norms, practices, routines – in other words, conventions – in order to make them work in practice. These conventions allow participants in markets to make assumptions about the value or quality of what they can buy, but they can change with time or circumstances. Some may be based on trust, others on trademarks, brands or packaging, and others on evaluations by price and commercial quality (Murdoch *et al.*, 2000).

the one for which it has been most completely worked out is artificial insemination (AI) in pigs.

The relationships between the technique and the authorities, discourse, and media (henceforth referred to as TADM) for pig AI are shown in Table 14.3.

Table 14.3. The TADM approach applied to pig Artificial Insemination

1. Dates and development stages	*2. Technique perceived by users*	*3. Authorities*	*4. Discourse*	*5. Media*
1940s pre-pig AI	Natural service + travelling boars	Traditional breeders and breed societies	Natural and traditional	Herdbooks and shows
early 1950s scientific investigations	Natural service+ AI by inseminators, low conception rates, untested boars	University scientists MAFF AIC	Scientific	Scientific literature
later 1950s commercial product available		CACAI CBCs MMB Vets Breed societies	Technical + Legal – approval, licensing and control	Agricultural textbooks Farming Press
1960s AI adopted by a few		PIDA MLC	Technical	
1970s DIY AI ncreasing	Natural service + DIY AI with effective equipment, high conception rates,proven boars			Extension activities - generic demonstration
1980s AI widely adopted				
1990		Breeding companies	Commercial branding	Adverts and websites

Source. Brassley (2007). For explanations of acronyms see text.

This shows that in the 1940s, before AI had been developed for pigs (it was already in use for cattle), sows were inseminated using natural service, usually by a boar kept on the farm. Although boars had to be licensed by the Ministry of

Agriculture, Fisheries and Food (MAFF)'s inspectors, their standards of what constituted a good or bad boar were formed by the traditional breeders and breed societies. Thus these authorities – the Ministry, and traditional breeders and breed societies – were the authorities. Their authority was legitimated by a traditional discourse based on natural mating of males and females of known parentage, tested by visual inspection through the traditional media, which were livestock shows and pedigree herd books. When scientists began work on AI for pigs in the 1950s these authorities, discourses, and media began to be contested, although for a time the traditional versions remained dominant. The emerging authorities were the university scientists who were working on pig AI and their supportive colleagues in MAFF (although not everybody in MAFF was initially supportive). The scientists used a scientific discourse very different from that of the traditional breeders, and the media through which they expressed it was initially the scientific and veterinary literature, although it was not long before they also wrote about their work in magazines read by farmers (Polge, 1954). There was then a period, between the later 1950s and the early 1970s, when the technology was developed, and its effectiveness, measured by conception rates, increased, through the joint efforts of members of the Agricultural Improvement Council (AIC) and the Central Advisory Committee on AI (CACAI), on both of which civil servants, scientists and traditional breeders met, to resolve potential conflicts. The cattle breeding centres (CBCs) and the Milk Marketing Board (MMB), both of which had experience in providing AI facilities for cattle, the Pig Industry Development Authority (PIDA), and the veterinary profession all made claims to have some authority over the process, as did the breed societies, but with less justification. Eventually, the Meat and Livestock Commission (MLC), which took over PIDA, emerged by the late 1960s as the principal provider of expertise and training in pig AI.

At the same time the discourse changed, from a scientific one, in which the search for understanding of the process was paramount, to a technical one that concentrated on what made for effective practice. The media used changed too, going from the scientific literature, through textbooks and the farming press, to extension activities, in which MLC officers used demonstrations and promotional materials to show farmers and workers how to engage with and carry out the process. As the technique became established, farmers and farm workers trained each other, so that by the 1980s what had been virtually a laboratory process carried out by expert scientists in white coats became a farmyard technique carried out by farm workers in muddy boots. By the 1990s the authorities had changed again, as pig breeding companies became the main sources of semen and advertised their branded (the new discourse) genetic bloodlines through press advertisements and websites, the new media[3].

[3] This is a rapid summary of the detailed account provided in BRASSLEY (2007).

Although the TADM approach to the analysis of technical change has been worked out in greatest detail for pig AI, it can also be applied to other technical changes. As the example of pig AI demonstrates, technical change can be a contested process, and the contests take place between the various authorities, discourses, and media, as well as within the minds of those upon whom these influencing agents operate. Many, if not all, technical changes have associated costs and benefits. Some of the benefits are obvious, as, for example, when a new crop variety offers higher yields, or a new pesticide deals with a source of animal or plant disease. Equally, the costs may be obvious too: seed of new varieties may be more expensive than that of established varieties; sprays are an additional cultivation cost; new machinery and improved buildings require capital. Other costs and benefits, or other considerations that affect the technology adoption decision, are less obvious. A new cereal variety may be used in much the same way as its predecessor, but a new precision drilling technique may make additional demands on the operator which imply additional training costs. Using additional nitrogenous fertilizer on cereal crops could potentially increase yields, but until short-strawed varieties were introduced it also created harvesting problems.

Analysing the way in which these difficulties were overcome, and the means by which potential adopters were made aware of their solution, is the purpose of the TADM approach. It can be applied, for example, to one of the most contested and long drawn-out technical changes of recent agricultural history: the introduction and adoption of silage. The replacement of hay by silage as a method of fodder conservation took nearly a century to become widespread in the United Kingdom, and over this period changing technologies were accompanied by corresponding changes in authorities, discourses, and media. To begin with, in the late nineteenth century, the authorities were prominent farmers and landowners, who compared methods with each other (the discourse might be termed 'exploratory'), and communicated through an Ensilage Society, agricultural shows and the *Journal* of the Royal Agricultural Society of England. But the virtues of the system were at best contested, for although silage offered the possibility of more reliable fodder conservation in a wet season, it proved unreliable in quality and high in both capital and labour costs. Although some interest persisted among landowners and farmers into the early twentieth century, it was the scientists who took over as the principal authorities in the interwar period. They became fascinated by the science of the process itself: what went on in the silo, chemically, biologically, or bacteriologically, how did it conserve the grass as useful fodder, and why, sometimes, did it all go so spectacularly wrong. They began to realise that good silage was produced when bacterial action rapidly produces lactic acid, whereas foul-smelling silage ('a smell as bad as ten motor cars' in the words of McConnell, 1906: 207) emerged when butyric acid had been formed. Thus the discourse was increasingly influenced by scientific terminology, and Amos and Woodman, who worked at Cambridge University, and S.J. Watson, at the research farm run by Imperial Chemical Industries (a firm manufacturing fertilizers among other chemicals) published their results in the *Journal of Agricultural Science*.

During the Second World War there was a further change. The university and college scientists were recruited into the Ministry of Agriculture, and the requirement was to maximise home-grown livestock feed, as cereal-based concentrates were no longer available. As civil servants, their discourse moved from one of science to national interest: 'twenty pounds of grass silage produces a gallon of milk and takes the place of 3.5 pounds of imported concentrated food. Thus every six tons of silage made has liberated one ton of shipping space', wrote Dr Ian Moore, formerly of the University of Leeds, in 1944 (Moore, 1944: 28). The media used changed too, from the scientific journal to the farming press, Ministry of Agriculture leaflets, and the advisory officers of the County War Agricultural Executive Committees (Riley, 2006).

Similarly, in the 1970s, when silage began to take over from hay, the discourse changed again, concentrating on conservation methodologies, waste minimisation, feed value, and profitability, with the Ministry advisers joined by leading farmers such as Richard Waltham (innovator of the Dorset Wedge system of silage making) and fertilizer company representatives as the authorities (Seddon, 1986: 30-31)[4]. Subsequently the technology was contested in yet another way, as wildlife conservationists became aware of its impact on floral diversity and fauna. In this case, conservation and production authorities came into conflict, using mutually unintelligible discourses and media espousing different scientific, economic and political values (Riley, 2005).

These are just two examples of the way in which the TADM technique can be used to tease out the complexities involved in the adoption of new technologies. Detailed studies of the use of pesticides, fertilizers, feeding stuffs, veterinary medicines, new crop varieties and new breeds of animals would show similarly more or less complex trajectories of development and adoption. Established technologies, such as fertilizers and feeding stuffs, responded rapidly to high post-Second World War product prices, and the relaxation of wartime restrictions. By 1950-1951 fertilizer usage was three times higher than it had been in 1935-1939, and by 1960 it was 50 per cent higher still, but thereafter the rate of increase slowed considerably. The same pattern is discernable for the use of concentrated feeding stuffs. Similarly the number of tractors on UK farms increased by 50 per cent between 1945 and 1950, and by a further 50 per cent by 1960, while the number of combine harvesters rose by five

4 The development of silage as a technique is covered in much greater detail in Brassley (1996).

times in the 1950s (Brassley, 2000). For some other technologies, as we have seen, the process was more complicated. Pesticides, for example, like pig AI, show how a science-based technique goes through various stages of authority and discourse, with corresponding changes of media, before it comes into every-day use in the fields. More recently, the argument over genetically modified organisms (GMOs) has illustrated the way in which contested authorities and discourses can lead to policy conflicts, and since the 1980s the requirement on the part of farmers to reduce inputs and outputs has illustrated that some discourses are more acceptable to farmers than others[5]. Similarly, the change in attitudes to organic farming could be analysed in terms of authorities, discourses and media, although there is insufficient space to do so here. These examples make it clear that technology is not just an exogenous variable, which implies that it is endogenous, in part at least, in the sense that the requirements of farmers, responding to price and cost changes, have some impact on the production of technology. Whether, and, if so, how, this happens requires further investigation. To what extent, for example, did individual farmers, or agricultural societies, or the National Farmers' Union, promote technical development or the adoption of new technologies?

The implication of these ideas is that the analysis of technical change in farming needs to look beyond agriculture itself. There is, for example, a wide range of potential authorities. In traditional systems, farmers themselves can be the source of authority, as they specify what makes a successful crop or a 'good-looking' animal, but by the nineteenth century, if no earlier, the presence of livestock judges at agricultural shows suggests that some farmers and land agents had more authority than others[6]. In the context of the UK in the post-war period, the list of authorities was longer, and included not only farmers but also agricultural scientists, agricultural educators, the Ministry of Agriculture, various commodity marketing boards, and, for many farmers, the representatives of firms in the ancillary industries. The centre of this network of authority was the Ministry of Agriculture. At first sight, the Ministry's principal contribution might be thought to be its expenditure on price support for agriculture, which in the years 1958-1964 averaged £181.6 million per year, or 11.7 per cent of the gross output of the agricultural industry, but in addition there were also input subsidies for fertilizers, lime, various capital improvements, drainage, rearing calves, and ploughing up grassland, on which annual expenditure averaged £101 million in the same period (Self & Storing, 1971: 83; Holderness, 1985: 22). Output expansion was thus encouraged through the market mechanism, although it should

[5] Note that some analysts have used the terms 'buy-in' and 'buy-out' for internalisation or rejection of policy goals and imposed norms of behaviour (van Woerkum & Aarts, 1998; Burton, 2004; Hall & Pretty, 2008).

[6] It could be argued that early agricultural writers might also be categorised as authorities, in which case authorities could be identified back to classical times. See in this regard Fussell (1972).

be remembered that to gain access to subsidies and grants farmers had to have the confidence to invest their own money. In addition, the Ministry had a network of paid employees and unpaid associates who were intimately involved in the various stages of the improvement process, from advising on the distribution of research expenditure to encouraging farmers to adopt new techniques. At the centre were bodies such as the Agricultural Research Council and the Agricultural Improvement Council, the latter bringing together scientists, bureaucrats, and representatives of various farming organisations. There were also more specialist committees, focussed on particular techniques or problems, which brought together a wide range of people from different organisations and parts of the country. One example of such a committee was the Central Advisory Committee on Artificial Insemination, which met through the 1950s. Its chair was the Earl Waldegrave, who at the time was Parliamentary Secretary to the Ministry of Agriculture, but who was also a farmer and landowner in the west of England, a former chairman of the Somerset Agricultural Executive Committee, governor of the Long Ashton research institute, trustee of the Bath and West Agricultural Society, and subsequently chairman of the Forestry Commission. The members included A.S. Cray, a pig breeder and member of the Pig Industry Development Authority, Dr Joseph Edwards, head of the Production division of the Milk Marketing Board, Sir John Hammond, a leading animal scientist from the University of Cambridge, L.E.A. Rowson, a vet, also from Cambridge University, one of the pioneer workers on the science of AI, Dr H.G. Sanders, formerly an agricultural academic at Cambridge and Reading universities but at that time Chief Scientific Adviser to the Ministry, and W.R. (later Sir Richard) Trehane, chair of the Milk Marketing Board (NA, MAF 124/100). Thus one relatively low-profile committee serves to show how the Ministry's activities brought together politicians, civil servants, scientists, members of commodity commissions and marketing boards, and farmers to exert authority over a specific field of technology. And, of course, the influence of these central Ministry committees would be transferred, through their members, to the scientists, advisers and, ultimately (and arguably), farmers and farm workers throughout the country[7].

The activities of agricultural scientists have been summarised by Blaxter and Robertson, who point out that a basic structure of agricultural research was in place before the Second World War. It included 32 separate research institutes, many, but not all, of them attached to universities, some of which also had agricultural faculties in which soil scientists, biochemists, geneticists, zoologists, engineers and economists carried out both basic scientific discovery work and also work on developing scientific findings for practical use (Blaxter & Robertson, 1995: 257). Their findings were

7 There were also local committees of the Ministry, known as Agricultural Executive Committees. See Self & Storing (1971: 139-156).

transmitted to students and those working on farms by a further cohort of agricultural educators working in universities and national and local agricultural colleges (Brassley, 2008). There were also numerous people and organisations concerned, directly or indirectly, in what is often called 'extension' work, or advisory work with farmers. Official provision came from the National Agricultural Advisory Service, established after the war in England and Wales (a different but comparable system operated in Scotland), with a system of district advisers organised on a county basis and supported by regional specialists in specific disciplines such as animal husbandry, crop husbandry, mechanisation, farm management, etc. The marketing boards were established to regulate the marketing of individual commodities, but the larger and more successful of them also became technical authorities, especially the Milk Marketing Board, through its production division (Martin, 2000: 77). Likewise there were also commodity commissions, such as the Home Grown Cereals Authority and the Meat and Livestock Commission, which exercised authority roles through their provision of technical advice and training (Blaxter & Robertson, 1995: 270). Even employees of commercial firms were seen as authorities by many farmers, and firms began to re-identify some of their sales staff as 'technical advisers'. The technical staff employed by one feed manufacturer, British Oil and Cake Mills (BOCM), for example, numbered over eighty people in the 1960s (Montague, 2000: 111). The point to be re-emphasised here, however, is that these various authorities were more or less influential at different times for different technologies. To cite a simple example: Ministry of Agriculture officials were probably the dominant authorities on concentrated livestock feeding stuffs when they were in short supply during and immediately after the Second World War; subsequently the feed company technical representatives were most influential with the majority of farmers.

As the authorities expanded in number, so the discourse of agriculture changed. Essentially, more scientific and technical thinking and language came into use. Instead of referring to putting 'nitre' or 'bag muck' on to a field of grass, farmers and farm workers might think in terms of putting a spring dressing of fifty units of straight nitrogen per acre on a ryegrass ley. Animal rations were calculated in terms of starch equivalents (and, later, metabolisable energy). Farm workers became practised in applying precisely calculated doses of herbicides, or injecting animals with veterinary medicines, including antibiotics. Instead of just looking at their bank balances, farmers began to use gross margin analysis on their accounts, and to compare their results with those produced by the national Farm Management Survey. This is not to say that this high-technology discourse was uncontested. Breed societies still judged champion animals on their appearance (although some of their more advanced members might talk about judging on phenotype rather than genotype), and the possession or otherwise of four white feet might determine whether or not a particular boar was thought to be a good example of his breed. The organic farming movement maintained a constant but, in the 1950s and 1960s, low-key and ineffective critique of modern farming methods, and their alternative discourse was dismissed as irrelevant by most authorities[8].

This new discourse was promulgated through a wide range of media. At the scientific end of the spectrum, the *Journal of Agricultural Science* was joined by other scientific journals, and at the more popular end, weekly farming journals such as *Farmers Weekly* and *Farmer and Stockbreeder* found their sales expanding, and they were subsequently joined by more specialist publications such as *Pig Farmer*, *Dairy Farmer*, and *Power Farming*. Radio, and later television, produced regular farming programmes, which mixed political and economic discussion with technical information (Cherrington, 1979: 142-148). The numerous meetings, discussion groups, farm visits, and training courses held by the advisory services, experimental husbandry farms and agricultural colleges should also be seen as part of these media.

V. Conclusion

In 1950, the farmer and writer A.G. Street observed that '[...] there is little doubt that the present prosperity of British farming is mainly due to one man, who is now dead. His name was Adolf Hitler' (Street, 1950: 224). The implication was that wartime and post-war shortages had produced agricultural policies that raised product prices. The analysis above shows that increased prices were found at the same time as increased output for the first ten years of the period, but thereafter output continued to rise as prices fell in real terms, even though government expenditure was used to maintain prices in money terms. Neither high prices nor the simple availability of new technologies are enough by themselves to explain output increases, because farmers have to respond to them, and are more likely to do so in some circumstances and less so in others. Much of the discussion therefore concerns the factors that persuade farmers to aim for high inputs, high outputs, and profit maximisation rather than low inputs, low outputs, and survival.

The argument put forward here is that these factors may be explained using the TADM-model, and examples of how the model can be applied to individual technical changes are provided. Overall, therefore, it is suggested that increased inputs explain the initial expansion of output before 1955, and the adoption of new technologies explains subsequent output increases after 1955. The implications of this might appear to be that authorities, discourse, and media did not change between 1945 and 1955,

8 This paragraph is based on the personal memories of the author, who first began work on farms in 1959 and became an agricultural student in 1964.

whereas changes in these variables explain all post-1955 changes. The first of these implications is almost certainly wrong. Although the immediate post-war years were dominated by the same problems of supply shortages as the wartime period (albeit for different reasons), investment in machinery and buildings increased significantly, and the prevailing discourse changed from one of production at any price to market-led expansion (Martin, 2000: 74-76). The pig AI and silage examples that have been analysed in most detail above turn out to shed most light on the post-1955 period rather than the immediate post-war decade. As has been pointed out above, it appears that it was the use of pre-existing technologies such as fertilizers, feeding stuffs, and mechanization that expanded most rapidly in the 1945-1955 period, but these have not yet been analysed using the TADM-model. It would be interesting to assess, for example, the extent of the contacts between the Agricultural Improvement Council and the fertilizer manufacturers in the post-war years, or the extent to which the National Agricultural Advisory Service, created in the post-war decade, encouraged the expansion of home-grown feeding stuffs or the use of purchased concentrate feed.

Thus it would be going too far to claim, from the limited analysis put forward here, that the TADM-model provides a complete and accurate explanation of all the changes in UK agriculture in the forty years after the end of the Second World War. It should be clear from the above discussion on authorities that much more work remains to be done on the way that authority worked, on how, for various techniques, it changed over time, and especially on the critical roles played by inter-connected individuals and organisations in constructing it. To an even greater extent, further work is also needed on changes in discourses and media. Given the quantity of material that survives, it is surprising that more has not been done on the analysis of media in particular[9]. Although it is easy to demonstrate, for example, that both print and broadcast media expanded their agricultural output from the 1950s onwards, very little work has been done in the UK on the effects that they had on the decisions made by farmers. An economist would see such changes as a reduction in transactions costs, and specifically in search and information costs. The availability of data on crop variety comparisons produced by the National Institute of Agricultural Botany (NIAB), for example, would reduce the risk of adopting a new variety or the cost of experimenting with one on the part of the ordinary farmer. But there is much more to it than this, for the historian also needs to ask why farmers would act on the information produced by the NIAB rather, for example, than that offered by local seed merchants, the editorial content of the farming press, advertisements, or the advice of their

9 John Martin devotes only about half a page to discussing the role of the farming press in technical change, and does not appear to mention radio or television (Martin, 2000: 2). Erin Gill (2011: 105-6, 122) has done some, work on the role of the BBC in the context of organic farming, and Nicola Verdon (2010) has examined the role of agricultural periodicals in changing rural domesticity and the farmhouse economy in the interwar period.

neighbours. Further work is also desirable on the relative importance of authorities, discourses, and media. Is it possible to identify, for example, technical changes that are promoted by, say, changes in the media, without any corresponding developments on the part of authorities, or vice versa? Do changing discourses *inevitably* produce media changes? Is it possible to argue that it is the change in authority that is always more significant, so that discourse and media developments simply flow from it, or are all three mutually supportive like the legs of a stool. Questions such as these require historical investigation of a much wider range of technical changes than has been possible so far.

There are thus two ways of using the TADM-model that could inform further research. One is to apply it to a wider range of individual technical changes than has so far been possible. The other is to concentrate on those aspects of it, across the whole range of agricultural technology, that have so far been neglected: the creation of authority, and the varying roles of the numerous people and agencies that had it; their influence in changing the dominant discourses employed in farming and its associated industries; and the media through which these discourses were promulgated. And finally, although this argument has used exclusively examples from the UK, there seems to be no reason why the same approach should not be used for any European country. The argument advanced here is that it provides a useful framework for the historical analysis of agricultural change. When applied to post-war changes in UK agriculture it appears to offer the possibility of a better explanation than the high price – supply response relationship suggested by simple conventional economics, for it requires the historian to enquire into the decisions made by farmers and the various forces that attempted to affect them. In so doing it becomes clear that the rapid output expansion of the post-war years identified at the beginning of this paper was the result of a complex interaction of economic, institutional, political, and social forces. What the TADM method might provide is a way of linking the effects of what might otherwise appear to be unrelated influences. Having said that, it is important to add that it still remains a methodology requiring much more testing against both the historical events in different countries and the different explanations offered by alternative models.

Archival sources

NA (UK National Archives, Kew), MAF 124/100, Central Advisory Committee on Artificial Insemination, minutes of meetings and papers, 1951-61.

Bibliography

BLAXTER, Kenneth & ROBERTSON, Noel (1995), *From Dearth to Plenty: the Modern Revolution in Food Production*, Cambridge University Press.

BRASSLEY, Paul (1996), 'Silage in Britain, 1880-1990: the Delayed Adoption of an Innovation', *Agricultural History Review*, 44/1, p. 63-87.

BRASSLEY, Paul (1997), *Agricultural Economics and the CAP: an Introduction*, Oxford, Blackwell.

BRASSLEY, Paul (2000), 'Output and Technical Change in Twentieth-century British Agriculture', *Agricultural History Review*, 48/1, p. 60-84.

BRASSLEY, Paul (2006), 'The Wheelwright, the Carpenter, Two Ladies from Oxford, and the Construction of Socio-economic Change in the Countryside between the Wars', in Paul BRASSLEY, Jeremy BURCHARDT & Lynne THOMPSON (eds), *The English Countryside between the Wars: Regeneration or Decline?*, Woodbridge, Boydell Press, p. 212-234.

BRASSLEY, Paul (2007), 'Cutting across Nature? The History of Artificial Insemination in Pigs in the United Kingdom', *Studies in the History and Philosophy of Biological and Biomedical Sciences*, 38, p. 442-461.

BRASSLEY, Paul (2008), 'Agricultural Education, Training and Advice in the UK, 1850-2000', in Nadine VIVIER (ed.), *The State and Rural Societies: Policy and Education in Europe 1750-2000*, Turnhout, Brepols, Rural Hisitory in Europe, 4, p. 259-278.

BURTON, Rob (2004), 'Seeing through the "Good Farmer's" Eyes: towards Developing an Understanding of the Social Symbolic Value of 'Productivist' Behaviour', *Sociologia Ruralis*, 44/2, p. 195-215.

CARDOZA, Anthony L. (1991), 'Commercial Agriculture and the Crisis of Landed Power', in Ralph GIBSON & Martin BLINKHORN (eds), *Landownership and Power in Modern Europe*, London, Harper Collins, p. 181-198.

CHERRINGTON, John (1979), *On the Smell of an Oily Rag*, London, Northwood Books.

EDGERTON, David (2006), *Warfare State: Britain, 1920-1970*, Cambridge, Cambridge University Press.

FEDERICO, Giovanni (2005), *Feeding the World: an Economic History of Agriculture 1800-2000*, Princeton NJ, Princeton University Press.

FUSSELL, George E. (1972), *The Classical Tradition in Western European Farming*, Newton Abbot, David and Charles.

GILL, Erin (2011), *Lady Eve Balfour and the British Organic Food and Farming Movement*, unpublished PhD thesis, Aberystwyth University.

GOLDMAN, Emily & ELIASON, Leslie C. (eds) (2003), *The Diffusion of Military Technology and Ideas*, Stanford CA, Stanford University Press.

HALL, Jilly & PRETTY, Jules (2008), *'Buy-in' and 'Buy-out': Linking Social Capital and the Transition to More Sustainable Land Management*, unpublished paper delivered at the conference on Rural Futures, University of Plymouth, April 2008.

HAYAMI, Yujiro & RUTTAN, Vernon W. (1985), *Agricultural Development: an International Perspective*, Baltimore, Johns Hopkins University Press.

HILL, Berkeley & RAY, Derek (1987), *Economics for Agriculture*, Basingstoke, Macmillan.

HOLDERNESS, Brian A. (1985), *British Agriculture since 1945*, Manchester, Manchester University Press.

HOLDERNESS, Brian A. (1996), 'Apropos the Third Agricultural Revolution: How Productive was British Agriculture in the Long Boom, 1954-73', in Peter MATHIAS & John A. DAVIES (eds), *Agriculture and Industrialisation from the Eighteenth Century to the Present Day*, Oxford, Blackwell.

MCCONNELL, Primrose (1906), *The Diary of a Working Farmer*, London, Cable Printing and Publishing Co.

MAFF [Ministry of Agriculture, Fisheries and Food] (1966), *A Century of Agricultural Statistics*, London, HMSO.

MARKS, Hilary F. & BRITTON, Denis K. (1989), *A Hundred Years of British Food and Farming: a Statistical Survey*, London, Taylor and Francis.

MARTIN, John (2000), *The Development of Modern Agriculture: British Farming since 1931*, Basingstoke, Macmillan.

MONTAGUE, Diane (2000), *Farming, Food and Politics: the Merchant's Tale*, Dublin, IAWS Group.

MOORE, Henry Ian. (1944), *Ploughing for Pasture*, London, Longmans.

MURDOCH, Jonathan, MARSDEN, Terry & BANKS, Jo (2000), 'Quality, Nature, and Embeddedness: Some Theoretical Considerations in the Context of the Food Sector', *Economic Geography*, 76/2, p. 107-126.

NISSEN, Morgens R. (2006), 'Danish Food Production in the German War Economy', in Frank TRENTMANN & Flemming JUST (eds), *Food and Conflict in Europe in the Age of the Two World Wars*, Basingstoke, Palgrave Macmillan, p. 172-192.

OUDSHOORN, Nelly & PINCH, Trevor (eds) (2003), *How Users Matter: the Co-construction of Users and Technologies*, Cambridge, Mass., MIT Press.

OVERTON, Mark (1996), *Agricultural Revolution in England: the Transformation of the Agrarian Economy 1500-1850*, Cambridge, Cambridge University Press.

OYLER, Philip (1950), *The Generous Earth*, London, Hodder and Stoughton.

POLGE, Cristopher (1954), 'AI May soon Help You to Breed Better Pigs', *Pig Farming*, November 27.

RILEY, Mark (2005), 'Silent Meadows: the Uncertain Decline and Conservation of Hay Meadows in the British Landscape', *Landscape Research*, 30/4, p. 437-458.

RILEY, Mark (2006), '"Silage for Self-sufficiency"? The Wartime Promotion of Silage and its Use in the Peak District', in Brian SHORT, Charles WATKINS, & John MARTIN (eds), *The Front Line of Freedom: British Farming in the Second World War*, Exeter, British Agricultural History Society, p. 77-88.

ROGERS, Everett M. (1995), *Diffusion of Innovations*, New York, Free Press, 4th ed.

SCHMIDT, Carl Theodore (1938), *The Plough and the Sword: Labour, Land and Property in Fascist Europe*, New York, Columbia University Press.

SEDDON, Quentin (1986), *The Silent Revolution: Farming and the Countryside into the 21st Century*, London, BBC Books.

SELF, Peter & STORING, Herbert J. (1971), *The State and the Farmer*, London, George Allen and Unwin.

STREET, Arthur George (1950), *Farmer's Glory*, Harmondsworth, Penguin. Re-issued with postscript; originally published in 1932 by Faber and Faber.

TRACY, Michael (1982), *Agriculture in Western Europe: Challenge and Response 1880-1980*, London, Granada.

VAN WOERKUM, Cees & AARTS, Noelle (1998), 'Communication Between Farmers and Government over Nature: a New Approach to Policy Development', in Niels G. RÖLING & Mana Anna Elisabeth WAGEMAKERS (eds), *Facilitating Sustainable Agriculture*, Cambridge, University Press, p. 59-82.

VERDON, Nicola (2010), '"The Modern Countrywoman": Farm Women, Domesticity and Social Change in Interwar Britain', *History Workshop Journal*, 70, p. 86-107.

WARRINER, Doreen (1939), *Economics of Peasant Farming*, Oxford, Oxford University Press.

WILMOT, Sarah (2007), 'From "Public Service" to Artificial Insemination: Animal Breeding Science and Reproductive Research in Early Twentieth-century Britain', *Studies in the History and Philosophy of Biological and Biomedical Sciences*, 38, p. 411-441.

YATES, Paul L. (1940), *Food Production in Western Europe: an Economic Survey of Agriculture in Six Countries*, London, Longmans Green and Co.

YATES, Paul L. & WARRINER, Doreen (1943), *Food and Farming in Post-war Europe*, London, Oxford University Press.